Anil K. Sharma, Poonam Bansal, Priti Duhan (Eds.)

Artificial Intelligence in Biotechnology

Also of Interest

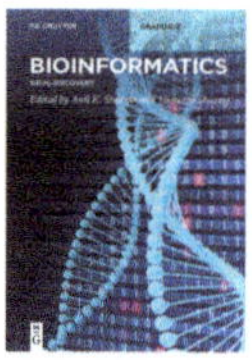

Bioinformatics.
Drug Discovery
A K Sharma, V Sharma (Eds.), 2025
ISBN 978-3-11-156788-4, e-ISBN (PDF) 978-3-11-156858-4,
e-ISBN (EPUB) 978-3-11-156882-9

Synthetic Biology for Therapeutics.
Engineering Cells for Living Drugs
Seker (Ed.), 2025
ISBN 978-3-11-132948-2, e-ISBN (PDF) 978-3-11-132949-9,
e-ISBN (EPUB) 978-3-11-132962-8

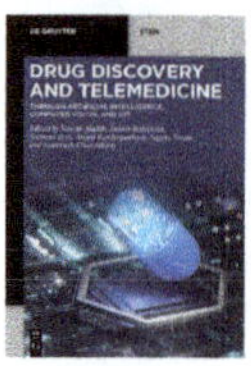

Drug Discovery and Telemedicine.
Through Artificial Intelligence, Computer Vision, and IoT
Mallik, Rahaman, Seth, Bandyopadhyay, Swain, Chakraborty (Eds.), 2025
ISBN 978-3-11-150398-1, e-ISBN (PDF) 978-3-11-150466-7,
e-ISBN (EPUB) 978-3-11-150476-6

Demystifying Artificial Intelligence.
Symbolic, Data-Driven, Statistical and Ethical AI
Gillain (Eds.), 2024
ISBN 978-3-11-142567-2, e-ISBN (PDF) 978-3-11-142614-3,
e-ISBN (EPUB) 978-3-11-142630-3

Artificial Intelligence and Computational Dynamics for Biomedical Research
Saxena, Brault (Eds.), 2022
ISBN 978-3-11-076199-3, e-ISBN (PDF) 978-3-11-076204-4,
e-ISBN (EPUB) 978-3-11-076208-2

Artificial Intelligence in Biotechnology

Biomanufacturing and Precision Medicine

Edited by
Anil K. Sharma, Poonam Bansal, and Priti Duhan

DE GRUYTER

Editors
Anil K. Sharma
Amity University Punjab
Mohali 140306, Punjab
India
anibiotech18@gmail.com

Poonam Bansal
Department of Bio-Sciences and Technology
Maharishi Markandeshwar (Deemed to be University)
Mullana, Haryana
India
poonambansal.biochem@gmail.com

Priti Duhan
Department of Biotechnology
K. L. Mehta Dayanand College for Women
Faridabad, Haryana
India

ISBN 978-3-11-914518-3
ISBN 978-3-11-221427-5 (PDF)
ISBN 978-3-11-221468-8 (EPUB)
DOI https://doi.org/10.1515/9783112214275

Library of Congress Control Number: 2026934667

Bibliographic information published by the Deutsche Nationalbibliothek
The Deutsche Nationalbibliothek lists this publication in the Deutsche Nationalbibliografie; detailed bibliographic data are available on the Internet at http://dnb.dnb.de.

De Gruyter and Walter de Gruyter GmbH are part of De Gruyter Brill.
www.degruyterbrill.com

Questions about General Product Safety Regulation:
productsafety@degruyterbrill.com

Cover illustration: BlackJack3D/E+/gettyimages

Preface

The convergence of biotechnology and artificial intelligence (AI) heralds a transformative era in scientific innovation. Biotechnology with its profound ability to manipulate biological systems for the betterment of humanity and AI, with its unparalleled capacity to process, analyze, and predict patterns in vast datasets, is revolutionizing our understanding and interaction with the living world. This book explores this dynamic intersection, offering a comprehensive examination of the tools, applications, and future potential of these integrated disciplines. The primary objective of this book is to illuminate how AI is reshaping biotechnology across diverse domains, including genomics, proteomics, healthcare, agriculture, and environmental sustainability. By harnessing AI, researchers and industry professionals are tackling some of the most pressing challenges in biotechnology – from accelerating drug discovery and optimizing production processes to personalizing medical treatments and addressing global environmental issues.

AI's integration into genomics has revolutionized genome sequencing, annotation, and editing. Advanced tools such as AlphaFold have set new benchmarks in protein structure prediction, enabling groundbreaking advancements in drug discovery and disease understanding. In drug development, AI accelerates the traditionally time-consuming and expensive process by predicting molecular interactions, identifying drug targets, and optimizing clinical trials. In healthcare, AI is driving the realization of precision medicine through personalized treatment plans, early disease detection, and enhanced diagnostic tools. Meanwhile, in agriculture and environmental biotechnology, AI is transforming practices by improving crop yields, monitoring soil health, and addressing challenges such as waste management and bioremediation. While the rise of AI in biotechnology presents immense opportunities, it also brings ethical and regulatory challenges.

The future of biotechnology in the AI era holds extraordinary promise. As AI technologies become more sophisticated, their integration into biotechnology will catalyze unprecedented advancements. The advent of quantum computing is poised to further enhance AI's capabilities, enabling breakthroughs in complex biological problems such as protein folding and metabolic pathway optimization. AI-driven insights into genetic and phenotypic data will pave the way for ultra-personalized healthcare, where treatments are tailored to individual patients with unparalleled precision. Additionally, AI will play a pivotal role in addressing global challenges like climate change and food security by optimizing biomanufacturing processes and creating sustainable agricultural practices. These advancements will necessitate interdisciplinary collaborations across computer science, biology, chemistry, and ethics, fostering innovation while addressing complex societal challenges.

Despite its vast potential, the integration of AI and biotechnology is not without obstacles. One significant challenge lies in the quality and availability of data. AI systems require vast amounts of high-quality data to function effectively, yet in many

 | https://doi.org/10.1515/9783112214275-202

areas of biotechnology, such data is scarce or fragmented. Robust regulatory frameworks are also essential to ensure that AI applications in biotechnology are safe, ethical, and equitable. Furthermore, the skill gap between AI and biotechnology expertise presents another hurdle. Addressing this gap will require transformative changes in educational curricula and training programs. This book is not merely a collection of chapters but a roadmap for understanding and navigating the rapidly evolving landscape of AI and biotechnology. It aims to inspire readers to think critically about the opportunities and challenges at this intersection and to contribute to the advancement of these fields. Each chapter has been carefully curated to provide a balanced mix of foundational knowledge, current trends, and future perspectives.

With this book, we invite you to embark on a journey through the exciting and transformative world of biotechnology in the AI era. The future is bright, and the possibilities are endless. Together, let us explore, innovate, and build a better tomorrow.

Editors

Contents

List of Contributing Authors

Poonam Bansal
Department of Biosciences and Technology
MMEC
Maharishi Markandeshwar (Deemed to be University)
Mullana, Ambala
Haryana
India
poonambansal.biochem@gmail.com

Priti Duhan
K. L. Mehta Dayanand College for Women
Faridabad
Haryana
India

Kumkum Verma
ICAR-IARI
Regional Station
Karnal
Haryana
India
Kumkum.verma89@gmail.com

Anupam Sharma
Department of Physics
Guru Kashi University
Talwandi Sabo
Bathinda 151302
Punjab
India

Anil K. Sharma
Department of Biotechnology
Amity University
Sector 82A, IT City Road
Sahibzada Ajit Singh Nagar 140306
Punjab
India

Pooja Verma
ICAR-IIWBR
Karnal 132001
Haryana
India

Bhupesh Gupta
Department of Computer Science and Engineering
MMEC
Maharishi Markandeshwar (Deemed to be University)
Mullana, Ambala
Haryana
India

Narender Mohan
ICAR-IIWBR
Karnal 132001
Haryana
India
narendermohan93@gmail.com

Praveen Kumar
Faculty of Health and Allied Sciences
Guru Kashi University
Talwandi Sabo, Bathinda
Punjab
India

Imran Sheikh
UCRD
Chandigarh University
Mohali
Punjab
India

Pooja Saini
Department of Biotechnology
Eternal University
Baru Sahib 173101
Himachal Pradesh
India

Varruchi Sharma
Department of Biotechnology and Bioinformatics
Sri Guru Gobind Singh College
Sector 26, Chandigarh 160019
India
varruchi.sharma2013@gmail.com

 | https://doi.org/10.1515/9783112214275-204

Aaliya Ali
UCRD
Chandigarh University
Mohali
Punjab
India

Syed Irfan Yaqoob
AIT-CSE
Chandigarh University
Ghauran, Mohali 140413
Punjab
India

Vikas Kushwaha
Department of Biotechnology
Goswami Ganesh Dutta Sanatan Dharma College
Sector 32, Chandigarh
India

Nitin Sharma
Department of Biotechnology
Chandigarh College of Technology
CGC, Landran
Mohali 140307
Punjab
India

Ashwanti Devi
Department of Biosciences and Technology
MMEC
Maharishi Markandeshwar (Deemed to be University)
Mullana, Ambala 133207
Haryana
India

Vandana Sharma
Department of Physics
MMEC
Maharishi Markandeshwar (Deemed to be University)
Mullana, Ambala 133207
Haryana
India

Graeme Mukhopadhyay
University Institute of Biotechnology
Chandigarh University
Gharuan, Mohali 140413
Punjab
India

Vikas Kumar
University Institute of Biotechnology
Chandigarh University
Gharuan, Mohali 140413
Punjab
India

Amit Joshi
Department of Biotechnology and Bioinformatics
Sri Guru Gobind Singh College
Sector 26, Chandigarh 160019
India

Diksha Pathania
Department of Biosciences and Technology
MMEC
Maharishi Markandeshwar University
Mullana, Ambala
Haryana
India
dikshapathania198@gmail.com

Priyanka Chauhan
SRMIST
Delhi NCR Campus
Ghaziabad
Uttar Pradesh
India

Aditya Chauhan
Department of Agriculture
Maharishi Markandeshwar University
Mullana, Ambala
Haryana
India

Sonu
School of Advanced Chemical Sciences
Shoolini University of Biotechnology and Management Sciences
Solan
Himachal Pradesh
India

J. K. Sharma
Department of Physics
MMEC
Maharishi Markandeshwar (Deemed to be University)
Mullana, Ambala 133207
Haryana
India

Anil Panwar
Department of Bioinformatics and Computational Biology
CCS Haryana Agricultural University
Hisar 125004
Haryana
India

Amit
Division of Genetics and Plant Breeding
ICAR-IIWBR
Karnal
Haryana
India

Vikas Kushwaha
Department of Biotechnology
Goswami Ganesh Dutta Sanatan Dharma College
Sector 32, Chandigarh
India

Anuprabha
Department of Botany
Goswami Ganesh Dutta Sanatan Dharma College
Sector 32, Chandigarh
India

Ajay Sharma
Department of Chemistry
Career Point University
Tikker-Kharwarian, Hamirpur 176041
Himachal Pradesh
India
and
Centre for Nano-Science and Technology
Career Point University
Tikker-Kharwarian, Hamirpur 176041
Himachal Pradesh
India

Punit Puri
Department of Zoology
DAV College
Jalandhar
Punjab
India

Manoj Singh
Department of Bio-sciences and Technology
MMEC
Maharishi Markandeshwar (Deemed to be University)
Mullana, Ambala 133207
Haryana
India

Milan Bera
Department of Physics
MMEC
Maharishi Markandeshwar (Deemed to be University)
Mullana, Ambala 133207
Haryana
India

Shivani Saklani
School of Biological and Environmental Sciences
Shoolini University of Biotechnology and Management Sciences
Solan 173229
Himachal Pradesh
India

Shallu Saini
Department of Bio-Sciences and Technology
Maharishi Markandeshwar Engineering College
Maharishi Markandeshwar (Deemed to be University)
Mullana, Ambala 133207
Haryana
India

Manuj Saini
Department of Genetics and Plant Breeding
Chaudhary Charan Singh Haryana Agricultural University
Hisar 125004
Haryana
India
doc.manujsaini@gmail.com

Hardeep Singh Tuli
Department of Bio-Sciences and Technology
Maharishi Markandeshwar Engineering College
Maharishi Markandeshwar (Deemed to be University)
Mullana, Ambala 133207
Haryana
India

Tanya
Department of Bioinformatics and Computational Biology
College of Biotechnology
Chaudhary Charan Singh Haryana Agricultural University
Hisar 125001
Haryana
India

Avinash Godara
Department of Bioinformatics and Computational Biology
College of Biotechnology
Chaudhary Charan Singh Haryana Agricultural University
Hisar 125001
Haryana
India

Vivek Kumar
Department of Bioinformatics and Computational Biology
College of Biotechnology
Chaudhary Charan Singh Haryana Agricultural University
Hisar 125001
Haryana
India

Ajeev Kumar
Department of Agricultural Biotechnology
College of Biotechnology
Chaudhary Charan Singh Haryana Agricultural University
Hisar 125001
Haryana
India
ajeevsangwan@hau.ac.in

Sonali Sangwan
Department of Biotechnology
Guru Nanak Khalsa College
Yamuna Nagar 135001
Haryana
India

Jatin Saini
Department of Agriculture
Maharishi Markandeshwar (Deemed to be University)
Mullana, Ambala 133207
Haryana
India

Shreya Nagpal
Department of Genetics and Plant Breeding
Chaudhary Charan Singh Haryana Agricultural University
Hisar 125004
Haryana
India

Poonam Bansal*, Priti Duhan, Kumkum Verma, Anupam Sharma, Anil K. Sharma, Pooja Verma

Chapter 1
Bridging Biotechnology and Artificial Intelligence

Abstract: The convergence of biotechnology and artificial intelligence (AI) is transforming scientific research, healthcare, and industrial processes, ushering in a new era of innovation and precision. This chapter explores the interdisciplinary integration of these two powerful fields, highlighting how AI technologies – such as machine learning, deep learning, and natural language processing – are accelerating advancements in genomics, drug discovery, personalized medicine, and synthetic biology. By analyzing large-scale biological data with unprecedented speed and accuracy, AI enables researchers to uncover hidden patterns, predict biological behavior, and design more effective therapeutic strategies. This chapter also addresses the challenges involved in merging these domains, including data complexity, ethical considerations, and the need for interdisciplinary collaboration. Case studies and real-world applications are presented to demonstrate the practical impact of this integration, such as AI-driven CRISPR optimization, automated diagnostics, and bioinformatics tools powered by neural networks. Ultimately, this chapter aims to provide a comprehensive overview of how bridging biotechnology and AI not only enhances scientific discovery but also opens new frontiers for innovation in medicine, agriculture, and environmental sustainability. As the synergy between these fields continues to evolve, it holds immense potential for reshaping the future of life sciences and driving solutions to some of humanity's most pressing challenges.

Keywords: Biotechnology, artificial intelligence, genomics, proteomics, synthetic biology

***Corresponding author**: **Poonam Bansal**, Department of Biosciences and Technology, Maharishi Markandeshwar (Deemed to be University), Mullana, Ambala, Haryana, India,
e-mail: poonambansal.biochem@gmail.com
Priti Duhan, K. L. Mehta Dayanand College for Women, Faridabad, Haryana
Kumkum Verma, ICAR-IARI, Regional Station, Karnal, Haryana
Anupam Sharma, Department of Physics, Guru Kashi University, Talwandi Sabo, Bathinda 151302, Punjab, India
Anil K. Sharma, Department of Biotechnology, Amity University Punjab, Sector 82A, IT City Road, Sahibzada Ajit Singh Nagar 140306, Punjab, India
Pooja Verma, ICAR-IIWBR, Karnal-Haryana, India

 | https://doi.org/10.1515/9783112214275-001

1.1 Introduction

The convergence of biotechnology and artificial intelligence (AI) represents a transformative frontier in scientific research and industrial innovation. Biotechnology, encompassing disciplines such as genomics, proteomics, and synthetic biology, has traditionally relied on experimental methodologies to explore biological systems (Madhavan & Mustafa, 2023). However, the advent of AI, particularly machine learning (ML) and deep learning (DL), has introduced computational approaches capable of analyzing vast and complex biological datasets (Gupta et al., 2021). This synergy enables the extraction of meaningful patterns and predictions that were previously unattainable through conventional methods. For instance, AI-driven platforms have accelerated drug discovery processes, reducing timelines from years to mere months.

The integration of AI into biotechnology is not merely a technological enhancement but a paradigm shift that redefines the capabilities of both fields. In drug discovery, AI algorithms have facilitated the identification of novel therapeutic targets and the design of compounds with optimized properties, expediting the development of treatments for complex diseases (Tiwari et al., 2023). In genomics, AI models process extensive genetic data to uncover disease-associated variants, paving the way for personalized medicine approaches (Ali, 2023a). Moreover, in agriculture, AI applications in precision farming have led to increased crop yields and resource efficiency, contributing to food security (Sharma, 2023). Collectively, these advancements underscore the profound impact of AI on enhancing the scope, efficiency, and precision of biotechnological applications.

This review aims to elucidate the multifaceted integration of AI within biotechnological domains, focusing on its applications in drug discovery, genomics, agriculture, and environmental sustainability. We examined case studies that highlight successful collaborations between AI and biotechnology, analyze the methodologies employed, and discuss the outcomes achieved. Furthermore, this chapter addresses the challenges and limitations inherent in this integration, such as data quality, model interpretability, and ethical considerations. By providing a comprehensive overview, current review seeks to inform researchers, practitioners, and policymakers about the potential and pitfalls of merging AI with biotechnology.

1.2 Foundations of Biotechnology

1.2.1 Historical Development

Biotechnology's roots trace back to ancient civilizations, where fermentation processes were utilized in the production of bread, beer, and wine. However, the field began to take a more scientific shape in the late nineteenth and early twentieth centuries with advancements in microbiology and genetics. The discovery of the structure of DNA in 1953 by Watson and Crick laid the groundwork for modern molecular biology, leading to significant breakthroughs such as the development of recombinant DNA technology in the 1970s (Glick & Patten, 2022). This era marked the beginning of genetic engineering, enabling the manipulation of organisms at the molecular level. The subsequent decades witnessed the emergence of various biotechnological techniques, including polymerase chain reaction in the 1980s, which revolutionized genetic analysis by allowing the amplification of specific DNA sequences (Kadri, 2019). These foundational developments set the stage for the integration of AI in biotechnology, facilitating more sophisticated analyses and applications.

1.2.2 Key Disciplines and Techniques

Biotechnology encompasses several key disciplines and techniques that have evolved over time discussed in Table 1.1.

Table 1.1: Integration of AI in core biotechnology disciplines and techniques.

Discipline/ technique	Biotechnology applications	AI contributions
Genetic engineering	Direct manipulation of genomes; use of tools like CRISPR-Cas9 for precise gene editing	Supports target identification and design strategies for genome modification
Fermentation technology	Use of microorganisms to produce products like antibiotics, hormones, and biofuels	Predicts optimal growth conditions and product yields through data-driven algorithms
Cell culture technology	Growth of cells in controlled environments for production of vaccines and monoclonal antibodies	Monitors and controls environmental parameters to ensure cell health and maximize productivity
Bioprocess engineering	Design and scale-up of processes for manufacturing bioproducts	Employs simulations to model and optimize bioprocesses, reducing operational costs, and enhancing efficiency

These disciplines have been instrumental in advancing biotechnology, and their integration with AI, continues to drive innovation across various sectors.

1.2.3 Applications in Health, Agriculture, and Industry

Biotechnology has found transformative applications in several domains as represented in Table 1.2.

Table 1.2: Applications of biotechnology and AI in health, agriculture, and industry.

Domain	Biotechnology applications	AI applications
Health	Development of recombinant therapeutics (e.g., human insulin and monoclonal antibodies)	Accelerated drug discovery through analysis of large datasets to predict molecular interactions
Agriculture	Genetically modified crops with enhanced traits (e.g., pest resistance and stress tolerance)	Precision agriculture: analysis of soil health, crop yield prediction, and resource optimization
Industry	Engineering microorganisms to produce bio-based products (e.g., biofuels and biodegradable plastics)	Optimization of metabolic pathways in microbes to enhance production efficiency and reduce waste

1.3 Foundations of Artificial Intelligence

The rapid integration of AI into biotechnology is underpinned by foundational advances in computer science, mathematics, and data systems. Understanding these foundations is essential to appreciate the transformative role AI plays in modern biosciences. This section outlines the evolution of AI, introduces core methodologies like ML and neural networks, and highlights how AI is reshaping scientific inquiry.

1.3.1 Evolution of AI Technologies

AI originated as a formal academic discipline in the mid-twentieth century, with early aspirations focused on mimicking human reasoning. Initial AI systems, developed during the 1950s and 1960s, relied heavily on symbolic reasoning and rule-based logic, which proved insufficient for complex, dynamic environments like those encountered in biology (Soares, 2025). The resurgence of AI in the twenty-first century has been driven by exponential growth in computational power, the availability of large-scale data, and innovations in algorithm design (Gill et al., 2022).

The introduction of probabilistic reasoning in the 1990s laid the groundwork for modern statistical learning approaches. The shift from rule-based to data-driven models allowed AI systems to learn patterns from empirical observations – an essential capability for dealing with biological heterogeneity and uncertainty (Brunner et al., 2024). Since 2012, DL has redefined AI capabilities, particularly in image and language processing, which are now being successfully applied to microscopy image analysis, genomics, and biomedical literature mining (Shetty et al., 2025). AI has evolved from a largely theoretical pursuit into a practical tool with direct applications in diagnostics, drug discovery, and systems biology. Its trajectory mirrors that of biotechnology: an interdisciplinary expansion driven by data, computation, and complexity.

1.3.2 Machine Learning, Deep Learning, and Neural Networks

ML is a subset of AI that enables computers to learn from data without explicit programming. In biotechnology, ML models are widely employed to classify biological samples, predict disease outcomes, and model genetic interactions (Madhukar et al., 2015). Algorithms such as support vector machines, random forests, and k-nearest neighbors have shown success in bioinformatics applications, particularly where feature selection and model interpretability are important. DL, a specialized form of ML, uses neural networks with multiple layers (deep neural networks) to model highly nonlinear relationships (Wang et al., 2020a). Convolutional neural networks (CNNs) have become instrumental in biomedical imaging tasks, such as identifying cancerous cells or segmenting organs in radiological scans (Chen et al., 2019). Recurrent neural networks, and more recently, transformers, are increasingly applied to sequential data like gene expression time series and protein sequences.

Artificial neural networks, the backbone of DL, are inspired by the structure of the human brain but operate through mathematical abstraction. In biotechnology, neural networks are being integrated into workflows for predicting protein folding, simulating metabolic pathways, and modeling cell signaling dynamics (Kathirvel et al., 2024). Recent breakthroughs such as AlphaFold demonstrate the profound potential of DL in solving grand challenges in structural biology.

1.3.3 AI in Scientific Research

AI is revolutionizing how scientific research is conducted, analyzed, and interpreted. In the context of biotechnology, AI enables automation in hypothesis generation, experiment design, and data interpretation. Techniques like natural language processing (NLP) are used to mine vast scientific literature for trends and novel insights, helping researchers stay abreast of developments across disciplines (Sandu et al., 2024). AI-driven platforms now assist in high-throughput screening of drug candidates, real-time analysis of omics data, and integration of multi-modal datasets (Xu et al., 2024). In synthetic biology, AI algorithms facilitate the design of genetic circuits and optimization of metabolic pathways. Moreover, reinforcement learning approaches are being investigated for adaptive experiment planning, where the AI system refines protocols based on prior experimental outcomes. The integration of AI in scientific research not only accelerates discovery but also democratizes access to advanced analytical tools (Youvan, 2024). By enabling scalable, reproducible, and data-centric science, AI contributes to a paradigm shift from hypothesis-driven to data-driven research models, aligning well with the high-dimensional, complex nature of biotechnological data.

1.4 Points of Intersection: Biotechnology Meets AI

The convergence of biotechnology and AI is reshaping modern bioscience by facilitating the analysis, interpretation, and application of complex biological data. This union is enabling innovations in genomics, proteomics, drug development, and synthetic biology. The sections below highlight how AI enhances biotechnological research and applications by bringing in computational precision, scalability, and predictive capabilities.

1.4.1 Data-Driven Biology

Biology has become a data-intensive science, with high-throughput technologies like next-generation sequencing, single-cell RNA-seq, and CRISPR screening generating petabytes of data. AI plays a pivotal role in turning this raw data into actionable insights. Through ML algorithms, complex biological patterns – such as gene-gene interactions, disease biomarkers, or epigenetic modifications – can be uncovered with improved accuracy and speed. AI models also excel in feature selection from multi-omics data, helping identify the most informative variables among tens of thousands (Hussein et al., 2024). For example, supervised learning has been used to stratify cancer subtypes based on gene expression profiles, leading to more precise prognostic and therapeutic strategies. Unsupervised learning, such as clustering, helps reveal hidden structures in biological data, including cellular heterogeneity within tissue samples (Phillip et al., 2021). The shift toward data-driven biology, supported by AI, reflects a move from hypothesis-centric research toward discovery-driven exploration, allowing researchers to form new hypotheses grounded in data patterns rather than intuition alone.

1.4.2 Computational Genomics and Proteomics

AI has become indispensable in computational genomics and proteomics, enabling large-scale annotation, interpretation, and modeling of genetic and protein-related information. In genomics, DL models are used to predict gene expression levels from regulatory sequences, detect structural variants, and prioritize variants of uncertain significance in clinical genetics. One of the most transformative applications is in the prediction of protein structures. Tools such as DeepMind's AlphaFold have surpassed traditional modeling methods in accuracy, solving structures of previously intractable proteins (Wang et al., 2024). These models leverage vast databases of known protein sequences and structures, applying attention-based architectures similar to those used in NLP. In proteomics, AI assists in interpreting mass spectrometry data, identifying post-translational modifications, and mapping protein-protein interaction networks. AI tools also help integrate transcriptomic and proteomic data to build comprehensive models of cellular states, which is critical in disease modeling and therapeutic targeting (Ali, 2023b).

1.4.3 Drug Discovery and Development

AI is revolutionizing the traditionally costly and time-consuming process of drug discovery. By integrating molecular modeling, cheminformatics, and biological datasets, AI systems can predict bioactivity, toxicity, and pharmacokinetics of drug candidates before laboratory testing. This accelerates early-stage screening and reduces the attrition rate in clinical pipelines. Generative AI models, such as generative adversarial networks and variational autoencoders, have been employed to design novel molecular scaffolds with desired biological properties (Martinelli, 2022). Reinforcement learning has also been applied to optimize lead compounds based on multi-objective functions, balancing efficacy, toxicity, and solubility. AI further plays a role in drug repurposing. By analyzing electronic health records, literature, and clinical trial data, AI tools can uncover new indications for existing drugs – an approach that gained prominence during the COVID-19 pandemic. Moreover, AI assists in the optimization of clinical trial designs, including patient recruitment, endpoint selection, and adaptive trial frameworks.

1.4.4 AI in Synthetic Biology

Synthetic biology, which involves designing and constructing novel biological parts and systems, benefits immensely from AI's ability to model, simulate, and optimize complex biological functions. AI enables predictive modeling of gene circuit behavior, pathway optimization for metabolic engineering, and automated design of genetic constructs. For example, ML models have been used to predict promoter strength, ribosome binding site efficiency, and the performance of CRISPR-based gene editing tools (Wang et al., 2020b). In metabolic engineering, AI algorithms optimize gene knock-in or knock-out strategies to enhance production yields of valuable biomolecules such as biofuels, pharmaceuticals, or industrial enzymes (Gupta et al., 2024). Automated platforms, or "robotic biologists," leverage AI to guide iterative design-build-test-learn cycles, reducing human intervention and increasing throughput. Tools like Bayesian optimization and active learning further refine these cycles, prioritizing experiments that are most likely to yield informative results. Overall, AI not only accelerates the design and prototyping stages of synthetic biology but also introduces a level of precision and predictability that is difficult to achieve through traditional methods.

1.5 Case Studies and Applications

The practical integration of AI in biotechnology has yielded remarkable outcomes across various domains. From genome annotation to personalized healthcare and sustainable agriculture, AI-driven solutions are driving precision, efficiency, and innovation. This section highlights selected case studies and applications that exemplify the tangible benefits of combining computational intelligence with biological systems.

1.5.1 AI in Genome Sequencing and Annotation

Genome sequencing has transitioned from being a labor-intensive process to a high-throughput, data-rich enterprise. AI enhances this evolution by enabling real-time interpretation and annotation of genomic data. One prominent case involves the use of DL models for base calling in nanopore sequencing technologies. Tools like DeepVariant and Bonito apply CNNs to improve the accuracy of nucleotide identification, reducing sequencing errors and enhancing downstream analysis (Ahmad et al., 2025). In genome annotation, AI assists in identifying coding regions, non-coding RNAs, regulatory elements, and structural variations. Transformer-based architectures, inspired by models like Bidirectional Encoder Representations from Transformers (BERT) and Generative Pre-trained Transformer (GPT), have recently been repurposed for genomic language modeling, treating DNA sequences like text to learn biologically relevant motifs (Sarumi & Heider, 2024). These models have improved annotation in non-model organisms where experimental validation is limited. Additionally, integrative platforms now combine AI with comparative genomics and epigenetic data to predict functional consequences of mutations, aiding in variant interpretation for clinical genetics.

1.5.2 Predictive Models for Disease Diagnosis

AI-powered predictive models are revolutionizing disease diagnosis by detecting subtle patterns in clinical, genomic, and imaging data. In oncology, ML classifiers have been trained on multi-omics datasets to distinguish between tumor subtypes and predict progression-free survival. For example, support vector machines and ensemble methods have successfully been used to predict breast cancer recurrence from transcriptomic profiles. Radiomics, the extraction of high-dimensional features from medical images, is another domain where AI excels. In lung and brain cancers, CNNs have demonstrated superior performance in early tumor detection compared to traditional radiological assessments (Naseer et al., 2021). Furthermore, integrating imaging data with histopathology and genomics through multi-modal AI models offers a more holistic diagnostic view. AI also shows promise in infectious disease diagnostics. During the COVID-19 pandemic, models using X-ray and computed tomography images, along with patient metadata, enabled rapid triage and risk stratification. Similar frameworks are being explored for tuberculosis, malaria, and sepsis using data from wearable sensors, electronic health records, and pathogen genomics.

1.5.3 Personalized Medicine and Precision Health

AI's capacity to integrate and analyze individual-level data is central to the realization of personalized medicine. By mining electronic health records, wearable device outputs, and genomic profiles, AI models can predict individual responses to therapies, assess disease risk, and optimize treatment plans. Pharmacogenomics is one area where AI has shown significant impact. ML models are used to predict adverse drug reactions and optimal dosing regimens based on genetic polymorphisms in drug metabolism genes. Tools such as Deep PDR utilize neural networks to correlate patient genotypes with pharmacological responses across diverse populations (Kalinin et al., 2018). Precision health initiatives, such as the "All of Us Research Program," are increasingly incorporating AI to identify predictive biomarkers from vast longitudinal datasets. AI algorithms also support early intervention by flagging at-risk individuals using predictive scores derived from lifestyle, environmental, and genetic data. Furthermore, reinforcement learning is being tested in adaptive clinical decision support systems to dynamically recommend therapies based on patient evolution over time, demonstrating the potential of AI to deliver real-time, patient-specific interventions.

1.5.4 Agricultural Biotechnology and Smart Farming

Agricultural biotechnology has embraced AI to enhance crop productivity, improve pest resistance, and reduce environmental impact. AI models assist in trait selection by analyzing genomic data from plant breeding programs, identifying optimal allelic combinations for drought tolerance, nutrient efficiency, and disease resistance. In smart farming, AI-driven platforms process data from satellite imagery, drone surveillance, and in-field sensors to monitor crop health, soil moisture, and weather conditions in real time (Kumari et al., 2025). Computer vision systems, trained with DL algorithms, can detect plant diseases or nutrient deficiencies at early stages, enabling targeted interventions. CRISPR-based gene editing in crops is also being augmented with AI for guide RNA design and off-target prediction. ML models help streamline the identification of target genes and predict editing efficiency, thereby accelerating the development of improved plant varieties. A notable case is the use of AI in vertical farming systems, where environmental variables such as light, humidity, and nutrient flow are optimized using reinforcement learning algorithms to maximize yield and resource efficiency. These AI-enabled systems represent a shift toward sustainable, data-informed agriculture that can adapt to climate variability and food security challenges.

1.6 Challenges and Ethical Considerations

The integration of AI in biotechnology holds tremendous promise, but it also raises significant challenges, particularly in the realms of data privacy, security, and ethics. As AI systems are increasingly used to analyze sensitive biological and health data, ensuring their responsible application is critical. This section explores some of the primary challenges and ethical concerns that arise at the intersection of biotechnology and AI.

1.6.1 Data Privacy and Security

AI applications in biotechnology often require access to vast amounts of sensitive personal data, including genomic sequences, health records, and environmental data. The secure management of these datasets is crucial to maintain individual privacy and trust in AI-powered systems. Traditional methods of data encryption, secure multi-party computation, and anonymization of health data have proven essential, but they are not without limitations. The risk of re-identifying individuals from seemingly anonymized datasets remains a significant concern, particularly with advancements in AI and ML that can correlate seemingly unrelated data points.

Moreover, in the context of genomics, whole-genome sequencing can reveal sensitive information about not only the individual but also their family members. Unauthorized access or misuse of this information could lead to discrimination, particularly in employment or insurance sectors. Ensuring robust data security measures, such as federated learning (where data does not leave local servers) and blockchain-based data provenance, is essential for safeguarding sensitive genomic data. The emerging regulatory frameworks, such as the European Union's General Data Protection Regulation and the Health Insurance Portability and Accountability Act in the United States, offer initial guidelines (Terry, 2017). However, they must be continually updated to address the growing complexities of AI and genomic data use. For instance, AI systems trained on large-scale datasets may need transparent audit trails to verify how sensitive data is handled and to maintain accountability.

1.6.2 Ethical Implications of AI in Life Sciences

AI's capabilities in life sciences, particularly in healthcare, bring forward ethical questions about decision-making, consent, and autonomy. One of the primary ethical concerns is the extent to which AI models can be trusted in making decisions that directly affect human lives. For example, AI systems deployed for disease diagnosis or treatment recommendations must be able to explain their decisions in a way that aligns with human understanding and medical ethics. This concept is central to the idea of "explainable AI" (XAI), which ensures that AI decisions can be traced back to understandable, verifiable reasoning. Moreover, the use of AI in genetic editing and gene therapy presents ethical dilemmas surrounding "designer babies" and genetic enhancement. While AI-powered platforms can accelerate advancements in genome editing, questions arise about the regulation of these technologies (Dixit et al., 2024). Should AI be used to make decisions on editing the human germline, especially when it involves traits that go beyond disease prevention (such as intelligence or physical characteristics)? The potential for misuse in this regard is high, and the long-term consequences of altering human genetics could be profound. Another ethical issue is the potential exploitation of vulnerable populations in AI-driven biomedical research. With the rise of AI in clinical trials and personalized medicine, there is a risk that marginalized groups may not have equitable access to AI-powered treatments or be disproportionately represented in datasets. Ensuring inclusivity in AI training datasets and research studies is critical to avoid exacerbating health disparities.

1.6.3 Bias and Transparency in AI Models

A significant challenge in the application of AI to biotechnology and healthcare is the inherent risk of bias in AI models. Since AI systems learn from historical data, they can perpetuate existing biases present in the data they are trained on. In healthcare, for example, AI models developed from data predominantly collected from certain demographic groups may perform poorly for underrepresented populations (Norori et al., 2021). This is particularly problematic when it comes to disease diagnosis or treatment, where biased models could lead to misdiagnoses or suboptimal care for certain ethnic or socioeconomic groups. In genomics, bias can also emerge if the training data fails to represent the genetic diversity of global populations. AI systems trained predominantly on data from individuals of European descent, for example, may struggle to accurately predict genetic risks or therapeutic responses in individuals from non-European backgrounds. This has led to calls for more inclusive and diverse datasets in AI-driven biotechnology research.

The transparency of AI models is also a key issue. Many DL models, while highly accurate, operate as "black boxes," meaning that their decision-making processes are opaque and not easily interpretable by humans (Hassija et al., 2024). In fields like healthcare, where decisions have serious consequences, it is essential that AI systems are transparent and interpretable. Researchers and developers are exploring various techniques in XAI to make these systems more transparent and accountable, but challenges remain in balancing model performance and interpretability. To address these concerns, researchers have proposed several frameworks and approaches for reducing bias and improving transparency, including fairness-aware algorithms, audit trails, and better documentation of training datasets. Furthermore, ethical guidelines and regulatory oversight are needed to ensure that AI in biotechnology is developed and deployed in a manner that is fair, accountable, and transparent.

1.7 Future Trends and Opportunities

As AI continues to evolve, its convergence with biotechnology is poised to reshape a wide range of industries. The future promises not only advancements in bioinformatics, genomics, and drug discovery but also the creation of entirely new opportunities for improving human health, agricultural practices, and environmental sustainability. This section explores some of the most promising future trends and opportunities in the intersection of AI and biotechnology.

1.7.1 Convergence of AI, IoT, and Biotech

The convergence of AI, the Internet of Things (IoT), and biotechnology is one of the most transformative developments in modern science and technology. IoT-enabled devices, such as wearable health monitors, biosensors, and lab equipment, generate vast amounts of real-time biological data. AI's ability to analyze and interpret this data is allowing for unprecedented insights into human health, disease progression, and environmental factors. For example, wearables that continuously monitor vital signs, blood glucose levels, or genetic expression are already integrated with AI algorithms to provide actionable health insights for personalized medicine.

The combination of AI and IoT can further revolutionize the healthcare industry by enabling remote monitoring of patients, predictive analytics for disease outbreaks, and personalized treatment plans that are dynamically adjusted based on real-time data (Bhambri & Khang, 2024). In agriculture, IoT sensors embedded in crops or livestock can collect data on growth conditions, pest invasions, or disease markers. AI models analyzing these data can enable precision farming by recommending optimal intervention strategies that enhance yield and minimize waste. Moreover, the rise of AI-powered, IoT-enabled biomanufacturing processes is expected to usher in the next era of biotech production. AI algorithms can optimize production lines by analyzing sensor data in real time, ensuring higher quality control, reduced waste, and more efficient use of raw materials. This convergence could also have profound implications for sustainability in biotech, as AI-enabled IoT devices can help monitor environmental conditions and reduce the carbon footprint of biotechnological processes.

1.7.2 AI-Enhanced Biotechnology Startups

The growing intersection of AI and biotechnology is spawning a wave of innovative biotechnology startups. These startups are leveraging AI to create cutting-edge products and services that address unmet needs in healthcare, agriculture, and environmental protection. AI-enhanced platforms are being developed to accelerate drug discovery, automate genomic analysis, and design new therapeutic molecules with greater precision than ever before (Moingeon et al., 2022). A notable trend is the rise of AI-driven drug discovery platforms, which use ML algorithms to predict how new compounds will interact with biological systems. Startups such as "In Silico Medicine and Benevolent AI" are pioneering these approaches, demonstrating the capacity of AI to streamline the drug development pipeline and reduce the time and costs associated with bringing a new drug to market. These companies are using AI not only to predict the efficacy of drug candidates but also to design entirely new molecules that might not have been identified through traditional methods.

In the realm of personalized medicine, AI-enhanced startups are creating platforms that use patient-specific data to tailor treatment plans. For instance, AI systems trained on genetic, environmental, and lifestyle data can help identify the most effective therapies for individual patients, thereby advancing the shift from a one-size-fits-all approach to more targeted, individualized healthcare. AI applications in precision oncology, for example, are being used to match cancer patients with the most appropriate treatments based on the genetic profiles of their tumors (Xu et al., 2019). Furthermore, AI is enabling biotech startups in the agricultural sector to address global food security challenges. AI-based tools for crop management, pest detection, and soil monitoring are helping farmers optimize production and reduce reliance on harmful pesticides.. By integrating AI with IoT devices, these startups are driving the next generation of smart farming practices.

1.7.3 The Role of Quantum Computing in Bio-AI

Quantum computing, though still in its early stages, holds tremendous promise for enhancing the capabilities of AI in biotechnology. Traditional computing architectures face limitations when it comes to processing and analyzing the vast amounts of data generated by biotechnological systems, especially in fields such as genomics, molecular dynamics, and drug design. Quantum computing, with its ability to process vast amounts of information simultaneously through qubits, could enable the development of AI models capable of solving complex problems that are currently beyond the reach of classical computers. One of the most exciting opportunities lies in drug discovery. Quantum computing could revolutionize molecular modeling by simulating molecular interactions with much higher precision than is possible with current techniques (Lappala, 2024). This will be particularly valuable in understanding protein folding, designing novel therapeutic agents, and optimizing the interactions between small molecules and biological targets. Companies like IBM and Google are already exploring quantum computing in partnership with pharmaceutical companies to improve drug discovery and development processes.

In genomics, quantum algorithms could expedite the process of sequencing and analyzing large genomic datasets. Current algorithms for genome assembly and alignment can take a significant amount of time and computational resources. Quantum computing could significantly speed up these tasks, enabling more efficient large-scale genomic studies and personalized genetic assessments (Ali, H. 2023a). Moreover, quantum machine learning (QML) is an emerging field that combines quantum computing and AI to process data more efficiently. By harnessing the power of quantum computing to perform faster and more accurate AI training, QML could accelerate the development of more powerful AI models in biotechnology, from predictive disease models to optimization in biomanufacturing. While the full potential of quantum computing in biotechnology is still being explored, the opportunities it presents are vast. As quantum hardware and software continue to mature, it is expected that bio-AI applications will become exponentially more powerful, opening up new avenues for innovation in drug design, genomics, and personalized healthcare.

1.8 Conclusion

The integration of AI into biotechnology has fundamentally transformed the way we approach complex biological problems, leading to groundbreaking advancements in fields such as genomics, drug discovery, personalized medicine, and agriculture. AI's ability to analyze large-scale biological data, recognize intricate patterns, and predict outcomes with remarkable accuracy has accelerated scientific discovery and innovation. As we have explored throughout this review, the intersection of AI and biotechnology has the potential to reshape healthcare, enhance sustainability in agriculture, and drive more efficient biomanufacturing processes.

However, the rapid evolution of AI in biotechnology also introduces challenges that must be addressed to ensure responsible and equitable outcomes. Issues such as data privacy and security, ethical considerations in genetic manipulation, and biases in AI models require careful attention. As AI systems become more embedded in biotechnological applications, it is imperative that researchers, policymakers, and industry leaders work collaboratively to establish transparent, ethical, and robust frameworks to govern their use.

Looking forward, the convergence of AI with emerging technologies such as the IoT, quantum computing, and automated systems offers exciting opportunities for further advancements. The potential for AI to enhance precision medicine, optimize agricultural practices, and revolutionize drug discovery is immense. With continued research, investment, and the development of interdisciplinary collaborations, the future of biotechnology and AI is poised to unlock new frontiers in human health, environmental sustainability, and bioengineering. In conclusion, the fusion of AI and biotechnology is not merely a technological advancement but a paradigm shift that will redefine the boundaries of what is possible in science and medicine. While there are challenges ahead, the vast opportunities for innovation and progress make this intersection one of the most exciting and transformative areas of modern research.

References

Ahmad, S., Bashar, A., Khanna, K., Bano, N., & Raza, K. (2025). Deep learning in variant detection and annotation. In *Deep Learning in Genetics and Genomics* (pp. 383–396). Academic Press.

Ali, H. (2023a). Artificial intelligence in multi-omics data integration: Advancing precision medicine, biomarker discovery and genomic-driven disease interventions. *International Journal of Science and Research Archive*, *8*(1), 1012–1030.

Ali, H. (2023b). Quantum computing and AI in healthcare: Accelerating complex biological simulations, genomic data processing, and drug discovery innovations. *World Journal of Advanced Research and Reviews*, *20*(2), 1466–1484.

Bhambri, P., & Khang, A. (2024). Managing and monitoring patient's healthcare using AI and IoT technologies. In *Driving Smart Medical Diagnosis through AI-Powered Technologies and Applications* (pp. 1–23). IGI Global.

Brunner, D., Legat, C., & Seebacher, U. (2024). Towards next generation data-driven management. In *Collective Intelligence: The Rise of Swarm Systems and their Impact on Society* (p. 152).

Chen, Y. C., Hong, D. J. K., Wu, C. W., & Mupparapu, M. (2019). The use of deep convolutional neural networks in biomedical imaging: A review. *Journal of Orofacial Sciences*, *11*(1), 3–10.

Dixit, S., Kumar, A., Srinivasan, K., Vincent, P. D. R., & Ramu Krishnan, N. (2024). Advancing genome editing with artificial intelligence: Opportunities, challenges, and future directions. *Frontiers in Bioengineering and Biotechnology*, *11*, 1335901.

Gill, S. S., Xu, M., Ottaviani, C., Patros, P., Bahsoon, R., Shaghaghi, A., . . . & Uhlig, S. (2022). AI for next generation computing: Emerging trends and future directions. *Internet of Things*, *19*, 100514.

Glick, B. R., & Patten, C. L. (2022). *Molecular Biotechnology: Principles and Applications of Recombinant DNA*. John Wiley & Sons.

Gupta, A., Kang, K., Pathania, R., Saxton, L., Saucedo, B., Malik, A., . . . & Mayfield, S. P. (2024). Harnessing genetic engineering to drive economic bioproduct production in algae. *Frontiers in Bioengineering and Biotechnology*, *12*, 1350722.

Gupta, R., Srivastava, D., Sahu, M., Tiwari, S., Ambasta, R. K., & Kumar, P. (2021). Artificial intelligence to deep learning: Machine intelligence approach for drug discovery. *Molecular Diversity*, *25*, 1315–1360.

Hassija, V., Chamola, V., Mahapatra, A., Singal, A., Goel, D., Huang, K., . . . & Hussain, A. (2024). Interpreting black-box models: A review on explainable artificial intelligence. *Cognitive Computation, 16*(1), 45–74.

Hussein, A., Prasad, M., & Braytee, A. (2024). Explainable AI methods for multi-omics analysis: A survey. *arXiv Preprint arXiv:2410.11910.*

Kadri, K. (2019). Polymerase chain reaction (PCR): Principle and applications. In *Synthetic Biology-New Interdisciplinary Science* (p. 1–17).

Kalinin, A. A., Higgins, G. A., Reamaroon, N., Soroushmehr, S., Allyn-Feuer, A., Dinov, I. D., . . . & Athey, B. D. (2018). Deep learning in pharmacogenomics: From gene regulation to patient stratification. *Pharmacogenomics, 19*(7), 629–650.

Kathirvel, I., & Gayathri Ganesan, N. (2024). Computational strategies to enhance cell-free protein synthesis efficiency. *BioMedInformatics, 4*(3), 2022–2042.

Kumari, K., Mirzakhani Nafchi, A., Mirzaee, S., & Abdalla, A. (2025). AI-driven future farming: Achieving climate-smart and sustainable agriculture. *AgriEngineering, 7*(3), 89. https://doi.org/10.3390/agriengineering7030089.

Lappala, A. (2024). The next revolution in computational simulations: Harnessing AI and quantum computing in molecular dynamics. *Current Opinion in Structural Biology, 89*, 102919.

Madhavan, M., & Mustafa, S. (2023). Systems biology–the transformative approach to integrate sciences across disciplines: Systems biology: Integrating biological sciences. *Physical Sciences Reviews, 8*(9), 2523–2545.

Madhukar, N. S., Elemento, O., & Pandey, G. (2015). Prediction of genetic interactions using machine learning and network properties. *Frontiers in Bioengineering and Biotechnology, 3*, 172.

Martinelli, D. D. (2022). Generative machine learning for de novo drug discovery: A systematic review. *Computers in Biology and Medicine, 145*, 105403.

Moingeon, P., Kuenemann, M., & Guedj, M. (2022). Artificial intelligence-enhanced drug design and development: Toward a computational precision medicine. *Drug Discovery Today, 27*(1), 215–222.

Naseer, A., Yasir, T., Azhar, A., Shakeel, T., & Zafar, K. (2021). Computer-aided brain tumor diagnosis: Performance evaluation of deep learner CNN using augmented brain MRI. *International Journal of Biomedical Imaging, 2021*(1), 5513500.

Norori, N., Hu, Q., Aellen, F. M., Faraci, F. D., & Tzovara, A. (2021). Addressing bias in big data and AI for health care: A call for open science. *Patterns, 2*(10), 100347.

Phillip, J. M., Han, K. S., Chen, W. C., Wirtz, D., & Wu, P. H. (2021). A robust unsupervised machine-learning method to quantify the morphological heterogeneity of cells and nuclei. *Nature Protocols, 16*(2), 754–774.

Sandu, A., Cotfas, L. A., Stănescu, A., & Delcea, C. (2024). A bibliometric analysis of text mining: Exploring the use of natural language processing in social media research. *Applied Sciences, 14*(8), 3144.

Sarumi, O. A., & Heider, D. (2024). Large language models and their applications in bioinformatics. *Computational and Structural Biotechnology Journal*, 5:23:3498–3505. doi: 10.1016/j.csbj.2024.09.031.

Sharma, S. (2023). Precision agriculture: Reviewing the advancements technologies and applications in precision agriculture for improved crop productivity and resource management. *Reviews in Food and Agriculture, 4*(2), 45–49.

Shetty, S., Khandelwal, A., Gore, M., Behera, R., Shaheeda, A., Bairy, M., & Shama, P. K. (2025). Overview of Artificial Intelligence. In *Artificial Intelligence and Biological Sciences* (pp. 1–25). CRC Press.

Soares, M. (2025). *Symbiotic Relationships in Environments of Change: The Potential of Biological Interactions in Artificial Intelligence*. (Doctoral dissertation, Royal College of Art).

Terry, N. (2017). Existential challenges for healthcare data protection in the United States. *Ethics, Medicine and Public Health, 3*(1), 19–27.

Tiwari, P. C., Pal, R., Chaudhary, M. J., & Nath, R. (2023). Artificial intelligence revolutionizing drug development: Exploring opportunities and challenges. *Drug Development Research, 84*(8), 1652–1663.

Wang, J., Zhang, X., Cheng, L., & Luo, Y. (2020a). An overview and metanalysis of machine and deep learning-based CRISPR gRNA design tools. *RNA Biology*, *17*(1), 13–22.

Wang, L., Wen, Z., Liu, S. W., Zhang, L., Finley, C., Lee, H. J., & Fan, H. J. S. (2024). Overview of AlphaFold2 and breakthroughs in overcoming its limitations. *Computers in Biology and Medicine*, 108620 176:108620. doi: 10.1016/j.compbiomed.2024.108620.

Wang, X., Zhao, Y., & Pourpanah, F. (2020b). Recent advances in deep learning. *International Journal of Machine Learning and Cybernetics*, *11*, 747–750.

Xu, J., Yang, P., Xue, S., Sharma, B., Sanchez-Martin, M., Wang, F., . . . & Parikh, B. (2019). Translating cancer genomics into precision medicine with artificial intelligence: Applications, challenges and future perspectives. *Human Genetics*, *138*(2), 109–124.

Xu, X., Li, J., Zhu, Z., Zhao, L., Wang, H., Song, C., . . . & Pei, Y. (2024). A comprehensive review on synergy of multi-modal data and ai technologies in medical diagnosis. *Bioengineering*, *11*(3), 219.

Youvan, D. C. (2024). AI-Driven Democratization of Academic Publishing: Leveraging Preprints for Equitable Knowledge Sharing.

Kumkum Verma*, Priti Duhan, Poonam Bansal, Anupam Sharma, Anil K. Sharma

Chapter 2 Revolutionizing Genomics with Artificial Intelligence

Abstract: The use of artificial intelligence (AI) in genomics has expedited advances in biological research, clinical diagnostics, and customized treatment. This research looks into how AI-powered technologies, such as machine learning and deep learning, are altering the way genetic data is examined, interpreted, and predicted. Traditional genomic techniques can struggle to deal with the complexity and high dimensionality of biological datasets. In contrast, AI has the ability to uncover hidden patterns, model complex gene interactions, and predict disease risks with unprecedented precision. Variant calling, gene expression profiling, medication development, and epigenetic studies are all possible applications. Furthermore, AI accelerates genome-wide association studies and improves the functional annotation of noncoding regions. Despite its enormous potential, challenges remain in terms of data standards, model interpretability, and ethical concerns concerning genomic privacy. This study critically examines existing AI approaches in genomics, highlighting important discoveries and discussing future possibilities for incorporating AI into genomic science. As data grows exponentially, collaboration between AI and genomics has the potential to provide greater insights into the molecular mechanisms of health and disease.

Keywords: Genomics, artificial intelligence, machine learning, bioinformatics, personalized medicine

***Corresponding author: Kumkum Verma,** ICAR-IARI, Regional Station, Karnal, Haryana, India, e-mail: kumkum.verma89@gmail.com
Priti Duhan, K. L. Mehta Dayanand College for Women, Faridabad, Haryana, India
Poonam Bansal, Department of Biosciences and Technology, MMEC, Maharishi Markandeshwar (Deemed to be University), Mullana, Ambala, Haryana, India
Anupam Sharma, Department of Physics, Guru Kashi University, Talwandi Sabo, Bathinda 151302, Punjab, India
Anil K. Sharma, Department of Biotechnology, Amity University Punjab, Sector 82A, IT City Road, Sahibzada Ajit Singh Nagar 140306, Punjab, India

 | https://doi.org/10.1515/9783112214275-002

2.1 Introduction

Genomics is a branch of molecular biology that investigates the arrangement, functioning, evolution, and sequencing of genomes, which constitute the total set of genetic material in an organism. Since the completion of the Human Genome Project (HGP) in 2003, the field of genomics has progressed from a descriptive discipline to one that promotes innovation in healthcare, agricultural, and environmental research (Collins et al., 2003). Modern genomics now encompasses various subfields, including transcriptomics, epigenomics, and metagenomics, which together generate vast and complex datasets that require advanced computational tools for effective analysis (Goodwin et al., 2016). The development of high-throughput sequencing methods, such as next-generation sequencing (NGS), made it possible for the rapid capture of genomic data at unprecedented sizes and resolutions (Mardis, 2017). However, this exponential growth in data volume presents a dual challenge while it opens up new avenues for biological discovery, it also demands innovative methods to process, interpret, and extract actionable insights from the data. Traditional bioinformatics tools, although powerful, are often limited in their capacity to manage this complexity efficiently and adaptively (Libbrecht & Noble, 2015).

Artificial intelligence (AI), including the use of machine learning (ML) and deep learning (DL), has become a powerful tool for dealing with complex, data-intensive genomics problems. These technologies excel at detecting detailed patterns in large datasets and making data-driven predictions, outperforming traditional rule-based systems. AI is currently being integrated across the genomic workflow. Base calling, sequence alignment, variant calling, and genome annotation are all examples of applications for ML techniques (Angermueller et al., 2016). DL techniques, such as convolutional neural networks (CNNs) and transformer-based models, have been used successfully to forecast gene expression (Zhou & Troyanskaya, 2015), recognize regulatory elements (Kelley et al., 2016), and identify disease-associated variants in noncoding regions (Sundaram et al., 2018).

DeepMind's AlphaFold, which used DL to achieve extraordinary precision in protein structure prediction, is one of the most well-known accomplishments at the junction of AI and biology. This accomplishment not only solved a decades-old structural biology challenge, but also revealed AI's transformative potential for interpreting genomic data and helping downstream applications including drug discovery and synthetic biology. AI integration with genomics is also important for the further development of precision medicine. AI is bridging the disparity between genomic discovery and clinical application by allowing for precise diagnostics, risk classification, and individualized treatment recommendations.

This chapter aims to provide a detailed review of how AI is transforming the area of genetics, to investigate the underlying concepts of AI and their application to genomic data analysis, to investigate specific use cases in which AI has improved the accuracy, efficiency, and scope of genomic research and clinical applications, to examine current issues and limitations of AI in genomics, such as data quality, transparency, and interpretability, to evaluate the ethical and societal ramifications of incorporating AI into genetic science and healthcare, and to propose future directions and emerging phenomena at the confluence of AI and genomics.

2.2 AI and Genomics: A Synergistic Relationship

The union of AI and genomics has created a powerful alliance, catalyzing transformative progress in biological research, precision medicine, and biotechnology. While genomics aims to decipher the blueprint of life encoded within DNA, AI provides the computational sophistication required to analyze, model, and predict the intricate patterns embedded in massive genomic datasets. This synergy has not only accelerated discovery but also expanded the possibilities of what can be achieved in the life sciences.

2.2.1 Historical Context of Genomic Research

The HGP, an international venture to sequence and to map the entire human genome, marked the beginning of the genomics discipline in 1990. The HGP, completed in 2003, was a watershed moment in biology and medicine, serving as a fundamental basis for understanding genetic diversity and illness (Collins et al., 2003). Early genomic research concentrated on gene identification, linkage research, and genome-wide association studies (GWASs), which frequently employed statistical methods and rudimentary bioinformatics tools to evaluate genetic data. Following the HGP, the emergence of NGS technology has accelerated and reduced genome sequencing costs (Mardis, 2008). This democratized access to genomic data, allowing for large-scale research in population genetics, cancer genomics, and customized medicine.

2.2.2 Early Uses of AI in Genomic Studies

The application of AI to genomics began modestly in the early 2000s, primarily through the application of ML algorithms like support vector machines (SVMs), decision trees, and clustering techniques for gene expression analysis and microarray data interpretation (Baldi & Brunak, 2001). These early models helped identify patterns in gene regulation, classify disease subtypes, and predict gene function. As genomic datasets grew in size and complexity, more sophisticated AI approaches were adopted. Neural networks and ensemble methods became popular tools for classifying DNA sequences, detecting gene interactions, and predicting transcription factor binding sites (Libbrecht & Noble, 2015). Despite computational limitations at the time, these methods provided proof-of-concept for AI's potential to extract biological insights from complex datasets. A notable example is the usage of hidden Markov models for gene prediction, which became standard in genome annotation pipelines (Krogh, 1997). These early successes laid the foundation for integration of DL and other advanced AI techniques in later years.

2.2.3 The Need for AI in Handling Big Data in Genomics

Modern genomics functions in an era characterized by "big data." Researchers may now create terabytes of data each experiment using technologies like whole-genome sequencing, single-cell RNA sequencing, and high-throughput epigenomic profiling (Stephens et al., 2015). This massive amount of data has overtaken traditional bioinformatics' analytical powers, making AI not just beneficial but also necessary. AI, particularly DL, offers unique advantages for handling genomic big data, such as scalability, in which AI models can be trained on large datasets and deployed to efficiently process millions of sequences, pattern recognition, in which deep neural networks excel at discovering complex, nonlinear relationships that may be missed by human-designed features or statistical models, and automation, in which AI reduces the need for manual curation and parameter For example, CNNs have been used to predict chromatin accessibility, histone modifications, and transcription factor binding from raw DNA sequences (Kelley et al., 2016). Transformer-based models, such as Enformer, have expanded these capabilities by modeling long-range genomic interactions (Avsec et al., 2021). Moreover, AI has enabled integrative analysis across diverse data types like genomics, transcriptomics, proteomics, and clinical metadata facilitating systems-level understanding of biological processes and disease mechanisms (Eraslan et al., 2019).

2.3 Key AI Technologies in Genomics

AI has surfaced as an indispensable tool in modern genomics, addressing challenges ranging from high-dimensional data interpretation to prediction of functional genomic elements. Several AI technologies, each with distinct capabilities, are now commonly used to extract insights from genomic datasets. These technologies span from conventional ML to cutting-edge DL architectures and generative models.

2.3.1 Machine Learning and Deep Learning in Genomic Data Analysis

ML and DL are central to modern genomic data analysis. They enable automated pattern recognition, prediction, and classification across vast and complex biological datasets, like gene expression profiles, DNA sequences, and epigenomic marks.

2.3.1.1 Machine Learning Applications

- Traditional ML techniques such as SVMs, random forests (RFs), and logistic regression are commonly used for:
- **Gene expression analysis:** ML models help in classifying tissue types, cancer subtypes, and patient outcomes from RNA-seq data (Libbrecht & Noble, 2015).
- **Mutation impact prediction:** PolyPhen and SIFT use ML to assess the functional effect of amino acid changes in protein-coding genes.
- **Genome-wide association studies (GWASs):** ML enhances single-nucleotide polymorphism (SNP) prioritization and phenotype-genotype correlation, especially when dealing with multi-omics data (Botta et al., 2014).

2.3.1.2 Deep Learning Applications

DL, particularly CNNs and recurrent neural networks (RNNs), has advanced the analysis of raw genomic data. These models not only enhance accuracy but also generalize across datasets and discover latent biological features, often outperforming traditional ML in scalability and feature learning.

- **Sequence analysis:** CNNs (e.g., DeepSEA and Basset) can predict functional annotations of noncoding variants, such as enhancers or splicing sites, from raw DNA sequences (Zhou & Troyanskaya, 2015; Kelley et al., 2016).
- **Transcriptomics:** Autoencoders and generative models can reduce dimensionality and denoise single-cell RNA-seq data while preserving biologically meaningful patterns (Lopez et al., 2018).
- **Multi-omics integration:** DL frameworks can handle different data modalities (e.g., DNA methylation and histone marks) to provide integrative insights for disease biology (Eraslan et al., 2019).

2.3.2 Natural Language Processing (NLP) for Genomic Literature Mining

NLP is revolutionizing how researchers extract insights from the vast and rapidly growing body of biomedical literature. By automating text mining, NLP enables real-time synthesis of knowledge across thousands of scientific publications and databases. By making vast textual data machine-readable and searchable, NLP bridges the gap between experimental genomics and the biomedical literature, enabling smarter and faster scientific discovery. Its applications in genomics include the following:

- **Gene-disease association discovery:** NLP tools can automatically identify links between genes and diseases from unstructured literature (Wei et al., 2016).
- **Variant interpretation:** Systems like PubTator and LitVar extract and normalize genomic variant mentions from biomedical abstracts to support clinical variant annotation (Allot et al., 2018).
- **Knowledge base construction:** NLP pipelines power resources like SemMedDB and DisGeNET by mining full-text articles for gene-protein-phenotype interactions.

Recent advances include the following: transformers for biomedical text, which comprise models such as BioBERT and PubMedBERT, trained on biomedical corpora, have significantly improved identification of named entities, relation extraction, and question answering in genomics (Lee et al., 2020), and literature-based hypothesis generation, in which NLP can suggest novel gene functions or therapeutic targets by identifying latent semantic connections across diverse studies.

2.3.3 Reinforcement Learning in Drug Discovery and Genomic Applications

Reinforcement learning (RL) is a new AI paradigm that teaches agents to make decisions by trial and error in a predefined environment. While its usage in genomics is still in its early stages, RL has demonstrated promise in drug discovery and genotype-to-phenotype modeling. Its applications in drug development include: (a) molecular design, where RL agents may generate novel structures of molecules by learning to maximize chemical parameters, like binding affinity and toxicity, in silico (Olivecrona et al., 2017); (b) drug-target interaction prediction in which RL helps model complex interactions between proteins and small molecules, combined with graph neural networks (Zhou et al., 2019); and (c) drug repurposing in which RL is being explored for optimizing repurposing strategies by navigating biological networks or literature-derived knowledge graphs (Xie et al., 2017). Moreover, its applications in genomic applications comprise: (a) gene regulatory network modeling in which RL can model feedback loops and decision processes in gene regulation, simulating the behavior of synthetic gene circuits and (b) experiment planning in which RL algorithms have been used to prioritize which genomic experiments to run next, maximizing information gain while minimizing cost (Angermueller et al., 2020).

Although still in its early phases, RL has great potential for difficult decision-making tasks in genomics and personalized medicine, particularly when combined with other AI techniques.

2.4 Applications of AI in Genomics

AI has become an essential tool in genomic research, providing powerful ways for processing, analyzing, and interpreting the massive amounts of data produced by modern sequencing technology. From basic science to clinical applications, AI has made substantial contributions to our understanding of genome function, variation, and disease association (Figure 2.1).

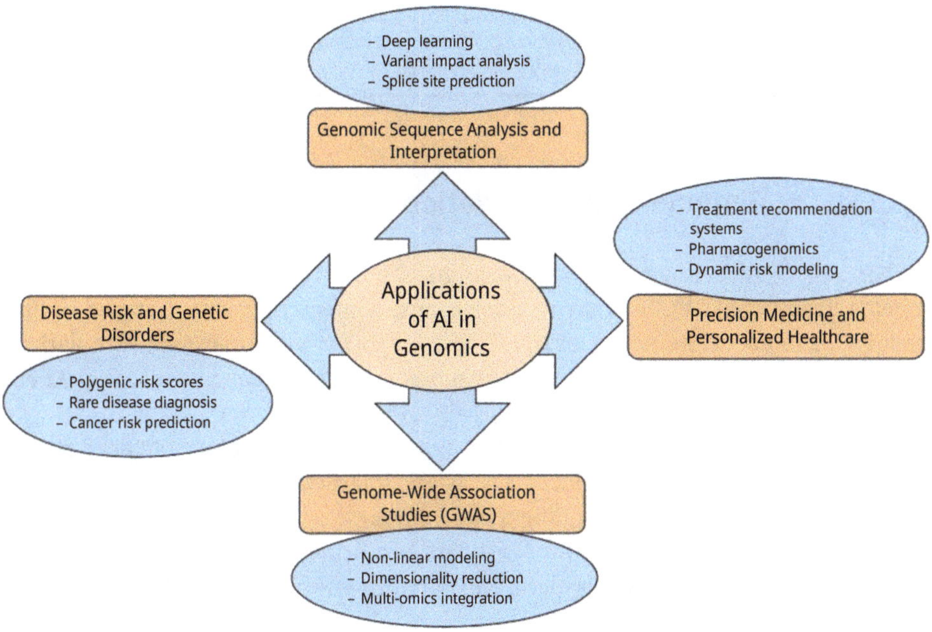

Figure 2.1: Artificial intelligence techniques used to analyze and interpret large-scale genomic data, enabling variant detection, functional annotation and disease prediction for precision medicine.

2.4.1 Genomic Sequence Analysis and Interpretation

The interpretation of raw genomic sequences, millions to billions of base pairs, requires sophisticated computational tools capable of discerning functional elements, variants, and regulatory structures. AI, particularly DL, is instrumental in automating and enhancing this process. By learning biologically meaningful patterns from large training datasets, AI models provide scalable, high-throughput solutions for genome annotation and variant interpretation as represented in Table 2.1.

Table 2.1: Applications of AI in genomics.

Application area	Key AI methods/tools	References
Deep learning for sequence prediction	DeepSEA and Enformer (CNNs and transformers for predicting chromatin features and gene expression)	Zhou and Troyanskaya (2015) and Avsec et al. (2021)
Variant impact analysis	Supervised learning using conservation, protein structure, and epigenetic data	Quang et al. (2014)
Splice site prediction	Recurrent neural networks (RNNs) for alternative splicing and cryptic splice site detection	Jaganathan et al. (2019)

2.4.2 AI for Genome-Wide Association Studies (GWASs)

GWASs aim to uncover genetic variations linked to attributes or diseases across the genome. Traditional GWASs often overlook complex interactions due to statistical limitations. To improve the power of GWAS to identify both common and rare variant associations, contributing to our understanding of complex traits and diseases, AI enhances this in the following ways:

- **Nonlinear modeling:** ML methods like as gradient-boosted trees and deep neural networks can discover complicated, epistatic interactions between SNPs that traditional GWAS may overlook (Mieth et al., 2016).
- **Dimensionality reduction:** Autoencoders and principal component-based models help to minimize feature space, which improves computational efficiency and interpretability.
- **Multi-omics integration:** AI models can combine GWAS data with transcriptomic and epigenomic profiles to prioritize causal variants and candidate genes (Chen et al., 2015).

2.4.3 Predictive Models for Disease Risk and Genetic Disorders

Among the most important uses of AI in genomics is disease risk prediction based on genetic information, which allows for early intervention and informed clinical decisions. Following predictive tools are increasingly being adopted in clinical workflows to screen at-risk individuals, inform genetic counseling, and guide preventive strategies represented in Table 2.2.

Table 2.2: Application of genomics in predictive models for disease risk.

Application area	Key AI methods/tools	References
Polygenic risk scores (PRSs)	Penalized regression, ensemble learning for variant weighting	Khera et al. (2018)
Rare disease diagnosis	Deep learning to find de novo or compound heterozygous mutations	Smedley et al. (2015)
Cancer risk prediction	Integration of germline, somatic variants, and family history via AI	Kourou et al. (2015)

2.4.4 AI in Precision Medicine and Personalized Healthcare

Precision medicine seeks to customize treatment and preventive methods to individuals' genetic, environmental, and lifestyle characteristics. AI is a major enabler of this paradigm because it can make sense of vast amounts of data from many sources. Furthermore, AI advancements are hastening the transition from one-size-fits-all healthcare to tailored prevention, diagnosis, and treatment based on an individual's genetic makeup:

- AI-powered platforms, such as Watson for Genomics, use genomic variants, clinical data, and biomedical literature to identify tailored treatments for cancer and rare disorders (Ferrucci, 2012).
- Pharmacogenomics involves using ML to anticipate patient responses to medications based on genetics, allowing for dose modifications and preventing side events (Relling & Evans, 2015).
- Dynamic risk modeling: AI integrates longitudinal genomic and clinical data to update disease risk profiles in real time, a step toward truly adaptive medicine.

2.5 AI-Driven Tools and Platforms in Genomics

The exponential growth in genomic information has necessitated the modification of AI-enhanced tools and platforms that automate and improve accuracy in sequence analysis, variant calling, bioinformatics, and genome editing. These tools are integral in transforming raw genomic information into actionable biological insights and clinical applications. These tools are vital for gene therapy, functional genomics, and synthetic biology, where safety, precision, and efficiency are critical. AI-driven tools and platforms enable rapid, accurate, and cost-effective genomic analysis and manipulation. From variant calling to genome editing, AI empowers researchers with new levels of precision and scalability that are vital for advancing biology and medicine.

2.5.1 Genome Sequencing and Variant Calling

NGS technologies generate vast amount of raw sequence data. Accurate and quick detection of genetic variations is crucial for research and clinical use. DeepVariant (created by Google) employs DL, specifically CNNs, to accurately identify SNPs and indels, outperforming standard variant callers such as GATK (Poplin et al., 2018). Clair3 and PEPPER-Margin-DeepVariant enhance variant discovery in long-read sequencing data (e.g., PacBio, Oxford Nanopore) by incorporating multiple neural networks into a variant-calling process (Shafin et al., 2021). AI models can also reduce sequencing errors, detect somatic mutations in cancer, and handle low-depth or noisy datasets through probabilistic learning and signal enhancement (Luo et al., 2019). These tools allow for fast, scalable, and highly accurate genome interpretation across species and clinical contexts.

2.5.2 Bioinformatics Software and AI-Enhanced Algorithms

AI-enhanced bioinformatics platforms are essential for processing, analyzing, and visualizing multi-dimensional genomic data. These tools improve pattern recognition, automate data interpretation, and facilitate complex model building. DeepMind's AI algorithm, AlphaFold, revolutionized protein structure prediction by predicting 3D protein structures from amino acid sequences with near-experimental accuracy (Jumper et al., 2021). While not exactly genomic, it enhances genome annotation and variation effect prediction. DeepSEA and Basset employ DL to predict chromatin accessibility, transcription factor binding, and variant effects based on genomic sequences (Zhou and Troyanskaya, 2015; Kelley et al., 2016). Gene network inference methods use AI to detect regulatory links in gene expression data, including Bayesian networks, decision trees, and DL to discover gene-gene interactions. These platforms are frequently connected with cloud-based pipelines (such as Terra, DNAnexus, and Seven Bridges), which enable large-scale genomics and biomedical data processing.

2.5.3 AI in CRISPR and Gene Editing Technologies

CRISPR-Cas systems are potent genome editing tools, and AI is rapidly being utilized to optimize guide RNA design, forecast off-target effects, and improve editing results. CNNs and RNNs are used by DeepCRISPR, CRISPR-Net, and DeepSpCas9 to predict on-target efficiency and off-target binding of guide RNAs, allowing for the creation of extremely targeted and successful modifications (Chuai et al., 2018; Wang et al., 2019). sgRNA Designer and CRISPR integrate ML models with sequence and structural features to guide CRISPR experiments. In case of AI-based multiplex editing, advanced models are being used to design multi-target strategies and minimize unintended edits in multiplex CRISPR applications (Kim et al., 2016).

2.6 AI in Drug Discovery and Development

The incorporation of AI into drug discovery is reshaping the pharmaceutical industry, particularly when paired with genomic data. AI improves our ability to find therapeutic targets, develop biomarkers, and simulate drug-genome interactions at unprecedented speeds and precision. AI is changing drug discovery by making it faster, cheaper, and more precise, particularly when combined with genomic data. From discovering promising drug targets to modeling complicated drug-genome interactions, AI is ushering in a new era of data-driven pharmacology and individualized treatment plans (Figure 2.2).

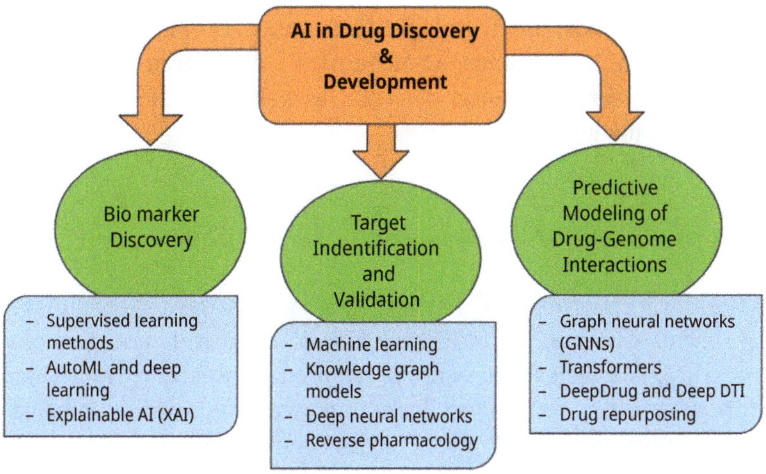

Figure 2.2: AI accelerates helps in biomarker discovery, target identification, lead optimization, and clinical decision-making by analyzing complex biological data and predictive modeling of drug- genome interactions.

2.6.1 Target Identification and Validation

Identifying and verifying the appropriate molecular targets is a critical stage in medication development. Traditional procedures are time-consuming and susceptible to failure in later stages. This technique is considerably improved by AI, namely ML, which analyzes multi-omics and biomedical literature data. To select novel targets, knowledge graph models such as BenevolentAI and IBM Watson for Drug Discovery search enormous databases of biological literature and genetics (Zeng et al., 2019). Deep neural networks can use gene expression, proteomics, and mutation data to identify disease-causing genes and pathways (Chen et al., 2016). AI is also used in reverse pharmacology, where algorithms infer potential targets from drug activity profiles in phenotypic screens. These tools dramatically reduce the target validation timeline and increase the likelihood of clinical success.

2.6.2 AI for Biomarker Discovery

Biomarkers are critical in diagnostics, treatment response prediction, and patient stratification. AI-driven approaches are enabling the identification of genomic, transcriptomic, and proteomic biomarkers that would be otherwise overlooked by traditional analyses. Supervised learning methods (RFs and SVMs) are widely used to classify disease states on the base of genomic and transcriptomic signatures (Xu et al., 2015). AutoML and DL enable biomarker discovery across high-dimensional datasets, identifying combinations of molecular features predictive of therapeutic outcomes (Libbrecht & Noble, 2015). Explainable AI is increasingly used to interpret biomarker models, helping researchers understand why specific features are relevant, which is critical for clinical adoption. For instance, AI models have discovered blood-based gene expression biomarkers for early cancer detection, as well as immunotherapy response predictors in genomics-guided oncology.

2.6.3 Predictive Modeling of Drug-Genome Interactions

AI aids in simulating the complicated interaction between genetic variants and therapeutic efficacy or toxicity, which is critical for precision pharmacogenomics and customized medicine. GNNs and transformers are being used for analyzing molecular structures, gene-drug networks, and protein-ligand interactions, providing information about efficacy and adverse effects (Yin et al., 2023). DeepDrug and DeepDTI are AI frameworks that predict drug-target interactions using genomic, transcriptomic, and structural data (You et al., 2019; Öztürk et al., 2018). AI is also applied in drug repurposing, predicting new indications for existing drugs by analyzing genomic signatures and disease-related pathways (Zhou et al., 2020). These predictive models not only enhance therapeutic efficacy but also support drug safety by identifying at-risk genotypes early in development.

2.7 Challenges and Limitations

Despite transformative potential of AI in genomics, several challenges and limitations hinder its widespread adoption and impact. These issues span from data quality concerns to technical, ethical, and infrastructural barriers.

2.7.1 Data Quality and Availability in Genomics

High-quality, diversified, and thoroughly annotated genomic datasets are essential for developing strong AI models. However, many available datasets suffer from inconsistencies, missing metadata, and lack of standardization. This is particularly true in clinical genomics, where privacy concerns often limit data sharing and accessibility across institutions and populations (Erlich & Narayanan, 2014). Furthermore, underrepresentation of diverse populations in genomic datasets introduces bias into AI models, which may lead to inaccurate predictions and health disparities (Popejoy & Fullerton, 2016).

2.7.2 Ethical Considerations in AI-Driven Genomic Research

AI applications in genomics raise pressing ethical questions. These include data ownership, consent for secondary use, the right to explanation of algorithmic decisions, and potential misuse of genetic data for nonmedical purposes. There is also concern that commercial interests could override patient rights or lead to exploitation of vulnerable populations. AI integration affects informed consent processes since patients may not fully comprehend how their genetic data will be utilized, particularly for AI model training and deployment (Shen et al., 2019).

2.7.3 Interpretability and Transparency of AI Models

Many AI models, in particular DL systems, are frequently derided as “black boxes” because of their complicated and opaque decision-making processes. This lack of interpretability creates a significant hurdle to clinical adoption, as explainability is required for trust, validation, and regulatory approval (Doshi-Velez & Kim, 2017). Without clear explanations, clinicians may be hesitant to act on AI-derived insights, even if they outperform traditional methods.

2.7.4 Integration with Existing Genomic Infrastructure

Most genomic analysis pipelines and healthcare IT systems are not designed to accommodate AI models. Integration challenges include differences in data formats, insufficient computational infrastructure, and lack of interoperability standards. Moreover, regulatory and data governance frameworks often lag behind technological advancements, complicating the execution of AI tools in clinical and research settings (Topol, 2019a). Overcoming these infrastructural barriers requires coordinated efforts among bioinformaticians, clinicians, data scientists, and policymakers.

2.8 Future Directions and Innovations

The convergence of AI and genomics is accelerating, offering unprecedented opportunities to transform biomedical research and healthcare. As the field matures, several emerging technologies, applications, and collaborative efforts are poised to reshape the future landscape.

2.8.1 Emerging AI Technologies for Genomics

Next-generation AI techniques, including as transformer-based architectures, graph neural networks and federated learning, are emerging as effective genomics tools. Transformers, which were originally designed for natural language processing, have recently been repurposed to assess genetic sequences with incredible precision (Ji et al., 2021). GNNs enable modeling of complex biological networks, including gene regulation and protein-protein interactions, providing insights into cellular processes. Federated learning offers a decentralized approach to model training across institutions, preserving privacy while expanding data access and model generalizability (Li et al., 2020).

2.8.2 AI in Single-Cell Genomics and Multi-Omics Integration

The emergence of single-cell sequencing technologies has created the need for AI methods capable of analyzing high-dimensional, sparse, and noisy data at cellular resolution. DL is now routinely applied to identify cell types, trajectories, and regulatory networks from single-cell transcriptomics (Lopez et al., 2018). Simultaneously, AI enables multi-omics integration by merging genomes, transcriptomics, proteomics, epigenomics, and metabolomics to provide holistic biological insights. This integrative approach is vital for unraveling complex diseases and identifying biomarkers that span molecular layers.

2.8.3 Collaborative Efforts: AI, Genomics, and Global Health Initiatives

Global consortia and interdisciplinary partnerships are essential to assure inclusive, ethical, and impactful development of AI in genomics. Initiatives such as the Human Cell Atlas, Global Alliance for Genomics and Health, and All of Us Research Program are incorporating AI to analyze vast, diverse datasets while prioritizing transparency, equity, and open science. Cross-sector collaborations, among academia, industry, governments, and nonprofits, are helping set data-sharing standards, ethical guidelines, and infrastructure support (Birney et al., 2017).

2.8.4 Prospects for AI in Revolutionizing Precision Medicine

AI is critical to the implementation of precision medicine, providing tools for tailored diagnostics, prognostics, and therapies. AI can produce personalized treatment plans by combining patient-specific genomic data with electronic health records, environmental factors, and lifestyle data. AI-powered real-time decision support tools are also emerging in clinical genomics, allowing for faster diagnosis and treatment optimization (Topol, 2019b). The ultimate goal is to move from reactive care to predictive, preventive, and participatory healthcare.

2.9 Conclusion

AI has quickly become an essential tool in genomics, changing the way we perceive and act on biological data. Its contributions span across the genomic research pipeline which include data processing and analysis, genomic interpretation, biomarker and drug target discovery, precision medicine, gene editing and CRISPR, literature mining, and knowledge discovery. The synergy between AI and genomics is only beginning to reach its full potential. Emerging innovations such as foundation models for DNA, real-time AI-assisted clinical genomics, and integrative multi-omics platforms promise to deepen our understanding of biology and disease. To realize the full potential of AI in genomics, continual attention must be devoted to data quality, ethical use, model transparency, and fair access. Interdisciplinary collaboration, inclusive data initiatives, and strong regulatory frameworks will be critical in managing these problems. Looking forward, AI will not only continue to revolutionize genomic science but also become a cornerstone of precision health, ultimately improving patient outcomes and global health equity.

References

Allot, A., Peng, Y., Wei, C. H., Lee, K., Phan, L., & Lu, Z. (2018). LitVar: A semantic search engine for linking genomic variant data in PubMed and PMC. *Nucleic Acids Research, 46*(W1), W530–W536. https://doi.org/10.1093/nar/gky355.

Angermueller, C., Pärnamaa, T., Parts, L., & Stegle, O. (2016). Deep learning for computational biology. *Molecular Systems Biology, 12*(7), 878. https://doi.org/10.15252/msb.20156651.

Angermueller, C., Belanger, D., Gane, A., Mariet, Z., Dohan, D., Murphy, K., . . . & Sculley, D. (2020, November). Population-based black-box optimization for biological sequence design. In International conference on machine learning (pp. 324–334). PMLR.

Avsec, Ž., Agarwal, V., Visentin, D., Ledsam, J. R., Grabska-Barwinska, A., Taylor, K. R., . . . Kelley, D. R. (2021). Effective gene expression prediction from sequence by integrating long-range interactions. *Nature Methods, 18*(10), 1196–1203. https://doi.org/10.1038/s41592-021-01252-x.

Baldi, P., & Brunak, S. (2001). *Bioinformatics: The Machine Learning Approach.* MIT Press.

Birney, E., Vamathevan, J., & Goodhand, P. (2017). Genomics in healthcare: GA4GH looks to 2022. *BioRxiv*, 203554.

Botta, V., Louppe, G., Geurts, P., & Wehenkel, L. (2014). Exploiting SNP correlations within random forest for genome-wide association studies. PloS one, 9(4), e93379.

Chen, L., Cai, C., Chen, V., & Lu, X. (2015). Trans-species learning of cellular signaling systems with bimodal deep belief networks. *Bioinformatics, 31*(18), 3008–3015.

Chen, Y., Li, Y., Narayan, R., Subramanian, A., & Xie, X. (2016). Gene expression inference with deep learning. *Bioinformatics, 32*(12), 1832–1839.

Chuai, G., Ma, H., Yan, J., Chen, M., Hong, N., Xue, D., . . . & Liu, Q. (2018). DeepCRISPR: Optimized CRISPR guide RNA design by deep learning. *Genome Biology, 19*, 1–18.

Collins, F. S., Morgan, M., & Patrinos, A. (2003). The human genome project: Lessons from large-scale biology. *Science, 300*(5617), 286–290. https://doi.org/10.1126/science.1084564.

Doshi-Velez, F., & Kim, B. (2017). Towards a rigorous science of interpretable machine learning. *arXiv Preprint arXiv:1702.08608*.

Eraslan, G., Avsec, Ž., Gagneur, J., & Theis, F. J. (2019). Deep learning: New computational modelling techniques for genomics. *Nature Reviews Genetics, 20*(7), 389–403.

Erlich, Y., & Narayanan, A. (2014). Routes for breaching and protecting genetic privacy. *Nature Reviews Genetics, 15*(6), 409–421.

Ferrucci, D. A. (2012). Introduction to "this is watson". *IBM Journal of Research and Development, 56*(3.4), 1–1.

Goodwin, S., McPherson, J. D., & McCombie, W. R. (2016). Coming of age: Ten years of next-generation sequencing technologies. *Nature Reviews Genetics, 17*(6), 333–351.

Jaganathan, K., Panagiotopoulou, S. K., McRae, J. F., Darbandi, S. F., Knowles, D., Li, Y. I., . . . Farh, K. K. H. (2019). Predicting splicing from primary sequence with deep learning. *Cell, 176*(3), 535–548.

Ji, Y., Zhou, Z., Liu, H., & Davuluri, R. V. (2021). DNABERT: Pre-trained bidirectional encoder representations from transformers model for DNA-language in genome. *Bioinformatics, 37*(15), 2112–2120.

Jumper, J., Evans, R., Pritzel, A., Green, T., Figurnov, M., Ronneberger, O., . . . & Hassabis, D. (2021). Highly accurate protein structure prediction with AlphaFold. *Nature, 596*(7873), 583–589.

Kelley, D. R., Snoek, J., & Rinn, J. L. (2016). Basset: Learning the regulatory code of the accessible genome with deep convolutional neural networks. *Genome Research, 26*(7), 990–999.

Khera, A. V., Chaffin, M., Aragam, K. G., Haas, M. E., Roselli, C., Choi, S. H., . . . Kathiresan, S. (2018). Genome-wide polygenic scores for common diseases identify individuals with risk equivalent to monogenic mutations. *Nature Genetics, 50*(9), 1219–1224.

Kim, D., Kim, S., Kim, S., Park, J., & Kim, J. S. (2016). Genome-wide target specificities of CRISPR-Cas9 nucleases revealed by multiplex digenome-seq. *Genome Research, 26*(3), 406–415.

Kourou, K., Exarchos, T. P., Exarchos, K. P., Karamouzis, M. V., & Fotiadis, D. I. (2015). Machine learning applications in cancer prognosis and prediction. *Computational and Structural Biotechnology Journal, 13*, 8–17.

Krogh, A. (1997). Two methods for improving performance of an HMM and their application for gene finding. *Center for Biological Sequence Analysis. Phone, 45*, 4525.

Lee, J., Yoon, W., Kim, S., Kim, D., Kim, S., So, C. H., & Kang, J. (2020). BioBERT: A pre-trained biomedical language representation model for biomedical text mining. *Bioinformatics, 36*(4), 1234–1240.

Li, T., Sahu, A. K., Talwalkar, A., & Smith, V. (2020). Federated learning: Challenges, methods, and future directions. *IEEE Signal Processing Magazine, 37*(3), 50–60.

Libbrecht, M. W., & Noble, W. S. (2015). Machine learning applications in genetics and genomics. *Nature Reviews Genetics, 16*(6), 321–332.

Lopez, R., Regier, J., Cole, M. B., Jordan, M. I., & Yosef, N. (2018). Deep generative modeling for single-cell transcriptomics. *Nature Methods, 15*(12), 1053–1058.

Luo, R., Sedlazeck, F. J., Lam, T. W., & Schatz, M. C. (2019). A multi-task convolutional deep neural network for variant calling in single molecule sequencing. *Nature Communications, 10*(1), 998.

Mardis, E. R. (2008). Next-generation DNA sequencing methods. *Annual Review of Genomics and Human Genetics, 9*(1), 387–402. https://doi.org/10.1146/annurev.genom.9.081307.164359.

Mardis, E. R. (2017). DNA sequencing technologies: 2006–2016. *Nature Protocols, 12*(2), 213–218.

Mieth, B., Kloft, M., Rodríguez, J. A., Sonnenburg, S., Vobruba, R., Morcillo-Suárez, C., . . . Müller, K. R. (2016). Combining multiple hypothesis testing with machine learning increases the statistical power of genome-wide association studies. *Scientific Reports, 6*(1), 36671.

Olivecrona, M., Blaschke, T., Engkvist, O., & Chen, H. (2017). Molecular de-novo design through deep reinforcement learning. *Journal of Cheminformatics, 9*, 1–14.

Öztürk, H., Özgür, A., & Ozkirimli, E. (2018). DeepDTA: Deep drug–target binding affinity prediction. *Bioinformatics, 34*(17), i821–i829.

Popejoy, A. B., & Fullerton, S. M. (2016). Genomics is failing on diversity. *Nature, 538*(7624), 161–164.

Poplin, R., Chang, P. C., Alexander, D., Schwartz, S., Colthurst, T., Ku, A., . . . & DePristo, M. A. (2018). A universal SNP and small-indel variant caller using deep neural networks. *Nature Biotechnology, 36*(10), 983–987.

Quang, D., Chen, Y., & Xie, X. (2014). DANN: A deep learning approach for annotating the pathogenicity of genetic variants. *Bioinformatics, 31*(5), 761–763.

Relling, M. V., & Evans, W. E. (2015). Pharmacogenomics in the clinic. *Nature, 526*(7573), 343–350.

Shafin, K., Pesout, T., Chang, P. C., Nattestad, M., Kolesnikov, A., Goel, S., . . . & Paten, B. (2021). Haplotype-aware variant calling with PEPPER-margin-deepVariant enables high accuracy in nanopore long-reads. *Nature Methods, 18*(11), 1322–1332.

Shen, J., Zhang, C. J., Jiang, B., Chen, J., Song, J., Liu, Z., . . . & Ming, W. K. (2019). Artificial intelligence versus clinicians in disease diagnosis: Systematic review. *JMIR Medical Informatics, 7*(3), e10010.

Smedley, D., Jacobsen, J. O., Jäger, M., Köhler, S., Holtgrewe, M., Schubach, M., . . . & Robinson, P. N. (2015). Next-generation diagnostics and disease-gene discovery with the exomiser. *Nature Protocols, 10*(12), 2004–2015.

Stephens, Z. D., Lee, S. Y., Faghri, F., Campbell, R. H., Zhai, C., Efron, M. J., . . . & Robinson, G. E. (2015). Big data: Astronomical or genomical?. *PLoS Biology*, *13*(7), e1002195.

Sundaram, L., Gao, H., Padigepati, S. R., McRae, J. F., Li, Y., Kosmicki, J. A., . . . & Farh, K. K. H. (2018). Predicting the clinical impact of human mutation with deep neural networks. *Nature Genetics*, *50*(8), 1161–1170.

Topol, E. (2019a). *Deep Medicine: How Artificial Intelligence Can Make Healthcare Human Again*. Hachette UK.

Topol, E. J. (2019b). High-performance medicine: The convergence of human and artificial intelligence. *Nature Medicine*, *25*(1), 44–56.

Wang, D., Zhang, C., Wang, B., Li, B., Wang, Q., Liu, D., . . . & Wang, Y. (2019). Optimized CRISPR guide RNA design for two high-fidelity Cas9 variants by deep learning. *Nature Communications*, *10*(1), 4284.

Wei, C. H., Peng, Y., Leaman, R., Davis, A. P., Mattingly, C. J., Li, J., . . . & Lu, Z. (2016). Assessing the state of the art in biomedical relation extraction: Overview of the BioCreative V chemical-disease relation (CDR) task. *Database*, *2016*, baw032.

Xie, L., Draizen, E. J., & Bourne, P. E. (2017). Harnessing big data for systems pharmacology. *Annual Review of Pharmacology and Toxicology*, *57*(1), 245–262.

Xu, Y., Dai, Z., Chen, F., Gao, S., Pei, J., & Lai, L. (2015). Deep learning for drug-induced liver injury. *Journal of Chemical Information and Modeling*, *55*(10), 2085–2093.

Yin, Q., Fan, R., Cao, X., Liu, Q., Jiang, R., & Zeng, W. (2023). DeepDrug: A general graph-based deep learning framework for drug-drug interactions and drug-target interactions prediction. *Quantitative Biology*, *11*(3), 260–274.

You, J., McLeod, R. D., & Hu, P. (2019). Predicting drug-target interaction network using deep learning model. *Computational Biology and Chemistry*, *80*, 90–101.

Zeng, X., Zhu, S., Liu, X., Zhou, Y., Nussinov, R., & Cheng, F. (2019). deepDR: A network-based deep learning approach to in silico drug repositioning. *Bioinformatics*, *35*(24), 5191–5198.

Zhou, J., & Troyanskaya, O. G. (2015). Predicting effects of noncoding variants with deep learning–based sequence model. *Nature Methods*, *12*(10), 931–934.

Zhou, Y., Hou, Y., Shen, J., Huang, Y., Martin, W., & Cheng, F. (2020). Network-based drug repurposing for novel coronavirus 2019-nCoV/SARS-CoV-2. *Cell Discovery*, *6*(1), 14.

Zhou, Z., Kearnes, S., Li, L., Zare, R. N., & Riley, P. (2019). Optimization of molecules via deep reinforcement learning. *Scientific Reports*, *9*(1), 10752.

Priti Duhan*, Kumkum Verma, Poonam Bansal, Anupam Sharma,
Anil Kumar Sharma, Bhupesh Gupta

Chapter 3
Unlocking Proteomics and Metabolomics Through AI

Abstract: Artificial intelligence (AI) is rapidly transforming the landscape of proteomics and metabolomics by permitting deeper, quicker, and more precise interpretation of complex molecular data. This review highlights how current developments in machine learning, deep neural networks, and generative modeling are redesigning the discovery pipeline from data acquisition and noise lessening to feature extraction, biomarker identification, and biological explanation. In proteomics, AI-driven algorithms now expand peptide-spectrum matching, enhance post-translational modification detection and support large-scale protein network reconstruction. Similarly, metabolomics profits from automated spectral annotation, refined compound identification, and predictive modeling of metabolic pathways. Across both fields, AI facilitates integration of multi-omics datasets, revealing functional relationships that were previously obscured by data heterogeneity and scale. Despite impressive progress, challenges remain, including limited training datasets, variable experimental standards, and the need for transparent, interpretable models. By synthesizing emerging methodologies and applications, this review underscores AI's growing role in accelerating biomarker discovery, refining disease classification, and advancing personalized medicine, while outlining future prospects for responsible and collaborative development in molecular sciences.

Keywords: Multi-omics, proteomics, AI, metabolomics, biomarkers

***Corresponding author: Priti Duhan,** K. L. Mehta Dayanand College for Women, Faridabad, Haryana, India, e-mail: pduhanbiotech@kuk.ac.in
Kumkum Verma, ICAR-IARI, Regional Station, Karnal, Haryana, India
Poonam Bansal, Department of Biosciences and Technology, MMEC, Maharishi Markandeshwar (Deemed to be University), Mullana, Ambala, Haryana, India
Anupam Sharma, Department of Physics, Guru Kashi University, Talwandi Sabo, Bathinda 151302, Punjab, India
Anil Kumar Sharma, Department of Biotechnology, Amity University Punjab, Sector 82A, IT City Road, Sahibzada Ajit Singh Nagar 140306, Punjab, India
Bhupesh Gupta, Department of Computer Science and Engineering, MMEC, Maharishi Markandeshwar (Deemed to be University), Mullana, Ambala, Haryana, India

 | https://doi.org/10.1515/9783112214275-003

3.1 Introduction

The rapid improvement in the field of biological sciences has accompanied an era of omics technologies, revolutionizing the way we study complex biological systems. Among such advancements, proteomics and metabolomics have appeared as prevailing disciplines that provide comprehensive perceptions into the protein complement and metabolic outline of a cell, tissue, or organism. These fields are crucial for understanding disease mechanisms, identifying biomarkers, and developing targeted therapeutic strategies. Recent developments in computational biology and data science have further accelerated progress in proteomics and metabolomics. As data generation becomes more efficient and voluminous, the integration of artificial intelligence (AI) has become indispensable. AI enhances the accuracy and speed of data analysis and also, uncovers patterns and relationships that are often imperceptible to traditional analytical methods. Proteomics and metabolomics are integral components of systems biology, providing comprehensive insights into the functional outputs of cellular processes. Proteomics emphases the large-scale study of proteins, including their expression, structures, functions, and interactions while metabolomics pertains to the systematic examination of small molecule and metabolites within a biological system (Patti et al., 2012; Aebersold & Mann, 2016). Together, these fields offer a dynamic view of cellular physiology, reflecting real-time biological activity and phenotypic changes. Technological advancements such as high-resolution mass spectrometry (MS), liquid chromatography (LC), and nuclear magnetic resonance (NMR) spectroscopy have significantly propelled these disciplines. However, the complication and volume of omics data demand innovative computational approaches for meaningful interpretation. AI encircling machine learning (ML), deep learning (DL), and natural language processing (NLP) has emerged as a transformative force in proteomics and metabolomics. AI tools can process high-dimensional datasets, uncover patterns, predict molecular interactions, and facilitate biomarker discovery with unmatched speed and accuracy (Libiseller et al., 2015; Webb-Robertson et al., 2015). In proteomics, AI supports protein structure prediction, post-translational modification (PTM) analysis, and protein-protein interaction (PPI) mapping. In metabolomics, AI assists in metabolite identification, spectral deconvolution, and pathway enrichment analysis (Chi et al., 2024; Grapov, 2018). Recent breakthroughs, such as AlphaFold for structure of protein prediction (Jumper et al., 2021) and profound learning models for metabolite annotation, highlight the growing influence of AI in solving long-standing challenges in life sciences.

This chapter explores the foundational concepts of proteomics and metabolomics, examines how AI is transforming these fields, and outlines the key objectives that guide this discussion. It will introduce the foundational concepts and technological advancements in proteomics and metabolomics. Moreover, it emphasizes on the current limitations, ethical deliberations, and the future potential of AI in advancing precision biology through proteomics and metabolomics.

3.2 Foundations of Proteomics and Metabolomics

Proteomics and metabolomics are two core pillars of systems biology that aim to capture the dynamic biochemical landscape of cells, tissues, and organisms. Proteomics deals with the large-scale study of the proteome (a complete set of proteins expressed in a biological system at a given time). Proteins are key effectors of cellular function, and their expression, modification (e.g., phosphorylation), and interactions provide insights into disease mechanisms, signaling pathways, and therapeutic targets. MS coupled with techniques like LC-MS/MS, remains the central analytical tool in proteomics. Metabolomics emphasizes the systematic profiling of metabolites, small molecules such as amino acids, lipids, sugars, and nucleotides that represent the end products of cellular metabolism. Metabolomics provides a real-time snapshot of physiological or pathological states and is often considered the most "downstream" omic layer. Analytical platforms include NMR spectroscopy and MS, often coupled with gas chromatography (GC) or LC.

3.2.1 Definitions and Scope of Proteomics

Proteomics refers to the large-scale, systematic study of the entire protein complement (proteome) expressed by a genome, cell, tissue, or organism under specific conditions. This discipline aims to characterize protein structures, functions, modifications, and interactions (Wilkins et al., 1996; Aebersold & Mann, 2016). In contrast to relatively static genome, the proteome is highly dynamic and responsive to physiological stimuli and environmental cues. Modern proteomics is often subdivided into various branches: (a) structural proteomics, which focuses on 3D structure determination; (b) functional proteomics, examining PPIs and pathways; and (c) clinical proteomics, geared toward biomarker discovery and disease diagnostics. Technological advancements, particularly in MS and LC, have enabled the high-throughput identification and quantification of thousands of proteins in a single experiment (Zhang et al., 2013).

3.2.2 Key Concepts in Metabolomics

Metabolomics is the wide-ranging analysis of metabolites (small-molecular-weight intermediates and products of metabolism) within a biological system. It offers a real-time portrait of the physiological state of a cell or organism and reflects both genetic and environmental influences (Patti et al., 2012; Nicholson et al., 1999). Metabolomics workflows generally involve (a) sample preparation and extraction, (b) analytical separation and detection, using MS or NMR spectroscopy, and (c) data preprocessing, normalization, and statistical analysis. Moreover, the two major approaches to metabolomics are (a) targeted metabolomics, which quantify known metabolites and (b) untargeted metabolomics, which aims to detect as many metabolites as possible for discovery-driven research (Scalbert et al., 2009).

3.2.3 The Interplay Between Proteomics, Metabolomics, and Genomics

Proteomics and metabolomics are complementary to genomics and transcriptomics, contributing to a systems-level understanding of cellular function. While genomics and transcriptomics reveal potential and regulatory control, proteomics and metabolomics reflect actual cellular activity and phenotypic outcomes (Hasin et al., 2017). Integrating these omics layers, often referred to as multi-omics, empowers (a) elucidation of molecular mechanisms underlying disease, (b) discovery of biomarkers aimed at diagnostics and prognosis, and (c) insights into drug response and metabolic adaptations (Karczewski & Snyder, 2018).

3.2.4 Traditional Approaches in Proteomics and Metabolomics

Historically, proteomics relied on 2D-gel electrophoresis coupled with MS for separation and identification of proteins. Over time, this has been largely supplanted by gel-free approaches, such as label-free quantification and isobaric labeling (e.g., Isobaric Tags for Relative and Absolute Quantitation (iTRAQ) and Tandem Mass Tags (TMT)) (Domon & Aebersold, 2006). In metabolomics, traditional workflows involved NMR spectroscopy, prized for its reproducibility and structural elucidation capabilities, but limited by its sensitivity. More recent efforts prioritize MS-based platforms using techniques such as GC-MS and LC-MS for broader metabolite coverage (Fiehn, 2002). While these traditional approaches laid the groundwork for omics-based research, they also presented challenges in data dimensionality, reproducibility, and standardization issues now being addressed with AI-enhanced methods.

3.3 AI Technologies Empowering Proteomics and Metabolomics

AI is reshaping proteomics and metabolomics by empowering the investigation of complex, high-dimensional data, revealing novel insights into molecular biology, disease mechanisms, and biomarker discovery. These technologies are particularly valuable in addressing the challenges of large-scale data processing, pattern recognition, and integration of multi-omics datasets.

3.3.1 Machine Learning Algorithms for High-Throughput Data Analysis

Proteomics and metabolomics generate vast, high-dimensional datasets from techniques such as MS and NMR. ML methods such as support vector machines (SVMs), random forests (RFs), and gradient boosting are essential for classifying samples, selecting biomarkers, and handling missing or noisy data. ML is especially useful for (a) sample classification (e.g., disease vs. control), (b) feature selection for reducing data dimensionality, and (c) batch effect correction and data normalization. For instance, SVMs are used effectively for cancer classification based on proteomic profiles, and RF models have demonstrated strong performance in metabolomics-based disease prediction (Gromski et al., 2015; Saccenti et al., 2014).

3.3.2 Deep Learning for Complex Data Interpretation

DL outshines at recognizing multifaceted, non-linear relationships in high-throughput omics data. In proteomics, convolutional neural networks (CNNs) can learn directly from raw MS spectra for peptide identification, while autoencoders are used in metabolomics for denoising and dimensionality reduction. DL has also enabled (a) de novo peptide sequencing from spectra (Tran et al., 2019), (b) prediction of protein functions and subcellular localizations, and (c) integration of multi-omics data for systems biology. DL models have significantly improved accuracy in PTM detection and disease-state classification in both proteomics and metabolomics studies (Sen et al., 2021).

3.3.3 Natural Language Processing for Literature and Data Mining

NLP helps to extract meaningful perceptions from the vast amount of biomedical literature and clinical notes. NLP replicas can perform (a) named entity recognition to categorize genes, proteins, metabolites, and diseases, (b) relation extraction to map associations among biological entities, and (c) topic modeling to uncover emerging trends in omics research. These tools assist in building curated biological databases, linking literature-based findings to experimental data, and uncovering new research directions.

3.3.4 Network-Based Approaches for Protein-Protein Interactions and Metabolite Pathways

AI and graph-based learning approaches are increasingly used to model biological networks such as PPIs, gene-metabolite interactions, and metabolic pathways (Figure 3.1). These approaches allow (a) prediction of unknown PPIs or metabolic links, (b) inference of functional modules and biological relevance, and (c) simulation of perturbation effects in metabolic networks. Graph neural networks (GNNs) and probabilistic graphical models can integrate diverse data types to detect crucial molecular hubs and trails involved in disease and drug reaction (Zitnik et al., 2018; Barabási et al., 2011).

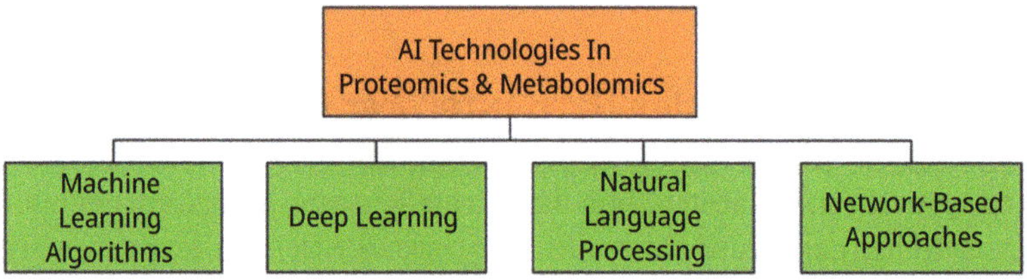

Figure 3.1: AI enables high-throughput analysis and integration of proteomic and metabolomic data to identify biomarkers, elucidate molecular pathways, and support disease diagnosis and therapeutic discovery.

3.4 Applications of AI in Proteomics

AI is transforming proteomics by improving the analysis, interpretation, as well as integration of multifaceted protein datasets. From MS to structural biology, AI algorithms are driving breakthroughs in protein identification, biomarker discovery, and interaction mapping, offering unprecedented insights into health and disease.

3.4.1 Protein Identification and Quantification

AI improves the speed and accuracy of protein identification and quantification, particularly from high-throughput data like tandem MS/MS. Traditional approaches such as database searching are now enhanced by ML algorithms that classify spectra, detect peptides, and reduce false discovery rates. Tools like Percolator, which applies semi-supervised learning, significantly improve peptide-spectrum match validation (Käll et al., 2007). DL-based implements can learn representations from raw spectral data, further enhancing protein inference and quantification accuracy.

3.4.2 Mass Spectrometry Data Analysis and AI-Driven Tools

AI facilitates the preprocessing, noise reduction, peak picking, and deconvolution of MS spectra. DL approaches like Prosit predict peptide fragmentation patterns to improve identification, while other tools use CNNs or transformers to interpret spectra directly. Automated systems driven by AI reduce dependency on expert-curated pipelines and adapt to variable data types, enhancing reproducibility and scalability in proteomic workflows.

3.4.3 Biomarker Discovery in Disease and Health

AI algorithms are extensively used to discover protein biomarkers associated with diseases, including cancer, cardiovascular, and neurodegenerative disorders. Supervised learning methods classify disease states and identify differentially expressed proteins, while unsupervised methods reveal hidden patterns. For example, RF and SVM classifiers have been used to predict Alzheimer's disease from cerebrospinal fluid proteomic profiles (Ray et al., 2007).

3.4.4 Predicting Protein Structures and Functions Using AI

Breakthroughs like AlphaFold2 have revolutionized structure prediction by using DL to accurately predict 3D structures of proteins from various amino acid sequences, even in the absence of homologous templates. In addition, AI models predict PTMs, subcellular localizations, and functional annotations using both sequence and structure information.

3.4.5 Protein-Protein Interaction Networks and Their Analysis

Understanding PPIs is fundamental for systems biology. AI-powered approaches, including GNNs and Bayesian models, are used to (a) predict unknown PPIs, (b) map disease-associated networks, and (c) integrate proteomic data with transcriptomics and interactomes. These models enable discovery of key molecular hubs and facilitate drug target identification in complex diseases.

3.5 AI in Metabolomics

AI is increasingly critical in metabolomics for managing complex, high-dimensional data, enabling accurate metabolite identification, discovering disease biomarkers, mapping metabolic pathways, and analyzing temporal changes in metabolism as represented in Figure 3.2. The integration of AI with metabolomics enhances interpretation, prediction, and discovery beyond the capabilities of traditional statistical methods.

3.5.1 Metabolite Identification and Quantification Using AI

Metabolite identification from MS or NMR data is a foremost challenge due to spectral complexity and database limitations. AI models including supervised classifiers, neural networks, and ensemble methods enhance annotation accuracy by learning spectral features, matching patterns, and predicting chemical properties. DL models such as CSI:FingerID and Spec2Vec have significantly improved metabolite annotation by learning from MS/MS fragmentation patterns (Dührkop et al., 2019; Huber et al., 2021). AI also supports automated quantification workflows, handling noise and variability more robustly than traditional algorithms.

3.5.2 Metabolic Pathway Mapping and Network Analysis

AI facilitates the reconstruction and interpretation of metabolic pathways by analyzing relationships among metabolites, genes, and enzymes. Network-based AI approaches, including graph convolutional networks and probabilistic models allow (a) pathway enrichment analysis, (b) prediction of missing metabolites or reactions, and (c) detection of pathway perturbations in disease contexts. These approaches enhance understanding of metabolic rewiring in cancer, aging, and metabolic disorders (Bai et al., 2024; Anh et al., 2024). AI tools can integrate data across omics layers (e.g., transcriptomics + metabolomics) to infer mechanistic insights.

3.5.3 AI in Identifying Disease Biomarkers in Metabolomics

AI excels in identifying metabolite signatures that distinguish disease states from healthy controls. Feature selection algorithms and classifiers such as LASSO, RFs, and deep neural complexes can identify minimal yet informative metabolite panels. The applications include (a) cancer biomarker discovery (e.g., lung, breast, and colorectal), (b) neurodegenerative disease profiling (e.g., Parkinson's and Alzheimer's), and (c) metabolic syndrome and diabetes diagnostics. AI models have shown better accuracy and generalization than conventional methods, especially when handling class imbalance and noisy clinical datasets (Hojjati et al., 2023; Bardanzellu et al., 2022).

3.5.4 Time-Series Analysis of Metabolomic Data

Longitudinal metabolomics data capturing temporal metabolic changes in response to stimuli (e.g., drug treatments, fasting, and circadian rhythms) requires specialized AI models for time-series analysis like recurrent neural networks, particularly long short-term memory networks, model temporal dependencies, and dynamic metabolic shifts. Time-aware clustering and dynamic Bayesian networks are also applied to infer causal metabolic events and predict future metabolic states (Hasin et al., 2017).

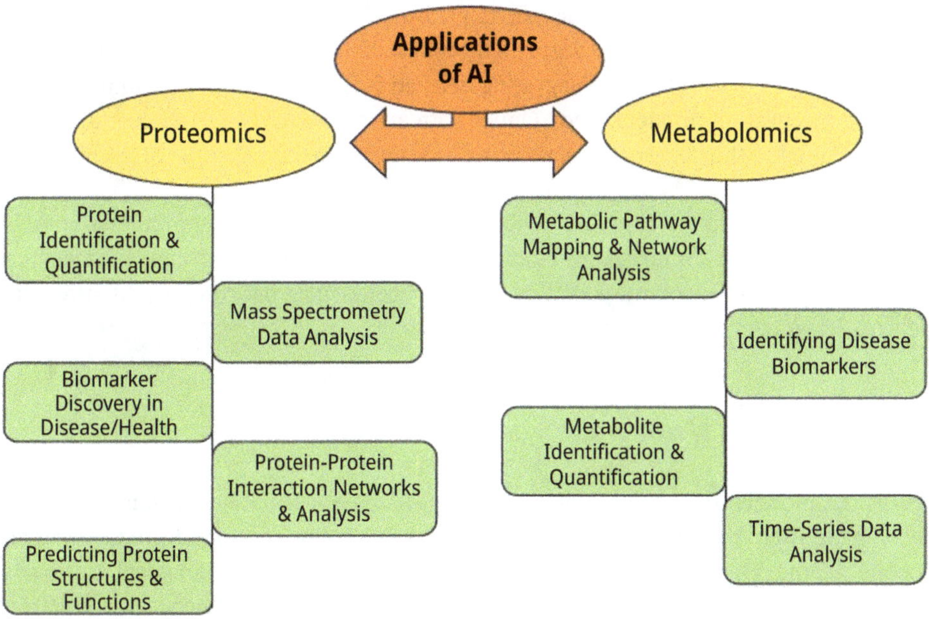

Figure 3.2: AI is applied to protein and metabolite identification, biomarker discovery, pathway analysis, and disease classification.

3.6 Integrating Proteomics and Metabolomics with AI

The integration of proteomics and metabolomics delivers a broad view of biological systems, capturing dynamic interactions between gene expression, protein function, and metabolic changes. AI offers powerful tools for deciphering these complex relationships, enabling automated data processing, feature selection, pattern recognition, and predictive modeling. ML techniques, such as SVM, RFs, and DL algorithms, have been working to interpret high-dimensional multi-omics datasets. These tools facilitate the identification of disease biomarkers and metabolic signatures connected with precise protein expressions, making systems biology more interpretable and actionable (Libiseller et al., 2015; Johnson et al., 2016).

3.6.1 Multi-omics Data Integration Techniques

Integration strategies can be broadly characterized into:

- Early integration (concatenation-based): Raw or normalized data from different omics are merged into a single matrix prior to analysis.
- Intermediate integration (transformation-based): Separate analyses of omics datasets are performed before integrating derived features or statistical summaries.
- Late integration (model-based): Omics data are analyzed independently and then combined at the decision level or using ensemble models (Misra et al., 2019).

AI-enhanced integrative frameworks such as MOFA (multi-omics factor analysis), DIABLO (data integration analysis for biomarker discovery using latent components), and DeepMO have improved interpretation and performance across complex disease datasets (Argelaguet et al., 2018; Singh et al., 2019).

3.6.2 AI-Driven Models for Systems Biology

AI-driven models in systems biology leverage DL, neural networks, and ensemble learning to infer causal mechanisms, simulate cellular responses, and predict system-level outcomes. Notable approaches include (a) GNNs to model biological pathways and protein-metabolite interactions (Zitnik et al., 2018), (b) autoencoders and variational autoencoders for dimensionality reduction and latent space modeling of multi-omics data, and (c) Bayesian networks and reinforcement learning for simulating dynamic systems and regulatory feedback loops (Li et al., 2023). These models enable the prediction of disease progression, drug response, and identification of novel therapeutic targets.

3.6.3 Correlation Between Protein Expression and Metabolic Profiles

Protein expression levels often directly impact metabolic fluxes and intermediate concentrations. Proteomics provides insight into enzyme abundances, while metabolomics reflects the actual physiological state of cells or tissues. AI algorithms can uncover nonlinear correlations between these layers, revealing functional modules and regulatory circuits. For illustration, ML models are used to correlate glycolytic enzyme levels with lactate production in cancer cells, offering insight into metabolic reprogramming (Patti et al., 2012). Multi-block partial least squares and canonical correlation analysis are also used for such integrative analyses (Cavill et al., 2016).

3.6.4 Applications in Personalized Medicine and Precision Health

The amalgamation of multi-omics data using AI is transforming modified medicine by enabling (a) patient stratification on the basis of molecular signatures, (b) prediction of drug efficacy and adverse effects based on metabolic and proteomic profiles, and (c) development of tailored therapeutic regimens targeting individual metabolic vulnerabilities. In cancer, for example, AI has helped identify patient-specific metabolic bottlenecks linked to proteomic alterations, guiding therapy choices (Hasin et al., 2017). In metabolic diseases, ML models have predicted disease onset by integrating serum proteomics with metabolic biomarkers (Wang et al., 2012).

3.7 AI in Disease Understanding and Drug Discovery

AI is increasingly recognized as a transformative tool in biomedical research, offering unparalleled capabilities in analyzing complex biological datasets, including those from proteomics and metabolomics. AI systems can model disease mechanisms, predict drug responses, find new therapeutic targets, and repurpose current medications by learning from massive multi-omics datasets.

3.7.1 AI for Disease Mechanism Exploration in Proteomics and Metabolomics

AI enhances the ability to map disease-related biological pathways by integrating proteomic and metabolomic profiles with systems biology models. Advanced ML algorithms and DL networks are used to (a) identify dysregulated metabolic pathways linked to disease progression, (b) model proteoform variation and PTMs, and (c) decode network-level interactions among enzymes, metabolites, and proteins. For example, CNNs and GNNs have been employed to understand how protein alterations contribute to metabolomics reprogramming in cancers and metabolic syndromes (Chicco & Jurman, 2020; Zitnik et al., 2018).

3.7.2 AI-Driven Approaches for Drug Target Identification

AI models facilitate drug target discovery by learning associations between disease phenotypes, molecular profiles, and protein functions. Notable techniques include (a) unsupervised learning for clustering disease subtypes based on omics profiles, (b) feature selection algorithms (e.g., LASSO and RFs) to prioritize candidate targets, and (c) deep generative models to simulate the effects of target perturbation. These methodologies are used to uncover innovative targets in diseases like Alzheimer's as well as triple-negative breast cancer by linking metabolic shifts with aberrant protein networks (Ebrahim et al., 2016; Du et al., 2024).

3.7.3 AI in Drug Repurposing Using Proteomic and Metabolomic Data

Drug repurposing with AI leverages existing proteomic and metabolomic data to match known drug mechanisms with new disease contexts. The approaches include (a) similarity-based learning: comparing omics-derived signatures between drugs and diseases, (b) knowledge graph embeddings: inferring relationships between molecules, diseases, and targets, and (c) transfer learning: applying insights from known therapeutic areas to emerging diseases. This methodology has been notably applied in rapidly identifying potential treatments during the COVID-19 pandemic using transcriptomic, proteomic, and metabolomic fingerprints (Zhou et al., 2020).

3.7.4 Toxicology and Safety Profiling Through AI Models

AI also accelerates preclinical research by predicting drug-induced toxicity and off-target effects from proteomic and metabolomic signatures. Key benefits include (a) early identification of hepatotoxicity, cardiotoxicity, and nephrotoxicity through AI analysis of metabolomic shifts, (b) integration of time-series omics data to track toxicity progression, and (c) adverse outcome pathway modeling using AI to simulate systemic toxicological responses. AI-powered platforms like DeepTox and REACHSafe have shown high accuracy in predicting toxicity profiles for both novel and existing compounds (Mayr et al., 2016; Lin & Chou, 2022).

3.8 Challenges and Limitations

In spite of the growing success of AI in biomedical research, integrating AI with proteomics and metabolomics comes with critical challenges and limitations. These issues span technical, biological, and ethical dimensions, influencing the reproducibility, trustworthiness, and practical deployment of AI-driven models in healthcare.

3.8.1 Data Quality, Variability, and Standardization

Proteomics and metabolomics datasets are often plagued by (a) instrumental and batch variability due to differences in sample preparation, instrumentation (e.g., LC-MS platforms), and acquisition protocols, (b) missing values arising from detection thresholds or stochastic sampling, and (c) inconsistent annotations and ontologies, making data integration and meta-analysis difficult. Poor data quality can lead to biased AI models, overfitting, and unreliable predictions. Standardization efforts like Human Proteome Organisation- Proteomics Standards Initiative (HUPO-PSI) (for proteomics) and metabolomics standards initiative aim to address these issues, but widespread adoption remains incomplete (Taylor et al., 2007; Salek et al., 2013).

3.8.2 Interpretability of AI Models in Biological Context

Many state-of-the-art AI approaches especially DL models are considered "black boxes" because (a) they do not offer clear explanations for predictions, (b) biological plausibility or causal inference is often lacking, and (c) their complexity hinders hypothesis generation and regulatory acceptance. This poses a barrier to clinical translation, where model transparency and explainability are critical. Emerging strategies like SHAP (SHapley Additive exPlanations), Local Interpretable Model-agnostic Explanations (LIME), and attention mechanisms are helping to increase interpretability, but they are still in early phases of adoption in systems biology (Tonekaboni et al., 2019).

3.8.3 Ethical Issues in AI-Based Proteomics and Metabolomics

AI applications in biomedicine raise several ethical concerns, including (a) data privacy and consent, especially when dealing with identifiable multi-omics profiles, (b) bias and fairness, as training datasets may underrepresent certain populations or diseases, and (c) accountability, where responsibility is unclear if an AI model makes a harmful or incorrect prediction. Moreover, the potential use of AI-generated biomarker profiles for insurance risk scoring or employment screening raises important societal questions (Topol, 2019; Mittelstadt et al., 2016).

3.8.4 Handling High Dimensionality in Proteomics and Metabolomics Data

Proteomics and metabolomics generate ultra-high-dimensional data (thousands of features vs. few samples), posing challenges such as (a) curse of dimensionality, leading to overfitting and poor generalizability, (b) feature redundancy and multicollinearity that confound model interpretation, and (c) limited sample sizes, which restrict the training of complex AI models. AI must be combined with robust statistical techniques (e.g., feature selection, dimensionality reduction via principal component analysis, t-distributed stochastic neighbor embedding, and autoencoders) and domain knowledge to guarantee valid and interpretable outcomes (Xu et al., 2021; Misra et al., 2019).

3.9 Future Directions and Innovations

As AI continues to evolve alongside experimental technologies, it holds immense promise for transforming proteomics and metabolomics research. Key innovations are shaping a future where AI models are not only more accurate and interpretable but also more integrated into real-world biomedical and clinical applications.

3.9.1 Advances in AI Models for Proteomics and Metabolomics

Future AI models are expected to surpass current limitations in prediction, inference, and scalability through:

- Foundation models and multi-modal learning (e.g., combining omics, imaging, and clinical data) for generalized biomolecular representations
- Transformer architectures (e.g., ProteinBERT and AlphaFold) skilled of catching long-range dependencies in protein sequences and omics time-series data
- Self-supervised learning on unlabeled omics data to improve feature learning and reduce dependence on annotated datasets

These innovations could revolutionize the ability to detect hidden patterns, predict protein function, and classify metabolic states across diseases and populations (Rives et al., 2021; Jumper et al., 2021).

3.9.2 The Role of AI in Emerging Technologies: Single-Cell Proteomics and Metabolomics

The rise of single-cell technologies permits the interrogation of cellular heterogeneity in unparalleled detail. However, this comes with data sparsity and noise. AI will be essential to:

- Denoise and impute missing data from single-cell MS or nanoPOTS assays
- Cluster and classify cell types and states based on proteomic/metabolomic signatures
- Infer lineage trajectories and intercellular signaling using graph-based AI approaches

Recent work has shown how DL can decode proteomic variations across single cells, aiding in tumor microenvironment analysis and immune profiling (Specht & Slavov, 2020).

3.9.3 Integrating AI with Real-Time Monitoring and Wearable Technologies

Wearable sensors are beginning to track metabolic biomarkers in real time, such as glucose, lactate, or electrolytes in sweat or interstitial fluid. Future AI models will:

- Analyze continuous metabolomic data streams from wearables using online learning algorithms
- Detect early warning signs of physiological imbalances (e.g., hypoglycemia and dehydration)
- Personalize health interventions in chronic disease management (e.g., diabetes and cardiovascular disease)

AI integration with wearables and lab-on-a-chip platforms will democratize metabolomic insights, extending beyond clinics to everyday environments (Kim et al., 2019).

3.9.4 AI for Enhancing the Reproducibility and Transparency of Omics Research

AI systems must support transparent, reproducible science, which is critical in omics research due to variability and complexity. Key strategies include:

- Automated documentation of workflows and parameters using tools like AutoML and reproducible notebooks
- Versioning of models and datasets for better traceability and reproducibility
- Explainable AI (XAI) to help understand which features drive predictions

Initiatives like the FAIR (Findable, Accessible, Interoperable, and Reusable) data principles, paired with AI transparency tools, are pushing the field toward open science (Wilkinson et al., 2016; Lundberg & Lee, 2017).

3.10 Conclusion

Over the past decade, the integration of AI with proteomics and metabolomics has led to transformative advances in biomedical research. AI has:

- Enabled multi-omics data integration, allowing researchers to decipher multifaceted biological networks and disease mechanisms with greater precision
- Improved biomarker discovery and target identification, particularly through the use of DL, feature selection, and unsupervised clustering approaches
- Accelerated drug discovery and repurposing, identifying novel therapeutic avenues based on proteomic and metabolic signatures
- Enhanced toxicology profiling, offering predictive models for drug safety assessment and side-effect profiling
- Facilitated personalized medicine, through accurate phenotyping, patient stratification, and individualized intervention strategies

These advances mark a shift from purely descriptive omics to predictive and prescriptive analytics, where AI provides actionable insights from high-dimensional biological data. Looking forward, AI is poised to become a cornerstone of systems biology and clinical decision-making. Key trends and innovations that will define the future include next-generation AI models (e.g., transformers and self-supervised learning) which may be capable of learning across heterogeneous and noisy omics datasets. The integration of real-time data from wearable biosensors and point-of-care devices may enable continuous metabolic monitoring and dynamic health interventions. XAI and ethical AI frameworks facilitate trust, fairness, and accountability in clinical and translational settings. Achieving greater reproducibility and interoperability is powered by open standards, FAIR data principles, and transparent model sharing. Ultimately, the synergy between AI and omics technologies will fuel a new era of precision health, where diagnosis, treatment, and prevention are tailored to individual biology. This paradigm will not only enhance patient outcomes but also redefine how diseases are understood, monitored, and treated on a global scale.

References

Aebersold, R., & Mann, M. (2016). Mass-spectrometric exploration of proteome structure and function. *Nature, 537*(7620), 347–355.

Anh, N. K., Lee, A., Phat, N. K., Yen, N. T. H., Thu, N. Q., Tien, N. T. N., . . . & Phuoc Long, N. (2024). Combining metabolomics and machine learning to discover biomarkers for early-stage breast cancer diagnosis. *Plos One, 19*(10), e0311810.

Argelaguet, R., Velten, B., Arnol, D., Dietrich, S., Zenz, T., Marioni, J. C., . . . & Stegle, O. (2018). Multi-Omics factor analysis – A framework for unsupervised integration of multi-omics data sets. *Molecular Systems Biology, 14*(6), e8124.

Bai, Y., Morita, K., Kokaji, T., Hatano, A., Ohno, S., Egami, R., . . . & Kuroda, S. (2024). Trans-omic analysis reveals opposite metabolic dysregulation between feeding and fasting in liver associated with obesity. *Iscience, 27*(3), 109121.

Barabási, A. L., Gulbahce, N., & Loscalzo, J. (2011). Network medicine: A network-based approach to human disease. *Nature Reviews Genetics, 12*(1), 56–68.

Bardanzellu, F., & Fanos, V. (2022). Metabolomics, microbiomics, machine learning during the COVID-19 pandemic. *Pediatric Allergy and Immunology, 33*, 86–88.

Cavill, R., Jennen, D., Kleinjans, J., & Briedé, J. J. (2016). Transcriptomic and metabolomic data integration. *Briefings in Bioinformatics, 17*(5), 891–901.

Chi, J., Shu, J., Li, M., Mudappathi, R., Jin, Y., Lewis, F., . . . & Gu, H. (2024). Artificial intelligence in metabolomics: A current review. *TrAC Trends in Analytical Chemistry, 178*, 117852.

Chicco, D., & Jurman, G. (2020). Machine learning can predict survival of patients with heart failure from serum creatinine and ejection fraction alone. *BMC Medical Informatics and Decision Making, 20*, 1–16.

Domon, B., & Aebersold, R. (2006). Mass spectrometry and protein analysis. *science, 312*(5771), 212–217.

Du, P., Fan, R., Zhang, N., Wu, C., & Zhang, Y. (2024). Advances in integrated multi-omics analysis for drug-target identification. *Biomolecules, 14*(6), 692.

Dührkop, K., Fleischauer, M., Ludwig, M., Aksenov, A. A., Melnik, A. V., Meusel, M., . . . & Böcker, S. (2019). SIRIUS 4: A rapid tool for turning tandem mass spectra into metabolite structure information. *Nature Methods, 16*(4), 299–302.

Ebrahim, A., Brunk, E., Tan, J., O'brien, E. J., Kim, D., Szubin, R., . . . & Palsson, B. O. (2016). Multi-omic data integration enables discovery of hidden biological regularities. *Nature Communications, 7*(1), 13091.

Fiehn, O. (2002). Metabolomics – The link between genotypes and phenotypes. *Functional Genomics*, 48(1–2), 155–171.

Grapov, D., Fahrmann, J., Wanichthanarak, K., & Khoomrung, S. (2018). Rise of deep learning for genomic, proteomic, and metabolomic data integration in precision medicine. *Omics: A Journal of Integrative Biology, 22*(10), 630–636.

Gromski, P. S., Muhamadali, H., Ellis, D. I., Xu, Y., Correa, E., Turner, M. L., & Goodacre, R. (2015). A tutorial review: Metabolomics and partial least squares-discriminant analysis–a marriage of convenience or a shotgun wedding. *Analytica Chimica Acta, 879*, 10–23.

Hasin, Y., Seldin, M., & Lusis, A. (2017). Multi-omics approaches to disease. *Genome Biology, 18*, 1–15.

Hojjati, F., Roointan, A., Gholaminejad, A., Eshraghi, Y., & Gheisari, Y. (2023). Identification of key genes and biological regulatory mechanisms in diabetic nephropathy: Meta-analysis of gene expression datasets. *Nefrologia, 43*(5), 575–586.

Huber, F., Ridder, L., Verhoeven, S., Spaaks, J. H., Diblen, F., Rogers, S., & Van Der Hooft, J. J. (2021). Spec2Vec: Improved mass spectral similarity scoring through learning of structural relationships. *PLoS Computational Biology, 17*(2), e1008724.

Johnson, C. H., Ivanisevic, J., & Siuzdak, G. (2016). Metabolomics: Beyond biomarkers and towards mechanisms. *Nature Reviews Molecular Cell Biology, 17*(7), 451–459.

Jumper, J., Evans, R., Pritzel, A., Green, T., Figurnov, M., Ronneberger, O., . . . & Hassabis, D. (2021). Highly accurate protein structure prediction with AlphaFold. *nature*, *596*(7873), 583–589.

Käll, L., Canterbury, J. D., Weston, J., Noble, W. S., & MacCoss, M. J. (2007). Semi-supervised learning for peptide identification from shotgun proteomics datasets. *Nature Methods*, *4*(11), 923–925.

Karczewski, K. J., & Snyder, M. P. (2018). Integrative omics for health and disease. *Nature Reviews Genetics*, *19*(5), 299–310.

Kim, J., Campbell, A. S., De Ávila, B. E. F., & Wang, J. (2019). Wearable biosensors for healthcare monitoring. *Nature Biotechnology*, *37*(4), 389–406.

Li, X., Huang, W., Xu, X., Zhang, H. Y., & Shi, Q. (2023). Deciphering tissue heterogeneity from spatially resolved transcriptomics by the autoencoder-assisted graph convolutional neural network. *Frontiers in Genetics*, *14*, 1202409.

Libiseller, G., Dvorzak, M., Kleb, U., Gander, E., Eisenberg, T., Madeo, F., . . . & Magnes, C. (2015). IPO: A tool for automated optimization of XCMS parameters. *BMC Bioinformatics*, *16*, 1–10.

Lin, Z., & Chou, W. C. (2022). Machine learning and artificial intelligence in toxicological sciences. *Toxicological Sciences*, *189*(1), 7–19.

Lundberg, S. M., & Lee, S. I. (2017). A unified approach to interpreting model predictions. *Advances in Neural Information Processing Systems*, *30*, 4768–4777.

Mayr, A., Klambauer, G., Unterthiner, T., & Hochreiter, S. (2016). DeepTox: Toxicity prediction using deep learning. *Frontiers in Environmental Science*, *3*, 80.

Misra, B. B., Langefeld, C., Olivier, M., & Cox, L. A. (2019). Integrated omics: Tools, advances and future approaches. *Journal of Molecular Endocrinology*, *62*(1), R21–R45.

Mittelstadt, B. D., Allo, P., Taddeo, M., Wachter, S., & Floridi, L. (2016). The ethics of algorithms: Mapping the debate. *Big Data & Society*, *3*(2), 2053951716679679.

Nicholson, J. K., Lindon, J. C., & Holmes, E. (1999). 'Metabonomics': Understanding the metabolic responses of living systems to pathophysiological stimuli via multivariate statistical analysis of biological NMR spectroscopic data. *Xenobiotica*, *29*(11), 1181–1189.

Patti, G. J., Yanes, O., & Siuzdak, G. (2012). Innovation: Metabolomics: The apogee of the omics trilogy. *Nature Reviews Molecular Cell Biology 13*(4):263–9.

Ray, S., Britschgi, M., Herbert, C., Takeda-Uchimura, Y., Boxer, A., Blennow, K., . . . & Wyss-Coray, T. (2007). Classification and prediction of clinical Alzheimer's diagnosis based on plasma signaling proteins. *Nature Medicine*, *13*(11), 1359–1362.

Rives, A., Meier, J., Sercu, T., Goyal, S., Lin, Z., Liu, J., . . . & Fergus, R. (2021). Biological structure and function emerge from scaling unsupervised learning to 250 million protein sequences. *Proceedings of the National Academy of Sciences*, *118*(15), e2016239118.

Saccenti, E., Hoefsloot, H. C., Smilde, A. K., Westerhuis, J. A., & Hendriks, M. M. (2014). Reflections on univariate and multivariate analysis of metabolomics data. *Metabolomics*, *10*, 361–374.

Salek, R. M., Steinbeck, C., Viant, M. R., Goodacre, R., & Dunn, W. B. (2013). The role of reporting standards for metabolite annotation and identification in metabolomic studies. *Gigascience*, *2*(1), 2047–217X.

Scalbert, A., Brennan, L., Fiehn, O., Hankemeier, T., Kristal, B. S., van Ommen, B., . . . & Wopereis, S. (2009). Mass-spectrometry-based metabolomics: Limitations and recommendations for future progress with particular focus on nutrition research. *Metabolomics*, *5*, 435–458.

Sen, P., Lamichhane, S., Mathema, V. B., McGlinchey, A., Dickens, A. M., Khoomrung, S., & Orešič, M. (2021). Deep learning meets metabolomics: A methodological perspective. *Briefings in Bioinformatics*, *22*(2), 1531–1542.

Singh, A., Shannon, C. P., Gautier, B., Rohart, F., Vacher, M., Tebbutt, S. J., & Lê Cao, K. A. (2019). DIABLO: An integrative approach for identifying key molecular drivers from multi-omics assays. *Bioinformatics*, *35*(17), 3055–3062.

Specht, H., & Slavov, N. (2020). Optimizing accuracy and depth of protein quantification in experiments using isobaric carriers. *Journal of Proteome Research*, *20*(1), 880–887.

Taylor, C. F., Paton, N. W., Lilley, K. S., Binz, P. A., Julian, R. K., Jr, Jones, A. R., . . . & Hermjakob, H. (2007). The minimum information about a proteomics experiment (MIAPE). *Nature Biotechnology*, *25*(8), 887–893.

Tonekaboni, S., Joshi, S., McCradden, M. D., & Goldenberg, A. (2019 October). What clinicians want: Contextualizing explainable machine learning for clinical end use. In *Machine Learning for Healthcare Conference* (pp. 359–380). PMLR.

Topol, E. (2019). *Deep Medicine: How Artificial Intelligence Can Make Healthcare Human Again*. Hachette UK.

Tran, N. H., Qiao, R., Xin, L., Chen, X., Liu, C., Zhang, X., . . . & Li, M. (2019). Deep learning enables *de novo* peptide sequencing from data-independent-acquisition mass spectrometry. *Nature Methods*, *16*(1), 63–66.

Wang, Y., Eddy, J. A., & Price, N. D. (2012). Reconstruction of genome-scale metabolic models for 126 human tissues using mCADRE. *BMC Systems Biology*, *6*, 1–16.

Webb-Robertson, B. J. M., Wiberg, H. K., Matzke, M. M., Brown, J. N., Wang, J., McDermott, J. E., . . . & Waters, K. M. (2015). Review, evaluation, and discussion of the challenges of missing value imputation for mass spectrometry-based label-free global proteomics. *Journal of Proteome Research*, *14*(5), 1993–2001.

Wilkins, M. R., Sanchez, J. C., Gooley, A. A., Appel, R. D., Humphery-Smith, I., Hochstrasser, D. F., & Williams, K. L. (1996). Progress with proteome projects: Why all proteins expressed by a genome should be identified and how to do it. *Biotechnology and Genetic Engineering Reviews*, *13*(1), 19–50.

Wilkinson, M. D., Dumontier, M., Aalbersberg, I. J., Appleton, G., Axton, M., Baak, A., . . . & Mons, B. (2016). The FAIR Guiding Principles for scientific data management and stewardship. *Scientific Data*, *3*(1), 1–9.

Xu, C., Cao, S., & Alam, M. A. (2021). Application of novel statistical and machine-learning methods to high-dimensional clinical cancer and (multi-) omics data. *Frontiers in Genetics*, *12*, 739442.

Zhang, Y., Fonslow, B. R., Shan, B., Baek, M. C., & Yates III, J. R. (2013). Protein analysis by shotgun/bottom-up proteomics. *Chemical Reviews*, *113*(4), 2343–2394.

Zhou, Y., Hou, Y., Shen, J., Huang, Y., Martin, W., & Cheng, F. (2020). Network-based drug repurposing for novel coronavirus 2019-nCoV/SARS-CoV-2. *Cell Discovery*, *6*(1), 14.

Zitnik, M., Agrawal, M., & Leskovec, J. (2018). Modeling polypharmacy side effects with graph convolutional networks. *Bioinformatics*, *34*(13), i457–i466.

Pooja Verma, Priti Duhan, Poonam Bansal, Kumkum Verma, Narender Mohan*, Praveen Kumar

Chapter 4
AI-Enhanced Drug Discovery and Development

Abstract: Drug discovery has traditionally been a slow, expensive, and uncertain journey, often taking more than a decade and costing billions, with only a small fraction of drug candidates ever reaching patients. The process has long depended on labor-heavy experiments and occasional strokes of luck, making it increasingly difficult to respond to fast-moving global health challenges. Today, however, advances in artificial intelligence (AI) have reshaped this landscape. AI is helping researchers in identifying several promising targets, vast chemical libraries, and also, refining potential drug candidates at unprecedented speed by combining machine learning, deep learning, natural language processing, and generative models with rich biological and clinical datasets. These systems enable the virtual assessment of billions of chemical structures, the prediction of in vivo pharmacological behavior, and the de novo design of therapeutic candidates, dramatically accelerating the discovery process. AI is also transforming later stages of development in both preclinical research and clinical trials as it supports better biomarker discovery, improves patient selection, and enables innovations such as digital twins and synthetic control arms that make trials faster, more efficient, and often safer. In drug safety and post-market monitoring, AI can detect potential risks earlier by analyzing patterns across massive real-world datasets. Yet these breakthroughs come with challenges. Issues such as biased data, opaque model behavior, evolving regulatory expectations, and concerns about fairness and transparency must be addressed to ensure these tools are used responsibly and effectively. Looking ahead in this chapter, these advances signal a fundamental reimagining of the ways in which medicines should be created and utilized, which will open the door to faster, more precise, and more affordable therapies for patients worldwide.

Keywords: Drug discovery, de novo, artificial intelligence, drug safety

***Corresponding author: Narender Mohan**, ICAR-IIWBR, Karnal 132001, Haryana, India, e-mail: narendermohan93@gmail.com
Pooja Verma ICAR-IIWBR, Karnal 132001, Haryana, India
Priti Duhan, K. L. Mehta Dayanand College for Women, Faridabad 121001,Haryana, India
Poonam Bansal, Department of Bioscience and Technology, MMEC, Maharishi Markandeshwar (Deemed to be) University Mullana, Ambala 133207, Haryana, India
Kumkum Verma, ICAR-IARI, Regional Station, Karnal 132001, Haryana, India
Praveen Kumar, Faculty of Health and Allied Sciences, Guru Kashi University, Talwandi Sabo, Bathinda, Punjab

 | https://doi.org/10.1515/9783112214275-004

4.1 Introduction

The drug discovery and development process represents a complex, multi-stage journey from initial target identification to post-market surveillance, traditionally requiring over a decade and exceeding $2.5 billion per approved therapeutic (DiMasi et al., 2016). This extended timeline stems from high attrition rates, with fewer than 10% of candidates surviving clinical trials due to challenges including target validation failures, unpredictable pharmacokinetics, and unexpected toxicity profiles (Shen et al., 2019). The pharmaceutical industry faces immense pressure to accelerate this process, reduce costs, and improve success rates, especially given the rising global health challenges and the urgent need for novel therapeutics. The traditional paradigm, heavily reliant on laborious wet-lab experiments and serendipitous discoveries, is increasingly proving insufficient to meet these demands. Artificial intelligence (AI) has emerged as a transformative force in this landscape, fundamentally restructuring pharmaceutical research through advanced computational methods that accelerate target identification, predict molecular behavior, and optimize clinical trial design.

By integrating machine learning (ML) with biomedical big data, AI systems can now identify novel drug targets in weeks rather than years, predict compound efficacy with increasing accuracy, and design optimized molecular structures de novo (Zhavoronkov et al., 2019). This paradigm shift is not merely an incremental improvement but a fundamental reimagining of how drugs are discovered and developed. AI's ability to process and derive insights from vast, complex datasets ranging from genomic and proteomic information to real-world patient data enables a level of precision and efficiency previously unattainable. This chapter examines how AI technologies are being integrated across the entire drug development continuum, from initial computational screening through post-marketing surveillance. We will delve into the specific applications of AI in target identification and validation, hit discovery and screening, lead optimization, preclinical and clinical development, and drug safety and toxicology. Furthermore, we will address persistent challenges in data quality, model interpretability, regulatory alignment, and ethical considerations that must be resolved to fully realize AI's potential in creating safer, more effective medicines. Finally, we will explore future directions, including the advent of foundation models, quantum computing, and automated laboratories, which promise to further revolutionize therapeutic development through patient-specific medicine designed at molecular resolution.

4.2 The Drug Discovery Pipeline

The conventional drug development pipeline progresses systematically through target identification, hit discovery, lead optimization, preclinical testing, clinical trials, and post-approval monitoring, with each stage presenting distinct computational challenges that AI approaches are uniquely positioned to address (Hughes et al., 2011). Traditional methodologies remain heavily reliant on manual experimentation and serendipitous discovery, exemplified by high-throughput screening (HTS) processes that typically test 10,000–100,000 compounds at costs exceeding $1 million per screen while yielding few viable candidates (Macarron et al., 2011). This labor-intensive and time-consuming approach often leads to bottlenecks and high failure rates, particularly in the early stages. In stark contrast, AI-driven virtual screening can evaluate billions of compounds computationally through cloud-based platforms, dramatically reducing resource requirements while expanding chemical space exploration (Bettanti et al., 2024; Gangwal et al., 2024). This shift from physical to virtual screening significantly accelerates the initial phases of drug discovery.

The implementation of deep learning architectures has demonstrated particular promise in addressing historically "undruggable" targets such as KRAS mutants, where AlphaFold-predicted protein structures have revealed cryptic binding pockets enabling novel therapeutic approaches (Tunyasuvunakool et al., 2021). This capability is critical for tackling diseases with previously intractable targets, opening new avenues for drug development. This paradigm shift toward computation-first approaches has reduced hit-to-lead timelines from years to months while improving success rates in later development stages through enhanced prediction of pharmacokinetic and toxicological properties (Stokes et al., 2020). AI's predictive power allows for earlier identification and deselection of compounds with unfavorable profiles, thereby reducing late-stage attrition. Furthermore, AI can optimize experimental design, analyze complex biological data, and even automate laboratory processes, leading to more efficient and cost-effective drug development. The various stages of drug development like target identification, hit discovery, and optimization are discussed in Table 4.1.

Table 4.1: Stages of drug development.

Stage	Output	AI impact
Target identification	Validated disease target (e.g., protein)	NLP mining of omics/literature data, network analysis, and multimodal data integration (Guo et al., 2025)
Hit discovery	Initial active compounds	Virtual screening of billion-molecule libraries, active learning, high-throughput data analysis (Wu et al., 2018)
Lead optimization	Improved efficacy/ safety profile	Generative chemistry + ADMET prediction, molecular dynamics simulations, and explainable AI for SAR (Pun et al., 2025)
Preclinical testing	IND-enabling data package	AI-predicted toxicity/species translation, in silico modeling, and organ-on-chip data analysis (Schieferdecker et al., 2024)
Clinical trials	Phase I–III safety/ efficacy data	Patient stratification via AI biomarkers, synthetic control arms, real-world evidence generation, and automated patient recruitment (Aggarwal et al., 2025)

Table 4.1 (continued)

Stage	Output	AI impact
Post-market surveillance	Real-world safety and efficacy data	NLP for adverse event detection, drug-drug interaction prediction, and pharmacovigilance (Chakraborty & Tiwari 2025)
Traditional versus AI-driven approaches		
Aspect	**Traditional**	**AI-driven**
Hit identification	10,000–100,000 compounds screened physically	1 billion+ compounds screened virtually, de novo design (Zhavoronkov et al., 2020)
Timeline	2–3 years (hit-to-lead)	6–12 months, sometimes weeks (Fleming, 2018)
Attrition rate	>90%	Reduced by 30–50% (early toxicity prediction and better lead selection) (Paul et al., 2010)
Cost per drug	>$2.5 billion	Potentially reduced by 20–30% (Wouters et al., 2020)

4.2.1 Key Challenges Addressed by AI

- **Undruggable Targets** (e.g., KRAS): AlphaFold2-predicted structures enable cryptic pocket identification, opening new therapeutic avenues (Tunyasuvunakool et al., 2021).
- **Polypharmacology**: Reinforcement learning (RL) models design multi-target inhibitors (e.g., for oncology), addressing complex disease mechanisms (Vamathevan et al., 2019).
- **Data Overload**: AI efficiently processes vast multi-omics datasets, extracting meaningful insights from genomics, proteomics, and clinical data (Mansoor et al., 2024).
- **Time and Cost**: AI accelerates various stages, from virtual screening to clinical trial design, significantly reducing the overall time and cost of drug development (Liu et al., 2024).

4.3 AI Technologies Transforming Drug Discovery

AI, particularly ML and deep learning, forms the computational backbone of modern drug discovery. These technologies enable the analysis of vast and complex datasets, leading to more accurate predictions and accelerated processes across the drug development pipeline. The integration of advanced AI models has revolutionized how researchers identify, design, and optimize potential drug candidates. ML algorithms, such as random forest and support vector machine models, have long been instrumental in enabling quantitative structure-activity relationship (QSAR) predictions that guide early compound selection (Abbas et al., 2024; Xu, 2022). These models can learn complex relationships between molecular structures and their biological activities, allowing for the prioritization of compounds with desired properties. Deep learning architectures, including convolutional neural networks (CNNs) and graph neural networks (GNNs), have demonstrated remarkable capabilities in processing complex biomolecular data. They can accurately predict protein-ligand binding affinities directly from amino acid sequences and molecular graphs, often without requiring resolved crystal structures (Gawehn et al., 2016; Yang et al., 2022). This capability is particularly valuable for virtual screening and de novo drug design, where rapid and accurate predictions are crucial.

Natural language processing (NLP) systems, such as BioBERT and SciBERT, have revolutionized knowledge extraction from scientific literature. These systems can automatically identify potential drug targets by analyzing relationships across millions of publications and patents, accelerating the identification of novel therapeutic avenues (Guo et al., 2025). NLP also plays a crucial role in extracting insights from unstructured clinical notes and patient records, which can inform target validation and patient stratification. Generative adversarial networks (GANs) and RL frameworks are increasingly employed for de novo molecular design. Platforms like Insilico Medicine's Chemistry42 generate novel compounds that meet multiple optimization criteria simultaneously while predicting synthetic accessibility (Zhavoronkov et al., 2019). These generative models can explore vast chemical spaces and design molecules with desired properties from scratch, significantly reducing the time and effort required for lead identification. Transformer-based models, including Chemformer and Molecular Transformer, have further accelerated drug discovery by accurately predicting reaction pathways and retrosynthetic routes, enabling more efficient synthesis of designed compounds (Schwaller et al., 2021; Shukla, 2025). These models leverage the power of attention mechanisms to capture long-range dependencies in molecular sequences, leading to highly accurate predictions of chemical reactions. The leading AI tools in drug discovery, such as AlphaFold, DeepChem, GENTRL, and BioBERT/SciBERT, have been discussed in Table 4.2.

Table 4.2: Leading AI tools in drug discovery.

Tool	Function	Case study
AlphaFold	Protein structure prediction	Solved 200M+ protein structures, enabling targeting of previously "undruggable" proteins (DeepMind) (Tunyasuvunakool et al., 2021)
DeepChem	Molecular property prediction	Predicted CYP450 inhibition with high accuracy (AUC = 0.91), widely used open-source platform (Wu et al., 2018)
GENTRL	De novo molecule design	Designed novel TXK kinase inhibitor and DDR1 kinase inhibitors in record time (Insilico Medicine) (Zhavoronkov et al., 2019)
PathExplorer	Target-disease linkage	Identified 12 novel ALS targets, accelerating research into neurodegenerative diseases (BenevolentAI) (Kategaya et al., 2019)
Chemistry42	Generative chemistry platform	Generated novel compounds meeting multiple optimization criteria simultaneously (Insilico Medicine) (Zhavoronkov et al., 2019)
BioBERT/ SciBERT	Biomedical NLP	Extracts disease-gene associations and drug-repurposing insights from scientific literature (Guo et al., 2025; Mohapatra et al., 2022)
DeepTox	Toxicity prediction	Outperformed traditional animal testing in predicting hepatotoxicity, improving early safety assessment (Mayr et al., 2016)
Unlearn.AI	Synthetic control arms	Reduced required control group sizes in clinical trials by 30% using digital twins (Viele et al., 2020)

4.3.1 Machine/Deep Learning Applications

- **Virtual Screening**: GNNs predict binding affinities, significantly outperforming traditional docking methods in screening speed (e.g., GraphConv can be 100× faster) (Gorgulla et al., 2020). This allows for the rapid evaluation of billions of compounds, drastically expanding the chemical space explored.
- **QSAR Modeling**: CNN-based models achieve high accuracy (e.g., $R^2 > 0.85$) for activity prediction, enabling more precise selection of compounds with desired biological effects (Wu et al., 2018). These models can capture intricate patterns in molecular structures that correlate with biological activity.
- **Protein Structure Prediction**: Beyond AlphaFold, other deep learning models like RoseTTAFold and ESMFold continue to advance the accuracy and speed of protein structure prediction, which is fundamental for understanding protein function and designing targeted therapies (Jumper et al., 2021).
- **Drug-Target Interaction Prediction**: Advanced deep learning models are being developed to predict interactions between drugs and their targets with high fidelity, even for novel compounds or targets with limited experimental data (Öztürk et al., 2018).

4.3.2 Natural Language Processing (NLP)

- **Target Identification**: BioBERT extracts disease-gene associations from massive biomedical text corpora (e.g., 30M+ PubMed abstracts), accelerating the discovery of novel therapeutic targets (Guo et al., 2025).
- **Drug Repurposing**: IBM Watson identified baricitinib for COVID-19 by analyzing over 25,000 scientific papers, demonstrating NLP's power in identifying new indications for existing drugs (Mohapatra et al., 2022).
- **Clinical Trial Document Analysis**: NLP can automate the extraction of critical information from clinical trial protocols, patient records, and regulatory documents, streamlining trial design and patient recruitment (Sorbello et al., 2023).

4.3.3 Reinforcement Learning (RL)

- **De Novo Design**: GENTRL generated DDR1 kinase inhibitors in just 21 days, showcasing RL's ability to rapidly design novel molecules with specific properties (Zhavoronkov et al., 2019). RL agents learn to navigate chemical spaces and optimize molecular structures based on predefined reward functions, such as desired binding affinity or ADMET properties.
- **Synthetic Route Planning**: RL is being applied to optimize multi-step chemical synthesis pathways, identifying the most efficient and cost-effective routes for producing complex molecules (Segler et al., 2018).

4.3.4 Generative Models

- **Molecule Generation**: GANs and variational autoencoders create novel scaffolds with optimized properties (e.g., solubility, log P, and synthetic accessibility), expanding the diversity of drug candidates (Popova et al., 2018). These models can learn the underlying distribution of chemical compounds and generate new molecules that adhere to specific design criteria.
- **Synthesis Planning**: Chemformer, developed by IBM RXN, predicts reaction pathways with over 90% accuracy, significantly improving the efficiency of chemical synthesis (Schwaller et al., 2021). This capability is crucial for translating designed molecules into tangible drug candidates.

4.3.5 Emerging AI Technologies

- **Federated Learning**: This approach enables collaborative model training across multiple institutions without sharing raw data, addressing data privacy and security concerns. It allows for the leveraging of diverse, distributed datasets to build more robust AI models for drug discovery (Rieke et al., 2020).

- **Quantum Computing**: While still in its early stages, quantum computing holds immense potential for simulating molecular interactions with unprecedented accuracy, potentially revolutionizing drug design and materials science. Quantum algorithms could accelerate complex calculations that are intractable for classical computers, leading to the discovery of novel compounds with superior properties (Cao et al., 2019).

4.4 AI in Target Identification and Validation

AI has dramatically transformed target identification through integrative analysis of multi-omics datasets, revealing novel disease associations that evade conventional approaches (Das et al., 2021). The sheer volume and complexity of data generated from genomics, transcriptomics, proteomics, and metabolomics necessitate advanced computational tools for meaningful interpretation. Deep learning frameworks can identify therapeutic targets by detecting intricate patterns across these diverse data types, with GNNs particularly effective in modeling complex biological networks to pinpoint critical nodes for intervention (Mansoor et al., 2024; Liu et al., 2024). This network-based approach allows for the identification of not just individual genes or proteins, but entire pathways and regulatory mechanisms implicated in disease. For example, BenevolentAI's knowledge graph platform identified baricitinib as a potential COVID-19 treatment by analyzing over 25,000 scientific papers and clinical trial records, leading to emergency authorization within months of pandemic onset (Mohapatra et al., 2022). This case highlights AI's ability to rapidly synthesize information from disparate sources and identify novel drug repurposing opportunities. Similarly, other AI-driven platforms are leveraging vast repositories of biomedical literature and real-world data to uncover previously unrecognized connections between diseases and molecular targets.

Protein-drug interaction prediction has advanced significantly through models like DeepDTA and PADME, which accurately forecast binding affinities using only sequence data when structural information remains unavailable (Öztürk et al., 2018). These models are crucial for early-stage drug discovery, allowing researchers to prioritize compounds that are most likely to bind effectively to their intended targets. The integration of AI with structural biology, particularly with tools like AlphaFold, has further enhanced this capability by providing highly accurate protein structures, which can then be used to predict binding sites and design targeted therapies with greater precision. Functional validation has been enhanced through CRISPR-AI integration, where ML interpretation of genome-wide knockout screens identified Werner syndrome helicase as a synthetic lethal target in microsatellite instability-high cancers, enabling new precision oncology approaches (Kategaya et al., 2019). This synergistic approach combines the power of high-throughput experimental techniques with AI-driven data analysis to accelerate the validation of novel targets. AI can also predict the functional consequences of genetic variations, helping to prioritize targets that are most likely to be therapeutically relevant. Recent advancements also include the use of multimodal AI models that integrate various data types, such as patient clinical data, imaging, and molecular profiles, to provide a more holistic view of disease biology and identify targets that are predictive of patient response to therapy (Topol, 2019). This approach moves beyond single-omics analysis to capture the complex interplay of biological systems, leading to more robust and clinically relevant target identification. Furthermore, explainable AI (XAI) techniques are being developed to provide insights into the decision-making process of these complex models, increasing trust and facilitating regulatory acceptance of AI-identified targets.

4.5 AI in Hit Discovery and Screening

Virtual screening has undergone revolutionary changes through AI, with GNNs now capable of evaluating billions of compounds in cloud-based environments at speeds unimaginable through physical screening (Gorgulla et al., 2020). This capability significantly reduces the time and cost associated with identifying initial active compounds. Traditional HTS involves physically testing large libraries of compounds, which is resource-intensive and often yields a low hit rate. AI-driven virtual screening, in contrast, allows for the rapid computational assessment of vast chemical spaces, prioritizing compounds with the highest likelihood of binding to a target. DNA-encoded library (DEL) technology combined with deep learning has been particularly transformative (Figure 4.1). Ex Scientia demonstrated the discovery of a novel A2a receptor antagonist from a library of 4 billion compounds through computational prioritization followed by minimal experimental validation (Brenner & Lerner, 1992; Flood et al., 2020). This integration of DEL with AI allows for the rapid identification of potent hits from ultra-large chemical libraries, significantly accelerating the hit identification process. AI algorithms can analyze the vast amount of data generated by DEL screens, identifying patterns and predicting active compounds with high accuracy.

Active learning approaches have significantly reduced wet-lab requirements, with systems like Cornell's DeepDock achieving a 90% reduction in experimental screening while identifying potent kinase inhibitors through iterative prediction-validation cycles (Zhang & Wen, 2021). Active learning dynamically selects the most informative compounds for experimental testing, thereby optimizing resource allocation and accelerating the discovery of novel hits. This iterative process allows the AI model to continuously improve its predictive capabilities based on new experimental data. Toxicity prediction has seen remarkable improvements through deep learning models such as DeepTox, which outperformed traditional animal testing in predicting hepatotoxicity by identifying complex molecular patterns associated with adverse outcomes (Mayr et al., 2016). Early and accurate prediction of toxicity is crucial for reducing attrition rates in later stages of drug development. AI models can analyze molecular structures and predict potential toxic effects, allowing researchers to deselect problematic compounds early in the discovery process.

Drug repurposing has emerged as a particularly successful application of AI; platforms like the L1000 matching gene expression signatures to identify novel indications for existing compounds, as exemplified by the rediscovery of topiramate for inflammatory bowel disease through computational analysis (Cui et al., 2021). AI can analyze vast amounts of existing drug data, including clinical trial results, adverse event reports, and molecular profiles, to identify new therapeutic uses for approved drugs. This approach offers a faster and more cost-effective path to drug development, as the safety profiles of repurposed drugs are already well-established. Furthermore, advancements in AI-driven phenotypic screening are enabling the identification of compounds that modulate complex biological processes without necessarily knowing the specific target. ML algorithms can analyze high-content imaging data from cellular assays, and identify subtle morphological changes indicative of desired biological activity (Caicedo et al., 2019). This approach is particularly valuable for discovering first-in-class drugs for complex diseases where the underlying molecular mechanisms are not fully understood.

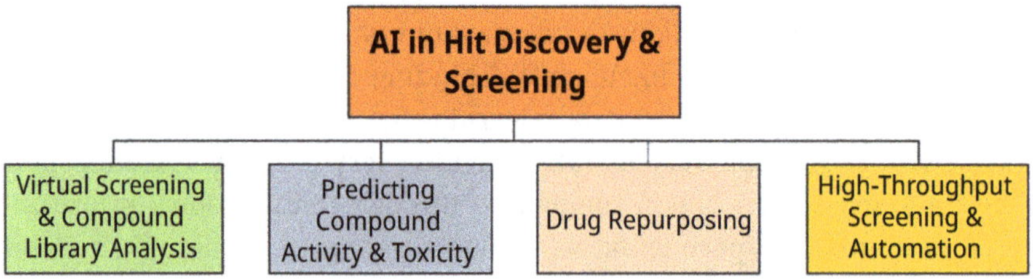

Figure 4.1: Artificial intelligence accelerates hit discovery and screening by rapidly analyzing vast chemical and biological datasets to predict active compounds in early drug discovery.

4.6 AI in Drug Optimization and Lead Discovery

Lead optimization has been dramatically accelerated through generative chemistry approaches, where RL systems design compounds meeting multiple property constraints simultaneously (de Souza et al., 2020; Wiegn et al., 2020). This represents a significant leap from traditional methods, which often involve iterative cycles of synthesis and testing. Insilico Medicine's GENTRL platform demonstrated this capability by designing novel discoidin domain receptor inhibitors in 21 days, with subsequent synthesis and validation confirming nanomolar potency against the target kinase (Zhavoronkov et al., 2019). Such platforms can rapidly explore vast chemical spaces, generating novel molecular structures with optimized properties like potency, selectivity, and ADMET (absorption, distribution, metabolism, excretion, and toxicity) characteristics. ADMET prediction has become increasingly sophisticated through models that incorporate quantum mechanical calculations and molecular dynamics simulations. Platforms like ADMET Predictor reduce preclinical attrition by 40% through early identification of problematic pharmacokinetic profiles (Cheng et al., 2013). AI models can predict how a drug will behave in the body, including its absorption rate, distribution to various tissues, metabolic pathways, excretion routes, and potential toxic effects. This early prediction helps in deselecting compounds with poor ADMET properties, saving significant time and resources in later development stages.

Structure-activity relationship modeling has been enhanced through XAI approaches, including SHapley Additive exPlanations analysis. XAI identifies critical molecular features driving biological activity to guide rational compound optimization (Pun et al., 2025). By providing insights into why a particular compound exhibits certain activity, XAI helps medicinal chemists make informed decisions about structural modifications, leading to more effective and targeted drug design. This transparency is crucial for building trust in AI-driven predictions and facilitating regulatory acceptance. Molecular stability optimization has benefited from transformer-based models that predict degradation pathways and suggest structural modifications to improve drug shelf-life, significantly impacting formulation development (Schwaller et al., 2021). Ensuring the stability of a drug molecule is paramount for its efficacy and safety during storage and administration. AI can analyze the chemical structure and predict potential degradation pathways, allowing for the design of more stable compounds and optimized formulations. Furthermore, AI is being used to optimize synthetic routes for lead compounds, ensuring that the designed molecules can be efficiently and cost-effectively synthesized. This involves predicting reaction yields, identifying suitable reagents, and optimizing reaction conditions, thereby streamlining the transition from design to synthesis (Coley et al., 2019). The integration of AI with automated synthesis platforms is paving the way for closed-loop drug discovery systems, where AI designs, synthesizes, and tests compounds with minimal human intervention.

4.7 AI in Preclinical and Clinical Development

Preclinical development has been transformed through ML models that predict interspecies translation, with XGBoost algorithms accurately forecasting human pharmacokinetics from rodent data to optimize dosing regimens (Schieferdecker et al., 2024). This capability is crucial for bridging the gap between animal studies and human trials, reducing the risk of unexpected outcomes in clinical phases. AI can also analyze complex preclinical data, including toxicology and efficacy studies, to identify potential issues early on, thereby improving the chances of success in later stages. AI-driven biomarker discovery has enhanced patient stratification in clinical trials, as demonstrated by PathAI's histopathology analysis platform identifying response signatures in oncology trials that improved patient selection (Aggarwal et al., 2025). By identifying specific biomarkers that predict patient response to a drug, AI enables the selection of patient populations most likely to benefit from a given therapy, leading to more efficient and successful clinical trials. This precision medicine approach minimizes exposure of non-responders to ineffective treatments and accelerates drug approval. Clinical trial design has been revolutionized through synthetic control arms created by companies like Unlearn. AI, whose digital twin technology reduced required control group sizes by 30% while maintaining statistical power through ML-generated virtual patients (Viele et al., 2020). Synthetic control arms leverage historical patient data and AI models to create a virtual control group, reducing the need for a large number of patients in the actual control arm. This not only accelerates trials but also reduces costs and ethical concerns associated with placebo groups.

Real-world monitoring during trials has advanced through wearable sensors integrated with deep learning algorithms, such as Apple Watch's arrhythmia detection in Pfizer's cardiac studies that enabled continuous safety assessment (Oren et al., 2021). Wearable devices collect continuous, high-fidelity data on patient health, providing a more comprehensive picture of drug effects in a real-world setting. AI algorithms can analyze this vast stream of data to detect subtle changes or adverse events that might be missed with traditional, intermittent monitoring. Patient recruitment has been accelerated through NLP analysis of electronic health records (EHRs), with companies like Deep 6 AI identifying eligible candidates 10× faster than manual methods by extracting relevant clinical features from unstructured physician notes (Sorbello et al., 2023). The role of AI in preclinical and clinical development has been shown in Figure 4.2. AI can sift through massive amounts of patient data to identify individuals who meet specific inclusion and exclusion criteria for clinical trials, significantly reducing the time and cost associated with recruitment. This is particularly beneficial for rare diseases or trials with very specific patient profiles. Furthermore, AI is being used to optimize trial sites, predict enrollment rates, and monitor trial progress in real-time, leading to more efficient and adaptive clinical trial management (Ghasemi & Khoshnevisan, 2020). The integration of AI throughout the preclinical and clinical development phases promises to streamline the entire process, bringing safe and effective drugs to patients faster.

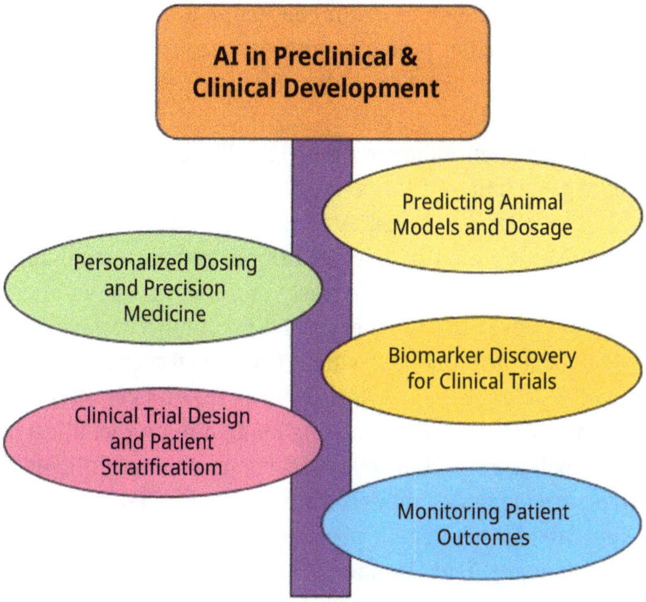

Figure 4.2: Artificial intelligence enhances preclinical and clinical development by predicting drug safety and efficacy, optimizing trial design and enabling data-driven decisions.

4.8 AI in Drug Safety and Toxicology

Computational toxicology has advanced significantly through deep learning models that predict adverse effects from chemical structure, with DeepTox identifying hepatotoxic compounds with 87% accuracy by recognizing complex molecular patterns associated with liver injury (Saikia et al., 2022). These AI models can analyze molecular descriptors and predict various toxicological endpoints, reducing the reliance on traditional, time-consuming, and often animal-intensive testing methods. This early prediction of potential toxicity allows for the deselection of problematic compounds, improving the safety profile of drug candidates even before preclinical studies. In silico approaches have reduced animal testing requirements through organ-on-chip technology integrated with computer vision algorithms that quantify cellular responses to drug exposure (Clarke et al., 2021). Organ-on-chip systems mimic human organ physiology in vitro, providing a more physiologically relevant model for toxicity testing. AI-driven image analysis of these systems can extract quantitative data on cellular health, morphology, and function, providing a high-throughput and ethical alternative to animal models. Drug-drug interaction (DDI) prediction has improved through transformer networks like DrugBERT, which analyzes pharmacological properties to forecast cytochrome P450 interference with high precision (Korotkov et al., 2022).

DDIs can lead to serious adverse events and are a major concern in drug development and clinical practice. AI models can predict potential interactions by analyzing the chemical structures and known pharmacological profiles of drugs, helping clinicians and researchers avoid harmful combinations.

Post-marketing surveillance has been enhanced through NLP analysis of social media and EHRs, enabling rapid detection of adverse events. This is demonstrated by the FDA's (Food and Drug Administration) Sentinel System that identified cardiovascular risks associated with rofecoxib years before traditional pharmacovigilance methods (Chakraborty & Tiwari, 2025). AI can continuously monitor vast amounts of real-world data, including patient-reported outcomes, social media discussions, and EHRs, to identify emerging safety signals much faster than traditional manual review processes. This proactive approach to pharmacovigilance allows for quicker intervention and improved patient safety. Furthermore, AI is being applied to predict idiosyncratic adverse drug reactions, which are rare but severe reactions that are difficult to predict based on chemical structure alone. ML models trained on large datasets of patient genomics and clinical data can identify genetic predispositions to certain adverse events, enabling personalized risk assessment (Menden et al., 2019). The integration of AI into drug safety and toxicology is not only accelerating the development of safer drugs but also improving patient outcomes by providing more comprehensive and timely safety information.

4.9 Challenges and Limitations

Despite substantial progress, AI-driven drug discovery faces significant hurdles that must be addressed to fully realize its potential. One of the most critical challenges lies in data quality and availability. Training datasets are often heavily biased toward well-studied target families and diseases, neglecting under-characterized proteins and rare diseases (Huang et al., 2021). The lack of high-quality, standardized, and diverse datasets can lead to biased AI models that perform poorly on novel targets or patient populations. Furthermore, proprietary data silos across pharmaceutical companies and research institutions hinder the development of more robust and generalizable AI models. Data governance, including data sharing agreements, standardization, and curation, remains a complex issue:

- **Model interpretability** remains a critical barrier to regulatory acceptance and widespread adoption. Complex deep learning systems often function as "black boxes," obscuring the reasoning behind their predictions (Rudin, 2019). Regulators, clinicians, and researchers require transparency to understand why an AI model makes a particular prediction, especially when it comes to patient safety and efficacy. Developing XAI techniques that provide insights into model decisions is crucial for building trust and facilitating regulatory approval.

- **Regulatory frameworks** are evolving but still present significant challenges. The rapid pace of AI innovation often outstrips the ability of regulatory bodies to develop comprehensive guidelines. While the FDA and other agencies have begun to issue discussion papers and guidance on AI/ML in drug development, there are still gaps in establishing standardized evaluation protocols, validation requirements, and post-market surveillance for AI-driven products (FDA, 2023). Clear regulatory pathways are essential for fostering innovation while ensuring patient safety.
- **Integration with existing infrastructure** presents practical obstacles, particularly in legacy systems not designed for AI workflows. Implementing AI solutions requires substantial investment in computational resources, high-performance computing infrastructure, and personnel training (Vayena et al., 2018). Many pharmaceutical companies have established traditional R&D pipelines, and integrating AI seamlessly into these workflows requires significant organizational change management and upskilling of the workforce.
- **Ethical considerations** around algorithmic bias necessitate careful attention. If training data reflects existing health disparities or biases, AI models can inadvertently perpetuate or even amplify these biases, leading to inequitable outcomes in drug development and patient care (Chen et al., 2021). Ensuring fairness, accountability, and transparency in AI algorithms is paramount to prevent exacerbating health inequalities. This includes addressing biases in data collection, model development, and deployment.
- **Validation and reproducibility** of AI models are also significant concerns. The performance of AI models can vary depending on the dataset, model architecture, and training parameters. Ensuring that AI-driven predictions are robust, reproducible, and generalizable across different settings and patient populations is crucial for their clinical utility. This requires rigorous validation studies and independent verification of AI model performance.

4.10 Future Directions

The landscape of AI-driven drug discovery is rapidly evolving, with several exciting avenues poised to further revolutionize the field:

- **Systems** like BioGPT demonstrate remarkable capabilities in generating biologically plausible hypotheses from billions of scientific tokens, enabling novel target identification and drug design with unprecedented scale and efficiency (Zhou et al., 2023). These large language models, adapted for biological and chemical data, can learn complex patterns and relationships, facilitating hypothesis generation and knowledge discovery.
- **Single-cell multi-omics integration** will advance precision medicine through AI systems that map therapeutic responses to individual patient biology, enabling

truly personalized treatment regimens. By analyzing data from individual cells across multiple omics layers (genomics, transcriptomics, proteomics, and metabolomics), AI can identify subtle cellular changes and predict patient-specific responses to therapies, leading to highly targeted and effective treatments (Stuart et al., 2019). This level of granularity will allow for the development of drugs tailored to an individual's unique biological profile.
- **Automated laboratories**, such as Carnegie Mellon's Chemputer platform, are pioneering closed-loop discovery systems where AI designs compounds, directs robotic synthesis, and analyzes results without human intervention (Burger et al., 2020). These autonomous labs promise to accelerate the entire drug discovery process, from hypothesis generation to experimental validation, by minimizing human error and maximizing throughput. The integration of AI with robotics and automation will create highly efficient and scalable drug discovery factories.
- **Federated learning** approaches are emerging to overcome data silos, allowing collaborative model training across institutions while preserving proprietary information through privacy-preserving techniques (Rieke et al., 2020). This is crucial for leveraging the vast amounts of biomedical data distributed across various organizations without compromising patient privacy or intellectual property. Federated learning enables the development of more robust and generalizable AI models by training on diverse datasets (Zhou et al., 2023).
- **Quantum computing**, while still in its nascent stages, holds immense potential for drug discovery. Its ability to simulate molecular interactions with unprecedented accuracy could revolutionize drug design by enabling the exploration of chemical spaces and the prediction of molecular properties at a level of detail currently impossible with classical computers (Cao et al., 2019). Quantum algorithms could accelerate complex calculations related to protein folding, drug-target binding, and reaction mechanisms, leading to the discovery of novel compounds with superior properties.
- **Digital twins** of patients, diseases, and even entire biological systems are also emerging as a powerful tool. These computational models, built from vast amounts of real-world data and AI, can simulate disease progression, drug response, and patient outcomes, allowing for in silico clinical trials and personalized treatment strategies (Viele et al., 2020). Digital twins can reduce the need for extensive physical experimentation and provide a platform for testing various therapeutic interventions in a virtual environment.

Finally, the increasing sophistication of XAI will be critical for the widespread adoption and regulatory acceptance of AI in drug discovery (Gunning and Aha, 2019). As AI models become more complex, the ability to understand and interpret their decisions becomes paramount. XAI techniques will provide transparency into the black-box nature of deep learning models, fostering trust among researchers, clinicians, and regulatory bodies (Arrieta et al., 2020).

4.11 Conclusion

AI has fundamentally transformed drug discovery and development, compressing decade-long timelines into years while reducing costs and failure rates through computational prediction and de novo design. The integration of deep learning with structural biology, exemplified by AlphaFold's revolutionary impact on protein structure prediction, has enabled targeting of previously "undruggable" proteins through identification of cryptic binding sites. AI-enhanced clinical trials have improved patient safety and recruitment efficiency while enabling precision medicine approaches through biomarker-guided stratification. From accelerating target identification and hit discovery to optimizing lead compounds and enhancing post-market surveillance, AI has demonstrated its immense potential across the entire drug development continuum. Despite significant progress, the field must address critical challenges in data quality, model interpretability, regulatory alignment, and ethical considerations. Overcoming these hurdles will require collaborative efforts from academia, industry, and regulatory bodies to establish standardized data practices, develop robust XAI techniques, and create adaptive regulatory frameworks. As foundation models, quantum computing, and automated laboratories mature, the convergence of these technologies promises a future where AI not only accelerates drug discovery but fundamentally reimagines therapeutic development through patient-specific medicine designed at molecular resolution. The ultimate goal is to deliver safer, more effective, and personalized medicines to patients faster than ever before, ushering in a new era of healthcare.

References

Abbas, M. K. G., Rassam, A., Karamshahi, F., Abunora, R., & Abouseada, M. (2024). The role of AI in drug discovery. *Chembiochem*, *25*(14), e202300816.

Aggarwal, A., Bharadwaj, S., Corredor, G., Pathak, T., Badve, S., & Madabhushi, A. (2025). Artificial intelligence in digital pathology – Time for a reality check. *Nature Reviews Clinical Oncology*, *22*(4), 1–9.

Arrieta, A. B., Díaz-Rodríguez, N., Del Ser, J., Bennetot, A., Tabik, S., Barbado, A., . . . & Herrera, F. (2020). Explainable artificial intelligence (XAI): Concepts, taxonomies, opportunities and challenges toward responsible AI. *Information Fusion*, *58*, 82–115.

Bettanti, A., Beccari, A. R., & Biccarino, M. (2024). Exploring the future of biopharmaceutical drug discovery: Can advanced AI platforms overcome current challenges?. *Discover Artificial Intelligence*, *4*(1), 1–16.

Brenner, S., & Lerner, R. A. (1992). Encoded combinatorial chemistry. *Proceedings of the National Academy of Sciences*, *89*(12), 5381–5383.

Burger, B., Maffettone, N., Gusev, V. V., Platunov, M., Aspuru-Guzik, A., & Green, W. H. (2020). A mobile robotic chemist. *Nature*, *583*(7815), 237–241.

Caicedo, J. C., Cooper, S., & Singh, S. (2019). Deep learning for cellular image analysis. *Nature Methods*, *16*(12), 1195–1201.

Cao, Y., Romero, J., Olson, J. P., Degroote, M., Johnson, P. D., Kiefer, P. M., . . . & Aspuru-Guzik, A. (2019). Quantum chemistry in the age of quantum computing. *Chemical Reviews, 119*(19), 10856–10915.

Chakraborty, S., & Tiwari, R. (2025). A clustering ensemble method for drug safety signal detection in post-marketing surveillance. *Therapeutic Innovation & Regulatory Science, 59*(1), 89–101.

Chen, I., Szolovits, P., & Ghassemi, M. (2021). Can AI be racist?. *Harvard Data Science Review, 3*(1).

Cheng, F., Li, W., Liu, G., & Tang, Y. (2013). In silico ADMET prediction: Recent advances, current challenges and future trends. *Current Topics in Medicinal Chemistry, 13*(11), 1273–1289.

Clarke, G. A., Hartse, B. X., Niaraki Asli, A. E., Taghavimehr, M., Hashemi, N., Abbasi Shirsavar, M., . . . & Hashemi, N. N. (2021). Advancement of sensor integrated organ-on-chip devices. *Sensors, 21*(4), 1367.

Coley, C. W., Green, W. H., & Jensen, K. F. (2019). Machine learning in computer-aided synthesis planning. *Accounts of Chemical Research, 52*(3), 762–774.

Cui, C., Ding, X., Wang, D., Chen, L., Xiao, F., Xu, T., . . . & Chen, K. (2021). Drug repurposing against breast cancer by integrating drug-exposure expression profiles and drug–drug links based on graph neural network. *Bioinformatics, 37*(18), 2930–2937.

Das, S., Das, S., & Das, S. (2021). Artificial intelligence in drug discovery: A review. *Journal of Applied Pharmaceutical Science, 11*(03), 001–010.

de Souza Neto, L. R., Moreira-Filho, J. T., Neves, B. J., Maidana, R. L. B. R., Guimarães, A. C. R., Furnham, N., . . . & Silva Jr, F. P. (2020). In silico strategies to support fragment-to-lead optimization in drug discovery. *Frontiers in Chemistry, 8*, 93.

DiMasi, J. A., Grabowski, H. G., & Hansen, R. W. (2016). Innovation in the pharmaceutical industry: New estimates of R&D costs. *Journal of Health Economics, 47*, 20–33.

FDA. (2023). *Artificial Intelligence/Machine Learning (AI/ML)-based Medical Devices: Discussion Paper and Request for Feedback*. U.S. Food and Drug Administration.

Fleming, N. (2018). How artificial intelligence is changing drug discovery. *Nature, 557*(7706), S55–S57.

Flood, D. T., Kingston, C., Vantourout, J. C., Dawson, P. E., & Baran, P. S. (2020). DNA encoded libraries: A visitor's guide. *Israel Journal of Chemistry, 60*(3–4), 268–280.

Gangwal, A., Ansari, A., Ahmad, I., Azad, A. K., & Sulaiman, W. M. A. W. (2024). Current strategies to address data scarcity in artificial intelligence-based drug discovery: A comprehensive review. *Computers in Biology and Medicine, 179*, 108734.

Gawehn, E., Hiss, J. A., & Schneider, G. (2016). Deep learning in drug discovery. *Molecular Informatics, 35*(1), 3–14.

Ghasemi, M., & Khoshnevisan, A. (2020). Artificial intelligence in clinical trials: A review. *Journal of Medical Systems, 44*(11), 1–11.

Gorgulla, C., Boeszoermenyi, A., Wang, Z. F., Fischer, P. D., Coote, P. W., Padmanabha Das, K. M., . . . & Arthanari, H. (2020). An open-source drug discovery platform enables ultra-large virtual screens. *Nature, 580*(7805), 663–668.

Gunning, D., & Aha, D. (2019). DARPA's explainable artificial intelligence (XAI) program. *AI Magazine, 40*(2), 44–58.

Guo, H., Xing, X., Zhou, Y., Jiang, W., Chen, X., Wang, T., . . . & Xu, J. (2025). A survey of large language model for drug research and development. *IEEE Access, 13*, 51110–51129.

Huang, K., Xiao, C., Glass, L. M., & Sun, J. (2021). Data-driven drug discovery and development: A review. *Journal of Chemical Information and Modeling, 61*(10), 4749–4763.

Hughes, J. P., Rees, S., Kalindjian, S. B., & Philpott, K. L. (2011). Principles of early drug discovery. *British Journal of Pharmacology, 162*(6), 1239–1249.

Jumper, J., Evans, R., Pritzel, A., Green, T., Figurnov, M., Ronneberger, O., . . . & Hassabis, D. (2021). Highly accurate protein structure prediction with AlphaFold. *Nature, 596*(7873), 583–589.

Kategaya, L., Perumal, S. K., Hager, J. H., & Belmont, L. D. (2019). Werner syndrome helicase is required for the survival of cancer cells with microsatellite instability. *Iscience, 13*, 488–497.

Korotkov, E., Zaytsev, K., & Fedorov, A. (2022). Use of 6 nucleotide length words to study the complexity of gene sequences from different organisms. *Entropy*, *24*(5), 632.

Liu, X., Shi, J., Jiao, Y., An, J., Tian, J., Yang, Y., & Zhuo, L. (2024). Integrated multi-omics with machine learning to uncover the intricacies of kidney disease. *Briefings in Bioinformatics*, *25*(5), bbae364.

Macarron, R., Banks, M. N., Bojanic, D., Burns, D. J., Cirovic, D. J., Garyantes, T., . . . & Sittampalam, G. S. (2011). Impact of high-throughput screening in biomedical research. *Nature Reviews Drug Discovery*, *10*(3), 188–195.

Mansoor, S., Hamid, S., Tuan, T. T., Park, J. E., & Chung, Y. S. (2024). Advance computational tools for multiomics data learning. *Biotechnology Advances*, 108447.

Mayr, A., Klambauer, G., Unterthiner, T., & Hochreiter, S. (2016). DeepTox: Toxicity prediction using deep learning. *Frontiers in Environmental Science*, *3*, 80.

Menden, M. P., Iorio, F., Garnett, M., McDermott, U., Benes, C. H., Weinstein, J. N., . . . & Saez-Rodriguez, J. (2019). Drug sensitivity prediction from genomic and epigenomic data using a deep learning model. *Cancer Research*, *79*(1), 129–141.

Mohapatra, S., Kumar, P. A., Farooq, U., Jain, P., Khan, R., Hassan, N., . . . & Iqbal, Z. (2022). COVID 19 pandemic challenges and their management: A review of medicines, vaccines, patents and clinical trials with emphasis on psychological health issues. *Saudi Pharmaceutical Journal*, *30*(7), 879–905.

Oren, O., Neilan, T. G., Fradley, M. G., & Bhatt, D. L. (2021). Cardiovascular safety assessment in cancer drug development. *Journal of the American Heart Association*, *10*(24), e024033.

Öztürk, H., Özgür, A., & Ozkirimli, E. (2018). DeepDTA: Deep drug–target binding affinity prediction. *Bioinformatics*, *34*(17), i821–i829.

Paul, S. M., Mytelka, D. S., Dunwiddie, C. T., Persinger, C. C., Munos, B. H., Lindborg, H. R., & Schacht, A. L. (2010). How to improve R&D productivity: The pharmaceutical industry's grand challenge. *Nature Reviews Drug Discovery*, *9*(3), 203–214.

Popova, M., Isayev, O., & Tropsha, A. (2018). Deep learning for de novo drug design. *Molecular Pharmaceutics*, *15*(10), 4337–4346.

Pun, Z., Xue, Q., & Zhang, Y. (2025). Enhancing ER α-targeted compound efficacy in breast cancer threapy with explainable AI and geneticAlgorithm. *PLoS One*, *20*(5), e0319673.

Rieke, N., Hancox, J., Li, W., Liu, H., Ma, J., Miller, C., . . . & Zhang, S. (2020). The future of digital health with federated learning. *NPJ Digital Medicine*, *3*(1), 1–7.

Rudin, C. (2019). Stop explaining black box models for high stakes decisions and use interpretable models instead. *Nature Machine Intelligence*, *1*(5), 206–215.

Saikia, S., Prajapati, J. B., Prajapati, B. G., Padma, V. V., & Pathak, Y. V. (2022). The role of artificial intelligence in therapeutic drug monitoring and clinical toxicity. In *Recent Advances in Therapeutic Drug Monitoring and Clinical Toxicology* (pp. 67–85). Cham: Springer International Publishing.

Schieferdecker, S., Rottach, F., & Vock, E. (2024). In silico prediction of oral acute rodent toxicity using consensus machine learning. *Journal of Chemical Information and Modeling*, *64*(8), 3114–3122.

Schwaller, C., Hiss, J. A., & Schneider, G. (2021). Molecular transformer for chemical reaction prediction and retrosynthesis. *Chemical Science*, *12*(1), 108–118.

Segler, M. H., Kogej, T., Tyrchan, C., & Waller, M. P. (2018). Generating focused molecule libraries for drug discovery with recurrent neural networks. *ACS Central Science*, *4*(1), 120–130.

Shen, J., Swift, B., Mamelok, R., Pine, S., Sinclair, J., & Attar, M. (2019). Design and conduct considerations for first-in-human trials. *Clinical and Translational Science*, *12*(1), 6–19.

Shukla, A. (2025). From code to cure: How generative AI is reinventing drug discovery. *Available at SSRN*, 5267158.

Sorbello, A., Haque, S. A., Hasan, R., Jermyn, R., Hussein, A., Vega, A., . . . & Ahadpour, M. (2023). Artificial intelligence-enabled software prototype to inform opioid pharmacovigilance from electronic health records: Development and usability study. *Jmir Ai*, *2*, e45000.

Stokes, J. M., Lopatkin, A. J., & Collins, J. J. (2020). Artificial intelligence for drug discovery. *Nature Reviews Drug Discovery*, *19*(12), 825–826.
Stuart, T., Butler, A., Hoffman, P., Hafemeister, H., Papalexi, E., Mauck III, W. M., . . . & Satija, R. (2019). Comprehensive integration of single-cell data. *Cell*, *177*(7), 1888–1901, e21.
Topol, E. J. (2019). High-performance medicine: The convergence of human and artificial intelligence. *Nature Medicine*, *25*(1), 44–56.
Tunyasuvunakool, K., Adler, J., Wu, Z., Green, T., Zielinski, M., Žídek, A., . . . & Hassabis, D. (2021). Highly accurate protein structure prediction for the human proteome. *Nature*, *596*(7873), 590–596.
Vamathevan, J., Clark, D., Blakeley, P., Chung, P., Danley, C. E., Gill, A., . . . & Emmerich, W. (2019). Applications of machine learning in drug discovery and development. *Nature Reviews Drug Discovery*, *18*(6), 463–477.
Vayena, E., Dzenis, Y., & Trachsel, M. (2018). Digital health and ethics: A systematic review of the literature. *Journal of Medical Internet Research*, *20*(11), e11576.
Viele, K., Saville, B. R., McGlothlin, A., & Broglio, K. (2020). Comparison of response adaptive randomization features in multiarm clinical trials with control. *Pharmaceutical Statistics*, *19*(5), 602–612.
Wiegn, P., Chan, R., Jost, C., Saville, B. R., Parise, H., Prutchi, D., . . . & Burkhoff, D. (2020). Safety, performance, and efficacy of cardiac contractility modulation delivered by the 2-lead optimizer smart system: The FIX-HF-5C2 study. *Circulation: Heart Failure*, *13*(4), e006512.
Wouters, O. J., McKee, M., & Luyten, J. (2020). Estimated research and development investment needed to bring a new medicine to market, 2000-2018. *Jama*, *323*(9), 844–853.
Wu, Z., Ramsundar, B., Feinberg, E. N., Gomes, J., Geniesse, C., Pappu, A. S., . . . & Pande, V. (2018). MoleculeNet: A benchmark for molecular machine learning. *Chemical Science*, *9*(2), 513–530.
Xu, J. (2022). *Evolving Drug Design Methodology: From QSAR to AIDD.*
Yang, C., Chen, E. A., & Zhang, Y. (2022). Protein–ligand docking in the machine-learning era. *Molecules*, *27*(14), 4568.
Zhang, L., & Wen, J. (2021). Active learning strategy for high fidelity short-term data-driven building energy forecasting. *Energy and Buildings*, *244*, 111026.
Zhavoronkov, A., Mamoshina, P., Vanhaelen, Q., Scheibye-Knudsen, M., Moskalev, A., & Aliper, A. (2019). Artificial intelligence for aging and longevity research: Recent advances and perspectives. *Ageing Research Reviews*, *49*, 49–66.
Zhavoronkov, A., Vanhaelen, Q., & Oprea, T. I. (2020). The future of drug discovery: AI-driven de novo design. *Trends in Pharmacological Sciences*, *41*(12), 923–930.
Zhou, Y., Guo, H., Xing, X., Jiang, W., Chen, X., Wang, T., . . . & Xu, J. (2023). BioGPT: A large language model for biomedical text generation and mining. *arXiv Preprint arXiv*, *2303*, 00868.

Imran Sheikh, Pooja Saini, Varruchi Sharma*, Aaliya Ali, Syed Irfan Yaqoob, Vikas Kushwaha, Nitin Sharma, Anupam Sharma, Ashwanti Devi, Poonam Bansal, Vandana Sharma, Anil Kumar Sharma*

Chapter 5 Navigating Ethics and Regulations in AI-Powered Biotechnology

Abstract: Artificial intelligence (AI) is also rapidly transforming biotechnology, biomanufacturing, and precision medicine to help appropriate the discovery process, clinical decision support by means of adaptive techniques, and simplify the pipelines of considerable complexity of production. This chapter discusses the transformational opportunities of AI integration, ethical, regulatory, and sustainability issues of AI integration in the biomedical ecosystem, and AI application in enhancing target discovery, drug design, digital twins, and real-time measures, and reveals threats, such as algorithm bias, unfair access, data governance issues, dual-use issues, and environmental impacts associated with large-scale models training. Responsible utilization is guided by principles that are analyzed through ethical aspects, including autonomy, human dignity, justice, equity, data stewardship, transparency, accountability, and biosecurity. International frameworks, such as those from the World Health Organization, OECD, and UNESCO, are complemented by regional regulations, for example, the EU AI Act, Food and Drug Administration guidelines in the United States, centralized regulations in China, and the General Data Protection Law (LGPD) in Brazil, with differences and gaps among them carefully noted. Examples of AI-enhanced develop-

***Correspondence authors: Varruchi Sharma,** Department of Biotechnology and Bioinformatics, Sri Guru Gobind Singh College, Sector 26, Chandigarh 160019, India, e-mail: Varruchi.sharma2013@gmail.com

***Correspondence authors: Anil Kumar Sharma,** Department of Biotechnology, School of Biological Sciences, Amity University, Mohali 140306, Punjab, India, e-mail: anibiotech18@gmail.com

Imran Sheikh, Aaliya Ali, UCRD, Chandigarh University, Mohali, India

Pooja Saini, Department of Biotechnology, Eternal University, Baru Sahib 173101, Himachal Pradesh, India

Syed Irfan Yaqoob, AIT-CSE, Chandigarh University, Ghauran, Mohali 140413, Punjab, India

Vikas Kushwaha, Department of Biotechnology, Goswami Ganesh Dutta Sanatan Dharma College, Sector 32, Chandigarh, India

Nitin Sharma, Department of Biotechnology, Chandigarh College of Technology, CGC, Landran, Mohali 140307, Punjab, India, e-mail: abhinitu30@gmail.com

Anupam Sharma, Department of Physics, Guru Kashi University, Talwandi Sabo, Bathinda 151302, Punjab, India

Ashwanti Devi, Poonam Bansal, Department of Biosciences and Technology, MMEC, Maharishi Markandeshwar (Deemed to be University), Mullana, Ambala 133207, India

Vandana Sharma, Department of Physics, MMEC, Maharishi Markandeshwar (Deemed to be University), Mullana, Ambala 133207, India

 | https://doi.org/10.1515/9783112214275-005

ment, such as COVID-19 vaccines and CRISPR control, as well as oncology diagnostics, demonstrate real-world conflicts between speed and safety, innovation and precaution, and openness and control. The new controversies surrounding the principles of intellectual property, open science, and green AI principles remind people of the necessity of striking a balance between commercial and noncommercial interests and the greater global good. Through establishing a combination of ethical foresight and strong foreign collaboration, AI in biotechnology can cease to be a question of efficiency benefits and create a more fair, sustainable, and patient-focused biomedical future.

Keywords: AI-driven biomanufacturing, precision medicine, algorithmic bias, ethical governance, regulatory frameworks, green AI

5.1 Introduction

Biotechnological studies and precision medicine are advancing faster than ever due to artificial intelligence (AI), which has been applied to tasks of detecting patterns and image interpretation in the initial stages of research but is now being used to identify targets, generate molecular design, perform high-throughput phenotyping, adaptive clinical decision support, and create biomanufacturing pipelines, among other tasks. These capabilities have the potential to bring about dramatic increases in speed, scale, and personalization of interventions, but they also raise new and challenging ethical and regulatory conflicts. The models that work well to achieve more rapid discovery can also be used to increase bias, obscure provenance, and create dual-use risks when applied to living systems. The current AI in the biotechnology spectrum is extensive in terms of technological level, spanning a range of models that operate under supervision, such as those used by radiologists or pathologists, at one end, and end-to-end models based on multimodal data integration, including genomic, imaging, and clinical data, at the other end. Every layer is burdened with specific issues of reliability, transparency, and governance. For example, due to the sudden popularity of large multimodal models, the World Health Organization (WHO) has given specific recommendations regarding the application of such models in the context of health (WHO, 2025). Since such systems must function in extremely high-stakes areas where mistakes, unrecognized assumptions, or malicious abuse can cause patient injury, disruption to the overall health of a population, or irreversible alterations of biological systems, ethical and regulatory systems do not serve as guardrails but rather as basic infrastructure. Ethics is the transformation of values into requirements on the design and implementation of privacy, fairness, responsibility, and control by humans. In contrast, regulation is the creation of enforceable guarantees of safety, effectiveness, and traceability to maintain public trust and align business interests with the welfare of society. Such stakes are indeed serious; in recent policy practice, jurisdictions are proceeding to codify regulations specific to AI. As an example, the Artificial Intelligence Act designed by the EU has proposed a new risk-based compliance regime, which has a categorical capture of health-proximate applications. Regulators, as well, revisit the current medical product pathways that consist of adaptive AI/ML systems. With the shifts in technology and changes in technology, multilateral bodies have re-examined the cross-border principles with renewed emphasis on the need to have globally consistent governance, which is also receptive to swift innovation (Van Kolfschooten & Van Oirschot 2024). The legislation and codes of ethics cannot be reactive but ought to be proactive and adaptive. Nonlinear, iterative AI-powered biotechnology development occurs, e.g., models trained on proprietary datasets can be fine-tuned, reused, or combined with wet-lab automation to form emergent behavior, causing static rulebooks to quickly fall out of step with practice. It is this strain that has seen the international institutions and standard-setting organizations shift toward pragmatic and principle-oriented advice. This initiative is reflected in new OECD and UNESCO guidance seeking to balance universal values, including human rights, transparency, and safety, with more flexible, risk-balanced administrative solutions that local regulators in their respective jurisdictions are free to adopt under their respective legal framework. One need not look far to discern the aspirational advantages of such multilateral arrangements; such arrangements have the potential to bring the goals of data stewardship and model validation together, restrict regulatory fragmentation that otherwise will invite arbitrage, and foster interoperable technical standards, including documentation practices that consume model cards and data provenance logs, which will render oversight, as well as international collaboration feasible. Nevertheless, it is hard to reach a common decision on the international level because of the variation in regulatory powers of the nations, social priorities, legal traditions, and commercial risks tolerance. An international governance model that is effective all over the world, in turn, requires stratified international principles and mechanisms of coordination, regional regulatory tools that are specific to legal frameworks, and local implementation tools that take into account health-system preparedness and socio-ethical background (Van Kolfschooten & Van Oirschot 2024). In addition to the jurisdictional design, several substantive ethical fault lines in AI-driven biotechnology warrant close consideration. First, the promise of precision medicine is based on large, heterogeneous, and high-quality data. However, aggregated biomedical data create privacy, consent, and equity issues, as historical under-representation or biased data collection may perpetuate inequities, affecting access to care or dose recommendations through the algorithm's results. Second, most high-performing models are black box in nature, and once these models are used to influence clinical decisions or manufacturing decisions, their opaqueness makes them difficult to hold responsible, obtain informed consent, and challenge the recommendations made by such systems. Third, generative models that suggest sequences, constructs, or experimental protocols increase the risks of abuse or accidental escape and require governance to serve as a bridge between AI safety communities and biosafety/biosecurity entities. Fourth, proprietary models and data monopolies can enhance the speed of innovation, yet also create a concentration of power, which leads to price-related problems, access, and independent validation, the regulation of which regulators and ethicists need to take into account when creating market entry and post-market surveillance regimes. The concerns of contemporary scholarly and policy research are on these points. It needs to be accompanied by the joint controls encompassing the techniques of audit, the principles of data governance, and legally binding post-market surveillance to learn how AI systems behave dynamically after being exposed to the real clinical workflows. The processes of realizing such ethical and regulatory objectives need mechanisms to the ground. First of all, risk-based classification systems need to be in place since the application of AI does not harm to the same degree. This requires a compromise between the needs of the use, in the forms of documentation, testing, human oversight, and openness, against the extent of possible damage to individuals and the population. The emerging regulation of AI/ML Software as a Medical Device (SaMD) by the European Union through its risk-based approach and the emerging guidance by the US Food and Drug Administration (FDA) toward maintaining stricter controls of systems that directly impact clinical care are instances of how regulatory agencies are increasingly enacting more stringent controls to systems that have direct impact on clinical care (Van Kolfschooten & Van Oirschot 2024). Second, an oversight paradigm based on life cycles would recognize that AI systems are regularly updated and evolving. Therefore, the premarket assessment would be complemented by real-time supervision, determination of performance drift, and a clear change-management guideline to maintain high levels of safety and effectiveness throughout updates. Third, strong documentation and external auditable models, aka model cards, data sheets, provenance chains, and publicly benchmarked datasets, contribute to external review and address informational asymmetries between developers, deployers, patients, and regulators. Fourth, federated and privacy-preserving computation methods, along with transparent consent systems, can expand access to data and strengthen the rights of individuals. It is essential to build capacity among regulators, clinicians, and institutional review boards to enable human actors to exercise meaningful control, rather than referring decisions to autonomous systems. The mentioned operational pillars are frequently cited in recent literature and policy roadmaps as essential for establishing reliable AI in health and biotechnology. However, there are still significant problems. Validation and reproducibility standards are not yet fully developed in most sub-domains, such as those related to generative models that generate biological sequences. Enforcement systems are not keeping pace with innovation rates, and even intra-country coordination across health, agriculture, defense, and research sectors is inconsistent. In addition, thorny normative issues arise regarding whether to prioritize the acceleration of innovation over precaution when a new AI-designed therapeutic shows promise in clinical trials. There is also the question of how to ensure fair global access to the benefits of AI-enhanced biotechnology, particularly when data and computing capacities are concentrated in high-income settings. Such problems will not be resolved simply by better technical tools and more specific regulations; rather, distributive justice as a quantifiable goal will be operationalized by including different stakeholders, such as patients, clinicians, ethicists, experts in the field of public health, industry, and affected communities, in the decision-making process. Procedural fairness, participatory rulemaking, public deliberation, and transparent risk-benefit assessments are the recent calls by international organizations that are suggested to be complementary to technical regulation (Floridi 2023). This chapter has a threefold objective: first, to give a brief but rigorous summary of the current technological ecosystem where AI meets biotechnology and precision medicine, as well as the exemplary cases and the sources of the ethical and safety risks produced by the applications. Second, to map the new regulatory environment, it is important to point out international frameworks (OECD, UNESCO, and WHO), regional frameworks, specifically the EU AI Act, and key national frameworks, including the FDA AI/ML frameworks, and lessons learned in these regimes on risk-proportionate regulation, life cycle regulation, use of standards, and transparency (Figure 5.1 and Table 5.1). Third, to make a realistic recommendation for governance, which encompasses the ethical desiderata and enforceable policy, including scalable risk classification, required documentation and audit trails, adaptive post-market surveillance and performance monitoring, mechanisms to reduce the potential for dual-use, and avenues to provide equity and facilitate international collaboration. In this chapter, abstract principles are grounded in specific examples and case studies from AI-assisted diagnostics, AI-based drug discovery, and automated cell-therapy manufacturing, thereby demonstrating how the tension between theory and practice is applied and how divergent solutions can be operationalized. This chapter is structured in a way that starts with the basic concepts and then the practical side. It begins with a technology primer that positions AI strategies used in biotechnology and precision medicine, identifying error modes and ethical issues associated with each solution. The next part discusses the normative frameworks and international standards, providing an overview of the positions and policy tools of the WHO, OECD, UNESCO, and other organizations. It then explores the responses of regional and national regulatory authorities, including an in-depth discussion of the EU AI Act and recent FDA activities, as illustrative regimes (Figure 5.1). It also follows the mapping of regulations, introduces an implementation toolkit comprising technical and policy tools, such as document requirements, audit frameworks, monitoring protocols, privacy-protecting architectures, and stakeholder engagement strategies, which institutions can utilize to implement the regulations. It then evaluates these tools based on their feasibility, their enforceability, and distributive justice. The second last section deals with the emergent and controversial issues such as dual-use governance, intellectual property (IP) and access, and global capacity building. It is based on recent literature that advances integrated biosecurity, AI governance, and models of multi-stakeholder stewardship. This chapter is finished by a discussion of proposed policy directions and a research agenda that identifies gaps in standards, validation procedures, and coordination across the borders that must be addressed to ensure AI-assisted biotechnology offers full-scale social benefits without any unacceptable risks. This chapter will present a single system that will allow professionals, scholars, policymakers, and corporate stakeholders in the biotechnology sector to sail in the stormy but potentially promising waters of AI using a mixture of new policy enactments, technical data, and moral logic (Singh et al., 2025a).

5.2 Ethical Dimensions of AI in Biotechnology

5.2.1 Autonomy and Human Dignity

As important ethical principles that drive the use of AI in biotechnological and precision-medicine settings, autonomy and human dignity do not represent abstract ideals but rather tangible limitations that need to be applied in designing algorithms, validating them, and introducing them into clinical practice. Autonomy, which can be characterized as the ability of patients and clinicians to exercise informed, free, and meaningfully self-directed choices, can be undermined silently when the decision-support systems give high-confidence suggestions without explaining their rationale, when consent procedures do not explicitly explain the algorithm roles, or when institutional practices give preference to algorithm output rather than dialogic shared decision-making. As a result, researchers and ethicists emphasize that autonomy must be preserved by being explainable, allowing for the clarification of doubts and providing patients and clinicians with the ability to comprehend, argue, and prevail over AI recommendations when they are inconsistent with their values or circumstances (Hildt 2025).

Human dignity, the claim that every individual has an inherent value that should be respected with more than instrumental treatment, is also being threatened when algorithms run the triage, risk stratification, or treatment choice in a manner that results in individuals being characterized as cohorts of data points, probabilistic labels, or cost-effectiveness ratios. The existence of a real tension between population-level optimization and the maintenance of narrative, relational, and value-laden elements of care that characterize dignified medicine is created by the ability of precision medicine to stratify and customize care and, in doing so, potentially hide individuality behind genomic or multimodal profiles (Johnson et al., 2021). Ethical implications are practical and multifaceted. At the clinician-practice interface, the evidence suggests that physicians' professional autonomy can be diminished by opaque or authoritative AI tools to which clinicians become accustomed or pressured to adhere, thereby transferring moral responsibility and risking moral de-skilling unless systems are intentionally created to complement, rather than replace, human judgment. Based on this, to maintain clinician agency, specific training, delimitation of responsibility, and system affordances that reveal uncertainty and thinking patterns are necessary to ensure clinicians are the ultimate moral agents in patient care (Funer & Wiesing 2024). In the

Table 5.1: Ethical and regulatory frameworks guiding responsible AI in biotechnology and precision medicine.

Level	Framework/ regulation	Key focus areas	Representative ethical/technical principles	Relevance to biotechnology and precision medicine
Global	World Health Organization (WHO, 2025)	AI ethics and governance in health	Safety, transparency, accountability, and human oversight	Provides global ethics guidance for AI deployment in health and genomics; promotes equitable access and safeguards against algorithmic misuse.
Global	Organization for Economic Co-operation and Development (OECD, 2024)	Reliable AI systems	Human-centered values, robustness, and risk management	Encourages interoperable standards for responsible innovation in biotech and AI.
Global	UNESCO (2023) Recommendation on AI Ethics	Rights-based global AI framework	Human dignity, inclusion, and sustainability	Sets universal values supporting fairness and equity in AI-powered biotechnology.
National (USA)	FDA AI/ML Action Plan	Software as a medical device (SaMD)	Transparency, life cycle monitoring, and post-market surveillance	Provides premarket and post-market guidance for adaptive AI in precision medicine.
National (China)	AI and Data Security Regulations	Sector-specific AI control and compliance	Data localization, algorithmic transparency, and state oversight	Regulates data-driven biotech AI under strict information control and ethical review.
Regional (EU)	EU Artificial Intelligence Act (2024)	Legal regulation of high-risk AI	Risk-based classification, documentation, and conformity assessment	Classifies AI used in biotechnology and medical devices as "high-risk," ensuring compliance and traceability.

patient-clinician relationship, there is a moral imperative to protect against depersonalization. Listening therapy, trade-off negotiations in the face of patient values and existential concerns cannot and may not be assigned to an algorithm without cautious precautions, as dignity is practiced and enacted through relationships and practices of respect, which are not necessarily reflected in technically correct results (Bouderhem 2024). Practically, the balance between automation and human control implies the need to implement an ecosystemic approach, which should be characterized by regulatory principles, enforcing transparency and validation, institutional policies defining the conditions in which AI outputs can affect care, education programs that promote literacy about algorithmic strengths and limitations, and workflow frameworks that place the human actor at the top of the final decision-making. These schemes enable systems to offer augmentative intelligence, which enhances precision and fairness while maintaining final normative decisions in the hands of humans (Elgin & Elgin 2024). In addition, since algorithmic systems can reproduce or amplify structural biases inherent in data, the protection of autonomy and dignity gains an equity aspect. Unless remediator interventions are implemented, marginalized patients may be misclassified, undertreated, or devalued based on models trained on unrepresentative datasets, notwithstanding the allocation of resources or choice of treatment, regardless of correct models. Thus, to exemplify ethical stewardship of AI, technical solutions (such as bias auditing and fairness constraints) and governance (including stakeholder engagement and transparency to communities impacted) should be prioritized in the same list, in order to respect the individuals in heterogeneous populations (Dankwa-Mullan 2024). Finally, the normative question that ought to guide developers, clinicians, and policymakers does not simply limit itself to the pragmatic question of feasibility. However, the issue is whether the automation is proper in its way, purpose, and extent. The answer to this question should be based on the fact that preserving human dignity, substantive autonomy and maintaining trust in the clinician -patient relationship are fundamental ethical principles that must guide the design, deployment and governance of healthcare technologies. Practical measures such as mandating explainability thresholds in high-stakes choices, expressing consent in algorithmic involvement, mechanisms of recourse in case patients disagree with algorithm suggestions, and continuous empirical evaluation of the impact of automation on patient experience and outcomes are all ethical imperatives if AI succeeds in realizing its potential in biotechnology without subverting the human essence of medicine (Funer & Wiesing 2024).

5.2.2 Justice, Equity, and Access

Justice, equity, and access are among the primary ethical requirements for the global introduction of AI-based biotechnologies, as the disparities in who benefits, who is harmed, and who suffers are likely to be institutionalized within the prevailing frameworks unless addressed proactively. Recent empirical studies have also shown that AI health applications, in particular those designed in wealthy places (the Global North) and implemented or tested in resource-constrained settings (the Global South), are less effective, with inappropriate predictive accuracy and even harmful biases attributable to divergences in population genetics, disease predicaments, healthcare infrastructure, and socio-environmental determinants. As an example, Hussain et al. (2025) carried out an integrative scoping review of AI health applications in the Global South, finding that tools trained on high-income country data tend to fail to translate into any adequate performance or applicability in emerging economies, a failure not only due to infrastructural gaps but to nonrepresentativeness in the data as well.

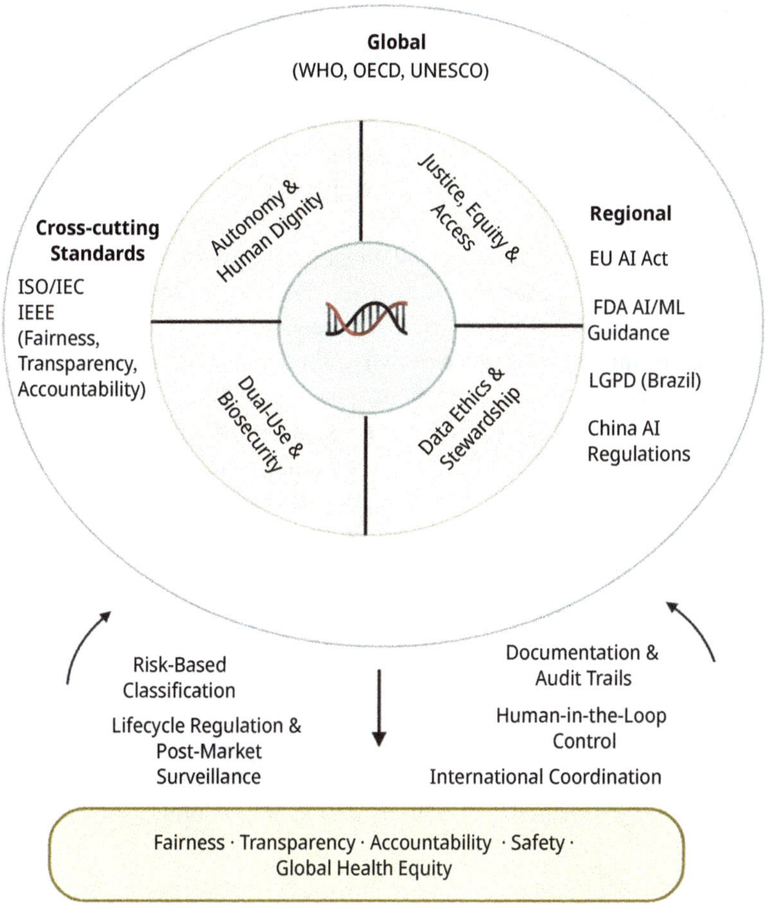

Figure 5.1: Integrated ethical-regulatory framework illustrating the convergence of ethical principles, regulatory instruments, and governance mechanisms for responsible AI in biotechnology and precision medicine.

Even in the clinical trials themselves, inequalities exist. Yang et al. (2024) examined 159 AI-enabled clinical trials, finding that they experienced high levels of geographic and gender disparities; of the trials, there were considerably fewer in low- and middle-income nations and underrepresentation of female participants, which increases the risk that the resulting AI models will be both less accurate, less safe, and more distrusted by such populations. The systematic underperformance of AI health algorithms, including both diagnostic and therapeutic decision-support systems, has also been reported as a manifestation of bias based on race, ethnicity, gender, and socioeconomic status, where models underperform with marginalized cohorts or suggest fewer or less aggressive treatments (Joshi 2024). Moreover, cultural, economic, and infrastructural barriers that are common in many countries of the Global South, such as the lack of a complete digital health record, limited internet access, limited local model development and validation capacity, and a lack of regulatory protections, contribute to such biases and accessibility to sophisticated AI-based treatments or accuracy diagnostics (Hussain et al., 2025). A growingly recognized phenomenon, which is currently named data colonialism, involves using Global South populations to extract data and then use it to train models, which in large part serve the interests of the Global North, with not much benefits being received by the origin communities, and in some cases, the models may pose risks to populations, such as misdiagnosis, or may be poorly suited to their disease burden (Hussain et al., 2025). Equity requires not just performance equality but also the equal sharing of opportunities. To fully enjoy the benefits of AI-based biotechnology, patients worldwide should have access to diagnostics, treatments, and biomanufacturing innovations. This means that underprivileged areas should be invested in and equipped with the necessary capacity, infrastructure, regulation, and workforce training. Based on this, the ethical implementation should include inclusive data curation (availability of diverse genetic backgrounds, environmental exposures, and lifestyles), continuous auditing of bias (to identify varying rates of error among different races, genders, and socioeconomic strata), and clear presentation of the model's performance in various subpopulations. These governance models empower communities in the Global South to amplify their voices and exert control over the development and utilization of AI tools. The only way AI-powered biotechnologies can uphold justice is by integrating the concept of justice at all levels, from data gathering to deployment, to ensure that AI-based technologies do not exacerbate existing disparities but maintain their potential to improve health outcomes for all people, rather than benefiting some at the expense of others.

5.2.3 Data Ethics

One of the underlying issues in the implementation of AI-enabled biotechnology and precision medicine is the area of data ethics, particularly in relation to the way genomic, proteomic, metabolomic, and other forms of omics data are gathered, disseminated, governed, and safeguarded, as a failure in this area can result in a lack of trust, infringement on individual rights, and exacerbation of disparities, as well as creation of a loophole to mishandle highly sensitive information. Informed consent is one of the main ethical demands in sharing genomic and biomedical data, as it is stated in the new principles of ethical data collection and sharing of human genomes provided by the WHO, individuals should be not only informed about the type of data (genomic, phenotypic, clinical) that will be collected but also of how and where this data will be used and what purposes will it serve and what period of time will it be used, as well as what are their rights to withdraw or restrict further use. Equally, the National Institutes of Health guidelines under the Genomic Data Sharing Policy mandate that consent forms list the types of data, future applications, and access permissions. The models of consent require additional adjustments in a variety of situations, particularly those prevalent in low- and middle-income countries. Qualitative empirical studies involving H3Africa groups have indicated that in most cases, participants anticipate feedback, information regarding the use of their data in downstream research, and continuous communication and appropriate measures, not just broad consent (Tindana et al., 2020). In addition to consent, the issues of ownership and stewardship are also at stake: individual-level health data cannot be merely accommodated under the binary owning vs. not owning framework because neither extreme is sufficient to uphold the ethical principles of autonomy, justice, fairness, or mitigation of risks. However, as Piasecki & Cheah (2022) present, governance structures like Data Access Committee and managed access, or data trusts and cooperatives, may provide the hybrid or distributed model of control and gain by having the patient and researcher. Recent suggestions, such as the Collaborative Health Data Ownership framework, attempt to conceptualize shared ownership, clear decision rights regarding data reuse, attribution, and sharing of benefits in the AI-powered integrative health care environment, but international inconsistency, undefined permission with regards to future use, and tension over who holds the keys to the data insights are real dangers (Liu & Guo 2024). The third area of criticality is in averting misuse of the sensitive multi-omics data. The enormous size, granularity, and interrelatedness of genomic, transcriptomic, epigenomic, proteomic, metabolomic, and environmental exposure data multiply the chances of reidentification, discrimination, group harms, and even the application of data in ways beyond the expectations of the participants. For example, the body of systematic review literature on exposome research has raised concerns that large multi-omics datasets can identify socially sensitive/heritable risk factors, which may implicate family members, communities, or entire populations in genetic/environmental exposures without their knowledge or consent (Safarlou et al., 2023), stewardship, data cooperatives, and democratized governance. Institutional forms of stewardship, data cooperatives, and other collective governance models provide mechanisms for individuals and communities to manage their data responsibly, ensuring equitable access, transparency, and accountability while safeguarding privacy and promoting shared benefits. A recent paper on Data Cooperatives outlines the models by which people can collectively control their data, make or delegate rights to manage access, and share in economic or other benefits without taking away sovereignty and ethical control. Additionally, data stewardship literature emphasizes the importance of role clarity (data stewards vs. controllers), standardization within the FAIR (Findable, Accessible, Interoperable, Reusable) principles, and AI-readiness in terms of data quality, interoperability, and governance capabilities. All these require striking a balance between conflicting values: how to ensure the scientific and social utility of large, integrative datasets without compromising individual consent, privacy, dignity, or allowing secondary abuse, whether in the form of privacy invasion, discrimination, or commercial exploitation. Biotechnology should be guided by ethical data stewardship, however, based on sound, open governance, tiered consent models, community consultation, secure and equitable stewardship, and effective legal/regulatory safeguards; only in these safeguards can precision medicine and biomanufacturing put AI-driven omics technology to work, in a manner that acknowledges persons as the origin of data and/or potential beneficiary, but not raw material to run algorithms.

5.2.4 Dual-Use Concerns and Biosecurity

Among the most urgent ethical aspects of AI in synthetic biology and biotechnology, more generally, dual-use issues and biosecurity issues intersect, since AI-based biodesign pipelines, generative modeling, and experimental automation implant changes create many opportunities to develop or modify biological agents, which increases risks that new technologies will be used in harmful ways. More recently, some examples have shown that large language models, when prompted accordingly, can be used to design new toxic proteins and small molecules with similarities to existing toxins, which highlights that biodesign pipelines may enable misuse (Hattoh et al., 2025). The merging of biological design tools with AI intensifies the alarm: with the use of AI-enabled synthetic biology, the design-build-test-learn cycle will be shortened to the extent that the actors with less wet-lab capacity may nevertheless produce dangerous agents, particularly when predictive models start to approximate pathogenesis or transmissibility (Science et al., 2025). Based on this, the ethical responsibility requires models of responsible innovation and risk management: the whack-a-mole governance challenge in the synthetic biology literature introduces the new models of policy that bridge command-and-control, stewardship, bottom-up, and more adaptive governance models to oversee and preempt the evolving biorisks instead of merely responding to their abuse (Undheim 2024). The most recent example is the Biosecurity Risk Assessment of the Use of Artificial Intelligence in Synthetic Biology, which provides valuable tools and prioritized risk measures that would map AI applications (gene editing, de novo gene design, automated lab experiments, etc.) on threats, vulnerabilities, likelihood, and mitigation options, thus providing a roadmap of how dual-use can be managed without unnecessarily crippling beneficial research (De Haro 2024). The open-source AI-biotech tools have particular ethical liabilities: although openness enhances innovation and democratization of access, by making global research possible, allowing small labs and underserved areas, it also opens such models or datasets to extensive abuse, especially when those are also open without security measures such as user validation, filtering of outputs, or restraints on specific features. In the article "Artificial intelligence and biological misuse: differentiating risks of language models and biological design tools," the author believes that open access to powerful tools should be balanced with the possibility of nonexpert abuse, and introduces differentiated access, gradual release, and prerelease assessments as the elements of the governance mix (Sandbrink 2023). In addition, the recent "A Call for Built-in Biosecurity Safeguards for Generative AI Tools" highlights the importance of built-in, AI-native protection systems, namely the use of watermarking, safety alignment, refusal or constraint mechanisms, anti-jailbreak design, and unlearning dangerous features, to complement regulatory and policy oversight. Meanwhile, worldwide collaboration and openness are required; regional biosecurity efforts, e.g., in Southeast Asia, are already experimenting with how to harmonize control, establish responsibility in screening DNA synthesis, and devise means of reporting and monitoring to verify that new AI-assisted synthetic biology studies do not fall into dangerous waters before being noticed (Wheeler 2025). The moral obligation will also imply that AI-biotech tools developers should anticipate the existence of dual-use opportunities during the initial phases of design, publish open-source tools, incorporate effective access controls, and perform ongoing risk evaluation and risk management. It also requires an insight into accountability, such as who takes which risk in the event of a misuse of a tool, where the liability is located, and what due diligence ought to resemble in research laboratories, industry, and government. Without such multilayered governance, transparency, and design safeguards, serious biosecurity issues may negate the potential of AI in biomanufacturing and precision medicine; ethical governance cannot be regarded as a hypothetical distant perspective of responsible innovation, but an important support pillar thereof.

5.3 Regulatory Frameworks and Guidelines

5.3.1 Global Landscape of AI Regulations

The global regulatory landscape on AI in biotechnology is rapidly developing into a multilayered structure that unites familiar medical products regulation with new AI-specific needs; national medical devices regulators, such as the US FDA, have revised existing SaMD regulatory frameworks to regulate AI/ML-enabled devices, released action plans, premarket expectations, and post-market surveillance recommendations, and have, concurrently, published lists of approved AI-enabled medical devices to increase transparency (Singh et al., 2025b). The European Medicines Agency (EMA) and the broader network of European medicines regulation in Europe are pursuing specific AI work plans and guidance to integrate AI throughout the medicinal product life cycle, and harmonize data-driven regulatory practice even as the European-wide Artificial Intelligence Act has created a risk-based legal framework that directly regulates high-risk systems using AI and obligates documentation and transparency, as well as conformity assessment (Madiega 2021). The United Kingdom Medicines and Healthcare products Regulatory Agency has published an AI regulatory strategy focusing on patient safety and innovation by 2030. Other larger regulators, including Japan Pharmaceuticals and Medical Devices Agency and India Central Drugs Standard Control Organization, have also published reports, work plans, and guidance documents that attempt to harmonize existing rules about their devices with new AI-related concerns, such as Predetermined Change Control Plans, good machine-learning practices, and the exceptional life cycle management of adaptive algorithms. Cross-border coordination through normative soft law at the international level is shaping expectations. The WHO has published ethics and governance guidance explicitly oriented toward AI in health (including with reference to large multimodal models), the OECD has revised its AI principles to provide interoperable standards of reliable AI, and the recommendation on the ethics of AI by UNESCO offers a global standards of rights-based AI that many states can draw on when harmonizing domestic policy with international standards (Morandín-Ahuerma 2023). In addition to these institutional initiatives, there are regionally oriented legal measures and political resolutions. The EU AI Act stipulates tangible responsibilities in risk-classification, transparency, and regulation of general-purpose and high-risk models, whereas the US policy up to now has preferred non-binding rights-based frameworks, the best-known being the White House Blueprint to an AI Bill of Rights that provides principles (safe and effective systems, discrimination protections, privacy, notice, and explanation) and informs federal procurement. Significantly to biotechnology and precision medicine, these regulatory strands overlap around a number of practical imperatives, including strong premarket validation and performance criteria for AI-based diagnostics and therapeutic decision-support, documentation, dataset provenance, and bias testing requirements to support equitable performance across populations; post-market monitoring and reporting of incidents to detect model drift and adverse outcomes; explicit allocation of responsibility among developers, deployers, and clinicians; and transparency requirements to support explainability, reproducibility, and patient-facing disclosure (Zanca et al., 2022). However, the torment of innovativeness and jurisdictional diversity generates tensions. Industry associations and confident nation leaders lament the potential compliance burden that may smother innovation, while patient champions and ethicists demand stricter regulations in high-stakes areas. Thus, global cooperation (WHO, OECD, UNESCO, and regional regulators) is playing a significant conciliating role, generating practical toolkits, mutual principles, and capacity-building to help low-resource regulators to access AI risks and benefits (Figure 5.1).

5.3.2 Data Protection and Privacy Regulations

Data protection and privacy regulations form the legal and ethical backbone for safe AI deployment in biotechnology and precision medicine, shaping what kinds of personal health and multi-omics data can be collected, how they may be processed, where they can travel, and what technical and organizational safeguards are required to protect individuals and populations. In Europe, the General Data Protection Regulation (GDPR) treats health and genetic data as special categories subject to heightened safeguards and prescribes legal bases, transparency obligations, data-minimization, and stringent rules for international transfers (adequacy decisions, standard contractual clauses, binding corporate rules), thereby profoundly shaping cross-border biomedical research and the design of AI pipelines that rely on pooled datasets. HIPAA imposes sectoral limits on the privacy and security of the so-called protected health information of covered entities, as well as their business associates, governing privacy and security, patient rights, and disclosure restrictions, and regulators are currently undergoing modernization to mitigate risks in AI models trained on electronic health records, including the problem of model inversion and mass reidentification. The General Data Protection Law (LGDP) of Brazil has proposed GDPR-like safeguards, such as special treatment of health and genetic data and an active supervisory authority (ANPD) to guide on the legal basis, as well as on cross-border flows, which show that non-EU jurisdictions are aligning their protection norms with those of GDPR, whilst also establishing the priorities of enforcement locally (de Lucena et al., 2024). The new Digital Personal Data Protection (2023) Act in India and the new underlying regulations will also be an initial step in the context of managing data in large and diverse populations. The provisions on lawful processing, the provisions on data fiduciaries, and the provisions on cross-border transfer will determine how the Indian research institutions, commercial laboratories, and multinational collaborations can conduct AI-driven research in biotechnology discovery and retain the rights of individual persons. All these national and regional frameworks, however, intersect with practical and ethical concerns, which are uniquely acute for biomedical data: the technical challenge of true anonymization of genomic and multi-omics data (where a small number of variants can reidentify individuals), the practicality of supporting data utility to facilitate discovery, and the mosaic of divergent rules across borders to transfer data, which can slow or thwart cross-border research projects, challenges noted in recent EU-US transfer analyses and multinational health-data cooperation more broadly (Lalova-Spinks et al., 2024). Privacy-preserving AI systems, such as federated learning, secure multiparty computation, homomorphic encryption, or differential privacy, are developing in response as viable strategies to enforce regulatory needs with scientific collaboration. Federated learning can be utilized to train models across distributed institutional silos, so it adheres to the constraints of data localization and data transfer, and systematic reviews of federated solutions in healthcare have shown that increasingly more applications are possible, to images, EHR, and genomic uses – albeit at the cost of the heterogeneity, governance, and composed privacy risks (Teo et al., 2024). Solutions to alternative cryptography, including multi-key homomorphic encryption, also allow computations on encrypted genomic data and are under development in bioinformatics to carry out privacy-preserving, aggregated analysis. But there is a trade-off between these methods in terms of computational cost, statistical noise, and implementation complexity, and which regulators and institutions should take into account when assessing privacy claims (Aherrahrou et al.,

2024). Besides the technologies, regulatory compliance in AI-driven biotechnology is required to be holistically governed: explicit data-sharing conditions and provenance logs to meet the demands of transparency under GDPR/HIPAA/LGPD/Digital Personal Data Protection, effective data-protecting impact analyses of high-risk AI-driven applications, contractual and liability models, business-to-business and business policies to create audit capabilities, and incident-response needs in the face of growing cyber threats and due to regulatory scrutiny of national regulating agencies. Notably, reconciling privacy protection with fairly researchable science is vital because the privacy protection should not inadvertently place groups of people in an underprivileged situation in the advancement of science. Prudent policymaking cannot afford to restrict safe and ethical access to diverse groups to the fruits of AI-related developments, rather than adding a border to the further concentration of AI-driven advances around resource-rich jurisdictions. Due of the ever-evolving nature of legal regulations and difference in priorities between enforcement, stakeholders in the biomedical manufacturing and precision medicine industry must adopt adaptive compliance, such as maintaining regulatory monitoring, investing in privacy-oriented engineering, documenting technical and organizational controls, and participating in global standard-setting so that the radical potential of AI is realized without impinging on individual rights, general confidence, or collaboration in biomedical advancement that forms the basis of global cooperation (Eden et al., 2025).

5.3.3 Regulatory Challenges in AI-Powered Biomanufacturing

The promise of AI-assisted biomanufacturing, ranging from process analytical technology, digital twins, predictive maintenance, and real-time quality control to optimization of upstream and downstream operations, carries with it a suite of regulatory challenges that stretch the boundaries of existing GMP (good manufacturing practice) regimes, quality assurance mechanisms, and oversight models. Regulatory agencies such as the FDA, EMA, and Medicines and Healthcare products Regulatory Agency have begun to wrestle with how to ensure AI/ML integration adheres to GMP's core demands of consistency, traceability, validation, and risk control, even as AI introduces non-determinism, model drift, and adaptive behavior, which traditional batch-wise validation frameworks are ill-equipped to accommodate (Niazi 2025). To illustrate, the FDA FRAME project and the new draft reflection paper on AI use in the drug life cycle by EMA highlight that marketing authorization holders need to demonstrate that the system is based on algorithms and models that meet GxP requirements such as data integrity, functionality under diverse conditions of operation, documentation of training data provenance, and strong change management of models that change over time (Saha et al., 2023). The other relevant challenge here is quality assurance in AI-assisted bioproduction. Biologics and vaccines, which are produced using living systems, contain complex purification workflows and are sensitive to environmental and raw-material variation, require control over critical quality attributes, and AI systems to monitor or control production must be tested not only against historic data but also against perturbations, upstream, and downstream variation, and to avoid over fitting, bias, or drift that could result in loss of product quality, potency, or immunogenicity. These settings are becoming increasingly differentiated as multivariate analysis, reinforcement learning, and neural networks are increasingly used. However, there is a big divide between the algorithm's performance on a pilot scale and regulatory acceptability at a commercial GMP scale (Khuat et al., 2023). The example of vaccine production can educate. During the COVID-19 pandemic, the use of mRNA vaccines was scaled to new levels and put them on new delivery schedules, and predictive modeling using AI was used to predict yields, optimize lipid nanoparticle formulations, and track process deviations. But some regulatory agencies have since indicated that lots brought to the market via emergency use are mostly post hoc QA and do not incorporate a fully validated AI-supported control system, posing a tradeoff between speed and full GMP validation. Further, complex upstream cell culture, downstream filtration, and chromatography steps, often used in biologics production, are sensitive to micro-variation. AI applications to control processes must meet regulatory requirements of reproducibility, scalability, sensor and actuator validation, and sensor drift or calibration bias, and record all change management.

An example is digital twins, which are increasingly considered as potent simulators of vaccine adjuvant production and biologic formulation, yet a recent review indicates that despite the ability to reduce deviation, trajectory toward greater productivity, and a human-machine interface, the models still face regulatory uncertainties regarding the validation of their predictive fidelity, auditability, data integrity, and accountability when the suggestions of the twin result in an out-of-specification batch (Shahab et al., 2025). Life cycle management of AI tools, particularly when those tools are adapting or learning after deployment, is also a concern to regulatory authorities, as GMP typically assumes fixed, validated processes. Therefore, change control plans, revalidation, drift, environmental variability, and worst-case scenarios must be specified. The harmonization across jurisdictions is still in its infancy, as manufacturers with a presence in several areas must meet conflicting or redundant requirements for data processing, risk evaluation, explainability, audit logs, and performance analysis. In total, the regulatory issues of AI-based biomanufacturing require both reforming the structures and the addition of new expectations for adaptive models, explainable AI, adequate validation across scales, and case-by-case negotiation among developers, manufacturers, regulators, and quality assurance specialists to ensure that the advantages of AI (better efficiency, less waste, greater consistency, possibly better access) are achieved without compromising the safety of products, public trust, and regulatory integrity.

5.3.4 Regulatory Pathways for AI in Precision Medicine

Regulatory pathways for AI in precision medicine are evolving rapidly but remain both fragmented and foundationally important because approval of AI-enabled diagnostic tools and therapeutics must reconcile traditional medical-product routes with the unique life cycle dynamics of software-based, data-driven systems commonly regulated as SaMD, so regulators have adopted tailored guidance, premarket expectations, and inventor-friendly yet safety-focused frameworks to assess performance, robustness, dataset provenance, clinical validation, and intended use (Fraser et al., 2023). The FDA AI/ML SaMD Action Plan and related CDRH initiatives in the United States express plans of market authorization and introduce the concept of Predetermined Change Control Plans to control planned model changes, and consistently expect detailed documentation (training data, cross-population validation, bias testing) to accompany market authorizations (Khinvasara et al., 2024). The EMA and the wider European regulatory system are also publishing reflection papers and directions that incorporate AI aspects throughout the life cycle of medicinal products, prioritizing conformity, transparency, and life cycle management. Meanwhile, the overall AI regulatory framework provides a risk-based overlay to high-risk medical AI, further informing evidentiary and conformity requirements. Adaptive regulatory models, in turn, are being developed to respond to continuously learning systems: instead of regarding an algorithm as fixed at approval, regulators are exploring strategies that require manufacturers to provide strong change-control procedures, monitoring mechanisms of model drift, explicit process triggers of revalidation, and versioning/audit trails which would allow traceability and accountability in the situation of models adapting after approval. These evolutionary pathways present challenging trade-offs between innovation and patient safety; hence, regulatory science programs are actively constituting practical tools to identify input changes, measure performance across sites, and understand the causes of performance variation in actual clinical situations (Babic et al., 2025). The use of post-market surveillance and integration of real-world evidence is emerging as a core element of the regulatory approach to AI in precision medicine. Regulators now regularly require plans of post-market performance monitoring, adverse-event-reporting mechanisms, and application of real-world evidence based on electronic health records, registries, and clinical practice to determine ongoing safety, effectiveness, and equity, thus transforming a previously one-time evidentiary dossier into a continuous evidence stream, which regulators can use to inform regulatory decisions, change labels, or withdraw. The examples of the case highlight both potential and shortcomings. AI diagnostics have been given the green light based on retrospective evaluation and limited prospective data, but on-the -independent reviews show that reporting gaps and uneven reporting of diverse datasets and practical use and performance continue to be unaddressed, and that further post-market surveillance, uniform reporting standards, and the definition of accountability between developers, providers, and regulators is necessary (Goswami et al.,

2025). To reach strong regulatory tracks, international coordination, further investments in regulatory science to authenticate monitoring instruments and criteria for adaptive algorithms, and explicit regulation about responsibility, disclosure, and equity are required. AI-based precision medicine will develop safely, progressing beyond its alluring prototypes to trustworthy, constantly overseen clinical devices that may provide an advantage to populations without undermining patient safety or community trust (Babic et al., 2025).

5.4 Emerging Issues and Debates

5.4.1 Algorithmic Bias and Discrimination

The discussion of "Algorithmic Bias and Discrimination" in AI in biotechnology, biomanufacturing, and precision medicine is now a matter of practice, as well as theory, with machine learning models trained on genomic and clinico-genomic data capable of simultaneously accelerating the discovery process and reinforcing health inequities. The discussion of algorithmic bias and discrimination in the chapter focuses on understanding how biased data and model design can create systematic errors, how those errors are manifested in genomics-specific examples, and what technical and governance solutions to the problem are underway. The main cause of most issues is the historical lack of diversity in genomic reference datasets and genome-wide association study cohorts. Historical datasets are disproportionately sampled of people of European ancestry, which leads to genetic models and sets of variants with less generalizability for underrepresented groups. Through this bias, there can be reduced accurate diagnosis, risk assessment, and treatment hypotheses of these groups. Recent reviews and analyses have highlighted the continuation of a gap in diversity and its clinical implications. At the same time, even platforms like "All of Us" have been created to improve the representation, there are still tangible gaps in the content and practical limitations (Fitipaldi & Franks 2023). The concrete evidence of these abstract risks is evident in the performance differences in polygenic risk scores that are trained on mixed-genetic populations of European origin, in other words, based mostly on European-ancestry cohort studies. These scores show significantly lower predictive accuracy over the range of genetic ancestry, with very strong amplifications of the individual-level that would, uncorrected, increase disparities, not diminish them (Ding et al., 2023). Genomic predictors of variants, as well as others, have also been demonstrated to be universally biased by ancestry, exhibiting uneven performance to prognosticate across populations and irregularly extrapolating compatibility to pathogenicity at one's fingertips, detecting variants in understudied cohorts (Pathak et al., 2024). By taking such unequal error rates and translating them into real clinical practice and biomanufacturing pipeline impairments, diversity may be prioritized unequally, inclusion may be unequal in clinical trials, disease risk may be misjudged, and trust in biomedical research among underserved population groups may be compromised. A multidisciplinary approach viewing bias detection and mitigation both as data-engineering and social-technical problems has become the common ground of the technical literature and applied work. Countermeasures to address size-driven differences in the data include intentional cohort diversification, reweighting, and resampling to balance the size of training sets and principled generation of synthetic data to increase the size of rare subgroups, an intervention at the data level. However, caution is warranted when such interventions are being conducted to avoid artifact introduction (Corpas et al., 2025). Techniques during in-processing methods, such as adversarial debiasing, fairness-conscious loss functions, and multitask or transfer-learning, have been demonstrated to work well in model making. Current research shows that in-depth transfer-learning models and multi-ancestry can be used to offer improved predictive capabilities with data underrepresented groups without negatively affecting overall performance in better-represented groups (Gao and Cui 2024). In scenarios where retraining is not viable, post-processing protective measures like subgroup-level threshold adjustments, post-processing calibrations, and audits of an algorithm system to more effectively gauge disparate impact are operationally useful interventions. Detailed surveys and comparative analyses categorize these methods into families of preprocessing, in-processing, and post-processing, providing useful insights into trade-offs between definitions of fairness, utility loss, and interpretability (Hort et al., 2024; Siddique et al., 2023). In addition to technical solutions, the emerging consensus focuses on transparency, stringent assessment across ancestries and sociodemographic dimensions, and governance solutions: making the performance of models obligatory, stratified by ancestry, sex, age, and other similar dimensions, conducting regular audit of fairness, ensuring community involvement to make sure that priorities of data-collection are set based on populations that are likely to benefit, and implementing institutional and regulatory strategies to enforce equity checks prior to deployment. The current policy-focused commentary and reviews recommend standardized reporting guidelines and fairness-by-design development pipelines that integrate statistical practices with domain knowledge and ethical supervision (Alderman et al., 2025). The discipline recognizes that technical mitigation can never replace structural change. The long-term solution involves investing in a diverse range of biobanks, ensuring equal recruitment of participants, funding that incentivizes inclusion, and fostering long-term collaboration with historically disadvantaged populations, so that models accurately represent the entire population. With the increased adoption of genomics and AI into precision medicine and biomanufacturing, the chapter contends, practitioners must not consider bias detection and mitigation as optional adjuncts but rather as design requirements to specify the constraints of models, as well as the efforts to quantify and minimize disparate effects so that the benefits of biotechnology are distributed across all populations equally (Xiang et al., 2024).

5.4.2 Intellectual Property (IP) in AI-Biotech

IP issues in the age of AI-driven biotechnology are a field of most disputed and actively developing legal, ethical, and scientific debate, with issues of patentability, ownership, and openness defining the future of biomanufacturing and precision medicine. In contrast to conventional inventions, AI systems can independently produce new molecules, protein structures, genetic circuits, or biological insights, which raises the challenging question of whether results generated by AI systems should be considered new under the established patent law criteria of novelty, inventive step, and nonobviousness. As an example, the controversy surrounding debates about DABUS (Device for the Autonomous Bootstrapping of Unified Sentience) case in which AI was listed as an inventor during patent practices underscores just how uncomfortable present IP regimes are with the reality of algorithmic creativity, since most jurisdictions such as the United States and Europe have reiterated that inventorship must be a natural person, despite the courts and academics also recognizing that AI-generated chemical scaffolds or biosynthetic pathways can present a transformative application in therapeutic and industrial practice. Similar issues can arise over ownership of AI-generated datasets and predictive insights; a large corpus of genomic sequences, phenotypes, and cell culture information is usually aggregated out of public repositories, private laboratories, and consortia, yet the models trained on it generate proprietary results like new drug targets or fermentation optimization plans. The question of who owns rights is unresolved, and recent controversies in genomics and digital health highlight the inappropriateness of existing categories of IP to the value derived based on data in its form (Andanda 2019). This stress is intensified by the gap between open science and proprietary models: as open releases of AI predictions in the biological field, such as AlphaFold2, provided evidence of the immense value of free access to such predictions, companies are moving into proprietary models of large-scale biological foundation models, with access to such models controlled by licensing and trade secrets in order to maintain competitive advantage and ensure a return on investment (Yang et al., 2023). The supporters of open science argue that making AI-generated biological knowledge publicly accessible helps hasten innovation, democratize access, and prevent the monopolization of key biomedical resources. However, industry voices note that the high cost of developing models and scaling up biomanufacturing may not be sustainable without IP protection measures. Consequently, policymakers and academics are beginning to consider intermediate models, such as patent pools, data commons with graded access, and time-limited exclusivity models, that aim to strike the right balance between incentives for private innovation and the social need for fair access to biotechnological innovation (de Beer 2015). In the case of AI in biotechnology, then, the issue of IP is not just about the technicalities of the law; it is about the essence of the social definition of inventorship, ownership, and access in a world where machines make direct contributions to discovery. The path taken by future protection systems will have a significant impact on how AI positively affects biomanufacturing productivity, how AI can be used to practice personalized medicine, and how the benefits of biotechnological advancement can be shared across societies.

5.4.3 Ethical AI in Drug Discovery and Clinical Trials

AI has become a potent method of drug-discovery pipeline in recent years, as well as clinical trial conduct, having the potential to accelerate drug-discovery and deliver more cost-effective and more accurate clinical trial outcomes. It is used in early-stage target identification, lead optimization, predictive risk modeling, trial design, patient selection, real-time monitoring, and, even more fundamentally, in silico or in silico-like simulations, a technology that has changed the pharmaceutical drug discovery and testing models. However, these innovations wield along serious ethical dilemmas that presuppose questions of equity, nondiscrimination, transparency, and accountability. With the increasing concern about the application of AI to trial design, participant selection, and participant monitoring, it is time to assess how these technologies could be either readily recreating or exacerbating biases, specifically when algorithmic exclusion of underrepresented populations is used, and to find out what defensive strategies are necessary when using digital twins and virtual patients to do in silico testing. Technology AI may also be used to establish more effective and flexible protocol design. With an advantage to construct biomarkers that further narrow selection or stratification criteria, algorithms may execute many trial simulated arms or outcomes, enabling optimization of dosing, decreased sample sizes, and generation of biomarkers. These include the virtual patient, digital twins, and causal disease models, which are being utilized to make up or even replace entirely to reduce the cost and timeline to regulatory milestones (Tubbs & Vazquez 2025). In the meantime, participant selection based on AI is a potential solution: predictive analytics can be applied to select participants who will benefit the most from an intervention, making scientific research more valid and safe. Even trial-time monitoring is being reinvented. Using AI-enhanced image processing, sensor feeds, and real-time information streams, endpoints and adverse effects can be measured dynamically and on a more granular level. These mighty capabilities are not without their share of ethical hazards, however. Lack of a widely representative training dataset, such as race, sex/gender, age, socioeconomic status, or geography, may result in AI systematically failing to or improperly classifying underrepresented samples, thereby not only creating inequitable access to the trials but also inaccurately generalizing outcomes. Recent research records the existence of algorithm bias in health data and the offense that intervention based on fairness (such as the error rate capping, demographic parity) is not magic; it is also misaligned to various conceptualizations of fairness and does not have an automatic negative impact on background structural properties (Zhang & Zhang 2023). The next area of focus is the morality of AI-generated digital twins and in silico testing. Individual smart computers of different patients are computational models of the patient, which reasonably produce simulations of their physiology, molecular, and lifestyle worlds' electronic twins, which promise to predict how patients respond to drugs, virtualized trial treatments, or provide approximate control groups. This type of virtual testing will allow making it safer, reduce human risks of exposure, and speed up development (Iqbal et al., 2022). However, they also present major ethical issues. The case of informed consent gets more complicated when a digital twin is used. Privacy risks increase as larger and more complex data are involved. The responsibility is abdicated through opaque AI models (black-box risk prediction/simulation), and the regulatory landscapes have not yet kept pace with the latest technological developments. In addition, models of digital twins trained on biased or unbalanced datasets may contribute to exclusion instead of addressing it. As an illustration, a system heavily reliant on data about populations in high-income countries might overlook physiological or environmental differences when applied in different contexts, resulting in inaccurate simulations of underrepresented populations or even counterproductive downstream decision-making. Overall, even though AI promises to transform the process of drug discovery and clinical trials, making the process more effective, targeted, and monitored, or even simulated and replaced by some in silico procedures, ethical oversight is necessary. To ensure fairness, representative data are needed to ensure transparency, interpretable models, clear disclosure, and strong consent processes; to ensure justice, thoughtful consideration of the group whose needs are historically underrepresented in biomedical research is needed to ensure responsibility, and defining accountability in connection to the harms caused by AI predictions or simulations is required. This chapter will delve deeper into these new issues and debates, but will address in particular (1) the ways AI is influencing trial design, participant selection, and monitoring; (2) the dangers of algorithmic exclusion; (3) the ethical, regulatory, and social impacts of digital twins and in silico testing; and (4) suggestions of governance, oversight, and ethical design that can mitigate both innovation and equity in precision medicine.

5.5 Case Studies and Policy Perspectives

Through limitations to a pair of case studies and current policy reactions, a certain degree of rigor and practical ways of exploring them analytically demonstrate that AI can and will play a central role in biomedical innovation, particularly in areas where more disruptive technologies like AlphaFold, performing protein structure prediction systems, emerge. Meanwhile, the regulatory and even the clinical landscape has struggled with the trade-offs between faster delivery and safety that are predetermined by the acceleration of timelines caused by pandemics and the use of adaptive design of trials, not to mention ethics that are underpinned by the necessity to balance swift authorization with an in-depth long-term safety profile. Case study 1 thus forces us to question when and under what conditions of governance the public interest might be best served by algorithmic acceleration, an inquiry that has given rise to a vast amount of comment in the medical ethics and policy literature in support of prespecified adaptive trial safeguards, public distribution of model assumptions, and post-market surveillance obligations where AI substantially reduces pre-authorization evidence generation (Wibawa 2021). In CRISPR and genetic editing, AI has quickly become an inseparable design and risk-assessment aid that enhances the selection of guide-RNA, predicts off-target events, and facilitates the prioritization of therapeutic candidates, but this same feature increases the regulatory and ethical risk due to the reduced technical barriers to intervention at the expense of technical validation, combined with an increased awareness of the hereditary effects of any changes, the equitable access to cures, and a system of dual-use capabilities. Regulatory frameworks on human genome editing proposed by the WHO and others are therefore relevant. Case study 2 thus brings into focus the tradeoffs between rapid algorithmic optimization (which could be used to facilitate clinical translation) and binding, internationally harmonized guardrails for human genome editing that are responsive to consent, intergenerational risk, and distributive justice (Khoshandam et al., 2024). (3) In cancer care, the development of AI-based diagnostic tools not only through radiomics but also through pathology image-analysis applications to genomic risk stratification models has elicited specific regulatory responses regarding safety, transparency, and life cycle management. The US FDA has shifted away from informal scrutiny of Software as a Medical Device to draft life cycle guidance of AI-enabled devices and public lists of approved AI tools, new requirements on demonstrating validity, and suggested ways of ensuring ongoing learning and indicating expectation of monitoring and real-world performance. Case study 3 thus brings to the fore an important regulatory dilemma: whether to give adaptive, learning systems permission to increase diagnostic accuracy over time and still to have updates that do not introduce untested biases or efficacy decreases in populations that are not overrepresented in training sets. Collectively, these three case studies of vaccination acceleration, CRISPR optimization, and oncology diagnostics demonstrate complementary technical benefits and similar ethical failures: model development transparency, training data representativeness, prediction explicability, harm accountability, and proportional yet enforced post-implementation monitoring. Policymaking environments across jurisdictions exhibit comparative disparities in regulatory approaches, yet increasingly demonstrate convergence toward common solutions to shared challenges. The AI Act, the first global comprehensive AI law in the world, which the European Union plans to adopt, subjects high-risk AI systems (some of which address healthcare) to stringent conformity requirements, transparency, and regulation, which impose obligations directly on the life cycle documentation and risk management of biotech AI systems (Madiega 2021). The United States has developed medical-device-specific pathways (e.g., the evolving AI/ML software guidance and the Breakthrough Devices pathway at FDA) to put more emphasis on the requirements of evidence and post-market surveillance, while maintaining flexible channels of innovation, which has produced debates as to whether device-focused regulation is more than sufficient to regulate algorithmic behavior throughout research and development phases (Singh et al., 2025b). The AI regulatory landscape in China is characterized by sectoral regulation, ideological regulation (e.g., models are expected to adhere to the core socialist values), and approval mechanisms for large models implemented by state agencies, a system that promotes transparency, data regulation, and cross-border cooperation. Brazil, through its LGPD and guidance on emerging ANPD, offers a data-protective-based methodology that acknowledges rights in automated decision-making and in having them explained and reviewed, which has an impact on biomedical AI in processing personal health and genomic data. The principles of LGPD, as well as the rule-making of ANPD, also shape how biotech companies approach consent in research and productization (Dourado & Aith 2022). Comparative policy analysis illustrates some complementarities and gaps: the EU model is highly effective at pre-deployment requirements and standardized categorization of risks but needs to determine the costs of innovation to serve SMEs; the US approach is iterative and device-oriented in regulation but can yield fragmentation between agencies and uncertain expectations of premarket transparency; China has consistent state oversight that hastens deployment but may treat public accountability as defined in Western models; and Brazilian LGPD grounds patient-centric data rights that are crucial to ethical research but needs operational guidance to those driven by genomic. Politically, the way to go forward would be (a) harmonizing trans-jurisdictional expectations on model validation, data representativeness, and post-market surveillance to mitigate regulatory arbitrage; (b) instilling transparency and explainability obligations based on clinical risk; (c) enhancing data protection mechanisms (consent, de-identification, and cross-border transfer protections) to the principles of LGPD/GDPR; and (d) establishing interoperable international fora utilizing WHO governance work on genome editing and AI in health to negotiate cross-cutting. Briefly, these case studies and policy comparisons indicate that the technological potential in AI-biotech should be accompanied by multilayered governance that maintains clinical safety, social justice, and international coordination to ensure speed, maximization, and business acceleration do not overtake the social trust and governance that will ensure sustainable, equitable advantage.

5.6 Future Directions

In the future, the intersection of AI and biotechnology will require a proactive architecture of norms, institutions, and practices that ensures innovation proceeds safely, fairly, and with social approval. Some of the more recent frameworks include international and standards organizations, such as ISO/IEC JTC 1/SC 42, which recently released a set of standards on data quality in analytics and machine learning (e.g., ISO/IEC 5259-1, 5259-3, 5259-4), . System life cycle processes (ISO/IEC 5338/53 IEEE) are also making increased efforts alongside ISO, including IEEE 3198-2025, offering ways to assess machine learning fairness. The IEEE "Horizon Scanning" white paper (IEEE-SA p3995) specifies key areas of focus, such as the safeguards of the AI models, rights to their data, misuse, instability, and resilience in the supply chain, and the IEEE CertifAIEd program certifies the ethical and transparent AI commitment of organizations. Standardization efforts are key forms of scaffolding in the rewarding of fairness, transparency, and accountability in basic model development. The incorporation of multilateral cooperation is the most important. Possible applications in computational biology to which dual-use might apply involve cross-border efforts to oversee what might be considered, along with transparency in regulation and alliance, and coordinating the organization of risk-management to avert category arbitrage, safe global biotechnological innovation, and address emerging risks. The life sciences and AI communities need to invest in high-stakes applications using human-in-the-loop systems to ensure accountability, explainability, and protections against unintended consequences, so that clinicians, researchers, and ethicists can maintain control over AI decision-making, and critical decision-making is limited to human input rather than algorithmic response. At the same time, prospective and scenario planning would have to be a regular thing: the integration of biotechnology and AI is likely to create new hazards that are hard to foresee. In line with this, regulatory agencies and industry conglomerates, as well as standards organizations, must participate in proactive horizon scanning and be ready to respond to emergent risks instead of reacting to them. It is also required to involve the populace and involve policymakers, patient bodies, and affected communities by biomanufactured products, and the general population, in the process of considering acceptable risk, equitable access, privacy of data, and principles like benefit sharing. This involvement can bring trust, a sense of legitimacy, and more closely align AI innovation with the expectations of the community. Therefore, academic, industrial, clinical, and governmental bioethics committees and AI ethics boards should be empowered and sufficiently resourced to review and assess AI projects rigorously. Their role and scope should not be confined to scrutinizing research protocols to administer the management of AI implementation, ongoing audit, impact analysis of algorithms, redress processes, and application of corrective measures when bias, hype, or malicious effects of AI systems occur. Combining these aspects reversal of robust global standards, human agency, scenario planning, involvement of citizens, and effective ethical stewardship institutions are all directions to fulfilling the purpose of the future where the AI in biotechnology and biomanufacturing meets the precision-medicine potential without jeopardizing fairness, accountability, transparency, or societal trust.

5.7 Conclusion

The use of computer-assisted technologies is transforming the future of biotechnology, biomanufacturing, and precision medicine because it is significantly shortening the discovery processes, increasing productivity in the production process, and enabling targeted therapeutic interventions. However, the hope of AI would not exist in isolation from its various attendant challenges, such as ethical and regulatory, as well as sustainability dimensions. The areas of innovation discussed in this chapter include, but are not limited to, algorithmic bias, data governance, dual-use risk, IP, and environmental impact, which make innovation seem to need a balance between creativity and responsibility, equity, and a global mindset. Emerging regulatory models, such as the EU AI Act, FDA recommendations, Brazilian LGPD, and new Chinese AI regulations, demonstrate convergence with fragmentation and may be viewed as the need to develop risk-based harmonized regulation. Ethical frameworks should also protect autonomy, dignity, justice, and transparency, and build a lasting human control over key applications. The main focus of future work ought to be the implementation of global norms (e.g., ISO and IEEE), human-in-the-loop management, and foresight exercises, in conjunction with participatory policy development with empowered ethics institutions and multilateral cooperation. Finally, AI promises to transform biotechnology not only through technical progress but also through an embedded sense of fairness, stability, and trust in its usage by citizens. At that, AI can cultivate a robust, multidisciplinary, patient-centered concept of biomedical innovation that will promote scientific advancement, be compliant with social norms, and promote health equity in the world.

References

Aherrahrou, N., Tairi, H., & Aherrahrou, Z. (2024). Genomic privacy preservation in genome-wide association studies: Taxonomy, limitations, challenges, and vision. *Briefings in Bioinformatics, 25*(5), bbae356.

Alderman, J. E., Palmer, J., Laws, E., McCradden, M. D., Ordish, J., Ghassemi, M., et al. (2025). Tackling algorithmic bias and promoting transparency in health datasets: The STANDING together consensus recommendations. *The Lancet Digital Health, 7,* e64–e88.

Andanda, P. (2019). Towards a paradigm shift in governing data access and related intellectual property rights in big data and health-related research. *IIC-international Review of Intellectual Property and Competition Law, 50,* 1052–1081.

Babic, B., Glenn Cohen, I., Stern, A. D., Li, Y., & Ouellet, M. (2025). A general framework for governing marketed AI/ML medical devices. *Npj Digital Medicine, 8,* 328.

Bouderhem, R. (2024). Shaping the future of AI in healthcare through ethics and governance. *Humanities and Social Sciences Communications, 11,* 1–12.

Corpas, M., Pius, M., Poburennaya, M., Guio, H., Dwek, M., Nagaraj, S., et al. (2025). Bridging genomics' greatest challenge: The diversity gap. *Cell Genomics, 5*(1).

Dankwa-Mullan, I. (2024). Health equity and ethical considerations in using artificial intelligence in public health and medicine. *Preventing Chronic Disease, 21,* E64.

de Beer, J. (2015) Open Innovation Policy Frameworks. 95p,

De Haro, L. P. (2024). Biosecurity risk assessment for the use of artificial intelligence in synthetic biology. *Applied Biosafety, 29,* 96–107.

de Lucena, B. A., Neves, I., De Alcântara, J. B., Camarago, M. E., & Neto, A. T. M. (2024). Systematic review in the implementation of the general data protection law in Brazil. *Multidisciplinary Studies: Management and Legal Sciences, 20,* 20–28.

Ding, Y., Hou, K., Xu, Z., Pimplaskar, A., Petter, E., Boulier, K., et al. (2023). Polygenic scoring accuracy varies across the genetic ancestry continuum. *Nature, 618,* 774–781.

Dourado, DdA., & Aith, F. M. A. (2022). A regulação da inteligência artificial na saúde no Brasil começa com a Lei Geral de Proteção de Dados Pessoais. *Revista de Saúde Pública, 56,* 80.

Eden, R., Chukwudi, I., Bain, C., Barbieri, S., Callaway, L., de Jersey, S., et al. (2025). A scoping review of the governance of federated learning in healthcare. *Npj Digital Medicine, 8,* 427.

Elgin, C. Y., & Elgin, C. (2024). Ethical implications of AI-driven clinical decision support systems on healthcare resource allocation: A qualitative study of healthcare professionals' perspectives. *BMC Medical Ethics, 25,* 148.

Fitipaldi, H., & Franks, P. W. (2023). Ethnic, gender and other sociodemographic biases in genome-wide association studies for the most burdensome non-communicable diseases: 2005–2022. *Human Molecular Genetics, 32,* 520–532.

Floridi, L. (2023). *The Ethics of Artificial Intelligence: Principles, Challenges, and Opportunities.*

Fraser, A. G., Biasin, E., Bijnens, B., Bruining, N., Caiani, E. G., Cobbaert, K., et al. (2023). Artificial intelligence in medical device software and high-risk medical devices–a review of definitions, expert recommendations and regulatory initiatives. *Expert Review of Medical Devices, 20,* 467–491.

Funer, F., & Wiesing, U. (2024). Physician's autonomy in the face of AI support: Walking the ethical tightrope. *Frontiers in Medicine, 11,* 1324963.

Gao, Y., & Cui, Y. (2024). Optimizing clinico-genomic disease prediction across ancestries: A machine learning strategy with pareto improvement. *Genome Medicine, 16,* 76.

Goswami, S., Masurkar, P. P., & Kaul, A. (2025). A comprehensive overview of real-world evidence in medical device approvals in the United States. *Expert Review of Medical Devices,* 22(7), 767–773.

Hattoh, G., Ayensu, J., Ofori, N. P., Eshun, S., & Akogo, D.. (2025). Can large language models design biological weapons? Evaluating moremi bio. *arXiv Preprint arXiv:250517154.*

Hildt, E. (2025). What is the role of explainability in medical artificial intelligence? A case-based approach. *Bioengineering, 12,* 375.

Hort, M., Chen, Z., Zhang, J. M., Harman, M., & Sarro, F.. (2024). Bias mitigation for machine learning classifiers: A comprehensive survey. *ACM Journal on Responsible Computing, 1,* 1–52.

Hussain, S. A., Bresnahan, M., & Zhuang, J. (2025). Can artificial intelligence revolutionize healthcare in the global South? A scoping review of opportunities and challenges. *Digital Health, 11,* 20552076251348024.

Iqbal, J. D., Krauthammer, M., & Biller-Andorno, N. (2022). The use and ethics of digital twins in medicine. *Journal of Law, Medicine & Ethics, 50,* 583–596.

Johnson, K. B., Wei, W. Q., Weeraratne, D., Frisse, M. E., Misulis, K., Rhee, K., et al. (2021). Precision medicine, AI, and the future of personalized health care. *Clinical and Translational Science, 14,* 86–93.

Joshi, A. (2024). Big data and AI for gender equality in health: Bias is a big challenge. *Frontiers in Big Data, 7,* 1436019.

Khinvasara, T., Tzenios, N., & Shanker, A. (2024). Post-market surveillance of medical devices using AI. *Journal of Complementary and Alternative Medical Research, 25,* 108–122.

Khoshandam, M., Soltaninejad, H., Hosseinkhani, S., Saltanatpour, Z., Goudarzi, M. T. H., & Hamidieh, A. A. (2024). CRISPR/Cas and artificial intelligence to improve precision medicine: Future perspectives and potential limitations. *Human Gene, 42*, 201356.

Khuat, T. T., Bassett, R., Otte, E., Grevis-James, A., & Gabrys, B. (2023). Applications of machine learning in biopharmaceutical process development and manufacturing: Current trends, challenges, and opportunities. *arXiv Preprint arXiv:231009991.*

Lalova-Spinks, T., Valcke, P., Ioannidis, J. P., & Huys, I. (2024). EU-US data transfers: An enduring challenge for health research collaborations. *Npj Digital Medicine, 7*, 215.

Liu, S., & Guo, L. R. (2024). Data ownership in the AI-powered integrative health care landscape. *JMIR Medical Informatics, 12*, e57754.

Madiega, T. (2021). *Artificial Intelligence Act.*

Morandín-Ahuerma, F. (2023). *Ten UNESCO Recommendations on the Ethics of Artificial Intelligence.*

Niazi, S. K. (2025). Regulatory perspectives for AI/ML implementation in pharmaceutical GMP environments. *Pharmaceuticals, 18*, 901.

Organisation for Economic Co-operation and Development. (2024). OECD framework for the classification of AI systems: Companion to the OECD AI principles. Paris: OECD Publishing. https://doi.org/10.1787/cb6d9eca-en.

Pathak, A. K., Bora, N., Badonyi, M., Livesey, B. J., Consortium, S. K. H., Ngeow, J., et al. (2024). Pervasive ancestry bias in variant effect predictors. *bioRxiv:2024.2005. 2020.594987.*

Piasecki, J., & Cheah, P. Y. (2022). Ownership of individual-level health data, data sharing, and data governance. *BMC Medical Ethics, 23*, 104.

Safarlou, C. W., Jongsma, K. R., Vermeulen, R., & Bredenoord, A. L. (2023). The ethical aspects of exposome research: A systematic review. *Exposome, 3*, osad004.

Saha, G. C., Eni, L. N., Saha, H., Parida, P. K., Rathinavelu, R., Jain, S. K., et al. (2023). Artificial intelligence in pharmaceutical Manufacturing: Enhancing quality control and decision making. *Rivista Italiana Di Filosofia Analitica Junior, 14*, 2023.

Sandbrink, J. B. (2023). Artificial intelligence and biological misuse: Differentiating risks of language models and biological design tools. *arXiv Preprint arXiv:230613952.*

Science C, Board T National Academies of Sciences E, Medicine. (2025). AI-enabled biological design and the risks of synthetic biology. In *The Age of AI in the Life Sciences: Benefits and Biosecurity Considerations*. National Academies Press (US).

Shahab, M. A., Destro, F., & Braatz, R. D. (2025). Digital twins in biopharmaceutical manufacturing: review and perspective on human-machine collaborative intelligence. *arXiv Preprint arXiv:250400286.*

Siddique, S., Haque, M. A., George, R., Gupta, K. D., Gupta, D., & Faruk, M. J. H. (2023). Survey on machine learning biases and mitigation techniques. *Digital, 4*, 1–68.

Singh, R., Paxton, M., & Auclair, J. (2025a). Regulating the AI-enabled ecosystem for human therapeutics. *Communications Medicine, 5*, 181.

Singh, V., Cheng, S., Kwan, A. C., & Ebinger, J. (2025b). United states food and drug administration regulation of clinical software in the era of artificial intelligence and machine learning. *Mayo Clinic Proceedings: Digital Health, 3*, 100231.

Teo, Z. L., Jin, L., Liu, N., Li, S., Miao, D., Zhang, X., et al. (2024). Federated machine learning in healthcare: A systematic review on clinical applications and technical architecture. *Cell Reports Medicine , 5*(2), 101419.

Tindana, P., Depuur, C., De Vries, J., Seeley, J., & Parker, M.. (2020). Informed consent in genomic research and biobanking: Taking feedback of findings seriously. *Global Bioethics, 31*, 200–215.

Tubbs, A., & Vazquez, E. A. (2025). Digital twins in increasing diversity in clinical trials: A systematic review. *Journal of Biomedical Informatics*, 169, 104879.

Undheim, T. A. (2024). The whack-a-mole governance challenge for AI-enabled synthetic biology: Literature review and emerging frameworks. *Frontiers in Bioengineering and Biotechnology, 12*, 1359768.

United Nations Educational, Scientific and Cultural Organization. (2023). Recommendation on the ethics of artificial intelligence. Paris: UNESCO. https://unesdoc.unesco.org/ark:/48223/pf0000381137.

Van Kolfschooten, H., & Van Oirschot, J. (2024). The EU artificial intelligence act (2024): Implications for healthcare. *Health Policy, 149*, 105152.

Wheeler, N. E. (2025). Responsible AI in biotechnology: Balancing discovery, innovation and biosecurity risks. *Frontiers in Bioengineering and Biotechnology, 13*, 1537471.

Wibawa, T. (2021). COVID-19 vaccine research and development: Ethical issues. *Tropical Medicine & International Health, 26*, 14–19.

World Health Organization. (2025). Ethics and governance of artificial intelligence for health: Guidance on large multi-modal models. Geneva: WHO.

Xiang, R., Kelemen, M., Xu, Y., Harris, L. W., Parkinson, H., Inouye, M, et al. (2024). Recent advances in polygenic scores: Translation, equitability, methods and FAIR tools. *Genome Medicine, 16*, 33.

Yang, R., Nair, S. V., Ke, Y., D'Agostino, D., Liu, M., Ning, Y., et al. (2024). Disparities in clinical studies of AI enabled applications from a global perspective. *NPJ Digital Medicine, 7*, 209.

Yang, Z., Zeng, X., Zhao, Y., & Chen, R. (2023). AlphaFold2 and its applications in the fields of biology and medicine. *Signal Transduction and Targeted Therapy, 8*, 115.

Zanca, F., Brusasco, C., Pesapane, F., Kwade, Z., & Beckers, R. Avanzo M Regulatory aspects of the use of artificial intelligence medical software. In *Seminars in Radiation Oncology, 2022* (Vol. 4, pp. 432–441). Elsevier.

Zhang, J., & Zhang, Z.-m. (2023). Ethics and governance of trustworthy medical artificial intelligence. *BMC Medical Informatics and Decision Making, 23*, 7.

Nitin Sharma, Graeme Mukhopadhyay, Vikas Kumar,
Varruchi Sharma*, Imran Sheikh, Anil K. Sharma*, Anupam Sharma,
Poonam Bansal, Vandana Sharma, Amit Joshi

Chapter 6
AI in Biomedical Imaging: A New Era of Precision and Insight

Abstract: Biomedical imaging has been playing a pivotal role in modern diagnosis, treatment selections, and clinical monitoring, but conventional interpretation is time-consuming and subjective. Advancement in artificial intelligence (AI)-based developments are revolutionizing the entire imaging pipeline, including acquisition and reconstruction, quantitative analysis, and predictive modeling. The emerging deep learning architectures, such as the convolutional neural network, the transformer, and the generative model, are improving the quality of the images, making it possible to perform a low dose of the images, automatically reduce and diagnose cancer, and discover clinically relevant biomarkers. The integration of imaging with electronic health records and genomic data and continuously measured physiological signals into the multimodal approach can further contribute to a transition to personalized medicine that is more predictive. Advanced technologies like foundational models, federated learning, and explainable AI extend scalability, collaboration, and clinical interpretability. Despite its promises, AI in imaging brings challenges around data privacy, regulatory endorsement, algorithmic fairness, and smooth integration into clinical workflows. This chapter identifies the current trends, case studies, and further directions to show how AI can

***Corresponding authors: Varruchi Sharma,** Department of Biotechnology and Bioinformatics, Sri Guru Gobind Singh College, Sector 26, Chandigarh 160019, India, e-mail: Varruchi.sharma2013@gmail.com

***Corresponding authors: Anil K. Sharma,** School of Biological Sciences, Amity University, Mohali 140306, Punjab, India, e-mail: anibiotech18@gmail.com

Nitin Sharma, Department of Biotechnology, Chandigarh College of Technology, CGC, Landran, Mohali 140307, Punjab, India, e-mail: abhinitu30@gmail.com

Graeme Mukhopadhyay, Vikas Kumar, University Institute of Biotechnology, Chandigarh University, Gharuan, Mohali 140413, Punjab, India

Imran Sheikh, UCRD, Chandigarh University, Mohali

Anupam Sharma, Department of Physics, Guru Kashi University, Talwandi Sabo, Bathinda 151302, Punjab, India

Poonam Bansal, Department of Bio-sciences and Technology, MMEC, Maharishi Markandeshwar (Deemed to be University), Mullana, Ambala 133207, India

Vandana Sharma, Department of Physics, MMEC, Maharishi Markandeshwar (Deemed to be University), Mullana, Ambala 133207, India

Amit Joshi, Department of Biotechnology and Bioinformatics, Sri Guru Gobind Singh College, Sector 26, Chandigarh 160019, India

 | https://doi.org/10.1515/9783112214275-006

move biomedical imaging beyond descriptive visualization toward biologically sensitive, patient-specific decision-making. Overall, AI represents not just a technological supplement, but a paradigm shifts toward more accurate, fair, and opinionated healthcare.

Keywords: Artificial intelligence (AI), biomedical imaging, deep learning, medical diagnosis, precision medicine, clinical decision support

6.1 Introduction

Biomedical imaging is intrinsically linked with contemporary healthcare and biotechnology in that, it allows us to see the internal anatomical structures, physiological processes, and disease conditions without invasive methods (Pesapane et al., 2018). Medical imaging is a rapidly expanding field, and many techniques are currently in use including magnetic resonance imaging (MRI), computed tomography (CT), positron emission tomography (PET), X-ray, ultrasound, and other optical and microscopic imaging systems. These imaging modalities have had great advantages in the capability to assess illnesses in their early stages and advance clinical results (Pinto-Coelho, 2023). Nevertheless, traditional biomedical imaging is dependent on manualization by radiologists, clinicians, and pathologists; therefore, this renders imaging taking a long time and subject to subjectivity depending on experience and cognitive exhaustion of the individual doing the interpretation (Mello-Thoms & Mello, 2023). Moreover, the rising amount and complexity of medical imaging data in hospitals and research laboratories have necessitated an augmenting load on medical care systems to handle, examine, and synthesize images at an effective and satisfactory level in a regular fashion (Geroski et al., 2023).

The term "biomedical imaging" (or "bioimaging") is an equivalent of the visual foundation of modern life sciences and medical diagnostics (James & Gambhir, 2012). It encompasses a wide range of technologies that provide the possibility to visualize biological processes on the level of molecules, cells, tissues, and the entire organism (Massoud & Gambhir, 2003). Bioimaging allows scientists and clinicians to see the living systems real time through radiological scans to fluorescence microscopy models, which is very important in disease processes, cellular workings, and therapeutic effects (Chudakov et al., 2010). Recent changes in high-resolution microscopy, optical coherence tomography, and molecular imaging have transformed the concept of pathophysiology and were able to simplify early disease detection (Weissleder & Pittet, 2008). Moreover, bioimaging would be of critical importance in drug discovery, regenerative medicine, and neuroscience, which can apply the same techniques to trace the drug-target interactions, the migration of stem cells, and neural activity. Bioimaging offers a central platform to facilitate the advancement of research and clinical diagnostics of various diseases by integrating the imaging data with computational models and bioinformatics and connecting biology, physics, and computational science (Berg et al., 2022).

Artificial intelligence (AI) is a branch of computer science that considers the creation of a smart system that has the capacity to execute functions in cognitive domains, that are traditionally developed by human beings, including learning, reasoning, problem-solving, and decision-making (Russell & Norvig, 2021). Essentially, AI is meant to be very similar to human intelligence, in that it uses algorithms and computer modeling that are able to handle large amounts of data, including patterns, forecast functions, or reach informed conclusions on their own (Nilsson, 2014). Machine learning (ML) is a sub-discipline of AI that allows them to improve on their own with experience, whereas deep learning (DL) is based on artificial neural networks and their resemblance to the human brain and able to process complex data, such as images, speech, or genomic sequences (LeCun et al., 2015). Over the past years, AI has taken over many fields, such as healthcare, finance, robotics, and education (Jordan & Mitchell, 2015). In biomedical imaging, it is possible to use AI to process health-related images, detect and classify diseases, and predict pathological conditions, improve diagnostic reliability and reduce the amount of work done by clinicians, and promote personalized medicine (Esteva et al., 2019). Continuous advancement of AI would help make medical care more efficient, evidence-based, and patient-focused (Topol, 2019).

The new AI has also presented disruptive potentials within biomedical imaging. Algorithms based on ML and DL, specifically, convolutional neural networks (CNNs) and transformer-based models, have the capacity to extract spatial and structural information through large imaging datasets and, therefore, enhance image augmentation, image segmentation, image classification, and disease prediction tasks with a high degree of accuracy without human intervention (Najjar,2023; Litjens et al., 2017). The given AI-based methods do not only help clinicians to save time on interpretation but also improve diagnostic accuracy and diminish the inter-observer variability (Chen et al., 2024).

In addition, AI is capable of combining imaging data with genomic, proteomic, and clinical data, which could be useful in the creation of precision medicine, when the treatment plan is created based on the unique attributes of each patient (Lambin et al., 2017). With the further development of biomedical imaging and the advent of AI, the sphere is entering the new stage at which it is getting more possible to perform automated analysis of the image, make a prediction, and introduce personalized diagnosis. The objectives of this chapter include identifying the principles of biomedical imaging and AI, discussing the innovations in AI-based image retrieval and recognition, explaining application in clinical practice, addressing the current trends in AI implementation (foundation models and multimodal information aggregation), and discussing ethical issues, information privacy, and deployment in healthcare.

6.2 Foundations of Biomedical Imaging and AI Integration

6.2.1 Biomedical Imaging Modalities

Biomedical imaging is a variety of technologies (Table 6.1) that enables one to visualize the structure of the body, physiological processes, as well as molecular activity within the body, which is critically important in diagnosis, planning treatment, and research application (Jensen et al.,2010; Helpern et al., 2010). MRI is one of the most essential modalities that are used to create detailed pictures of soft tissues using strong magnetic fields and radiofrequency waves, so it is especially applicable in neurological, musculoskeletal, and cardiovascular imaging (Bernstein et al., 2004). The use of CT entails rotating the X-ray beams to create cross-sectional and three-dimensional images and is often used in emergency medicine to carry out fast assessment of trauma, internal bleeding, and tumor (Ginat & Gupta, 2014).

PET is a nuclear medical procedure that is employed to visualize the activity of metabolism and the functional activity by identifying the uptake of radiotracers, which are mostly used in oncology to detect tumors and measure treatment response (Cherry et al., 2018). Blasts that are tolled over the body, ultrasound imaging is a useful, non-ionizing, and convenient tool that is extensively used in obstetrics, cardiology, and abdominal diagnostics (Szabo, 2014). One of the oldest modalities, which is the simplest and the most common to use, X-ray imaging is still necessary in skeletal investigation, pulmonary analysis, dental analysis, etc. because of its availability and relatively low cost (Seibert, 2004). Also, cellular and sub-cellular structures can be visualized using optical imaging methods, such as fluorescence and bioluminescence, to study molecular pathways and pathogenesis of diseases (Ntziachristos, 2010). Light, confocal, and super-resolution microscopy is a microscopy that visualizes the cells and tissues with high resolutions in order to identify the research and diagnostic pathology (Schermelleh et al., 2019). Cryo-electron microscopy at the frontier of structural biology allows imaging biomolecules at near-atomic resolution in their native commodious position and has transformed the comprehension of protein and virus structures and speedy therapeutic discoveries (Nogales and Scheres, 2015).

6.2.2 AI Techniques for Imaging

The application of AI to biomedical imaging is mainly based on ML and DL solutions, which present different levels of possibilities in the analysis of medical images (Erickson et al., 2017). Techniques like ML rely on the handcrafted features provided by the experts and are commonly applied in classical classification and regression of imaging. Nevertheless, such techniques are restricted by the quality and comprehen-

Table 6.1: Comparison of major biomedical imaging modalities.

Imaging modality	Physical principle	Key clinical uses	Advantages	Limitations
X-ray	Differential absorption of ionizing radiation	Bone fractures and chest imaging	Fast, inexpensive, widely available	Ionizing radiation and limited soft tissue contrast
CT (computed tomography)	Multiple X-ray projections reconstructed into cross-sectional images	Trauma assessment, tumor localization, and vascular imaging	High spatial resolution rapid scanning	Higher radiation dose, requires contrast agents in some cases
MRI (magnetic resonance imaging)	Nuclear magnetic resonance of hydrogen atoms in a magnetic field	Neurological, musculoskeletal, abdominal, and cardiac imaging	Excellent soft tissue contrast and non-ionizing	Expensive, slower, and contraindicated in metal implants
PET (positron emission tomography)	Detection of gamma rays from radiotracer decay	Metabolic activity imaging and oncology staging	Functional/ metabolic insights	Ionizing radiation, very costly, and requires radiotracers
Ultrasound	High-frequency sound wave reflection patterns	Obstetrics, cardiac imaging, and abdominal organs	Portable, real-time imaging, and no radiation	Operator-dependent and limited depth and resolution in obese patients
Optical imaging	Light absorption/ emission interactions	Surgical guidance and cellular/tissue visualization	High resolution at cellular scale	Limited depth penetration
Microscopy	Direct magnification of cellular structures	Histopathology and cell-level research	Cellular and subcellular resolution	Requires tissue samples and limited field of view
Cryo-EM	Electron microscopy of frozen samples	Structural biology and protein complex visualization	Near-atomic structural detail	Expensive, highly specialized, and complex sample preparation

siveness of features that are manually generated (Suganyadevi et al., 2022). DL, in its turn, is an automatic learning of hierarchical images based on raw imaging data and thus can provide high-quality image recognition, image segmentation, and pattern discovery without predefined features (LeCun et al., 2015). CNNs are the most common types of DL architecture used in biomedical imaging because they are capable of identifying spatial patterns, and structural image information (Litjens et al., 2017). CNNs have proved to be very effective in identifying tumors, detecting organ boundaries, and identifying anomalies in various computing imaging modalities, such as MRI, CT, and microscopy. Transformer-based models that were originally designed to work with natural language processing have more recently been applied to medical imaging tasks. Transformers, as a special technology, employ self-attention to learn long-range spatial relationships, and are most useful in 3D volumetric image analysis and multimodal image integration (Chen et al., 2021).

Graph neural networks (GNNs) are another new sub-categorization of imaging AI that can be thought of as image structures where nodes and edges represent the graph. GNNs are particularly convenient when it comes to anatomical and cellular relationship modeling in radiomics and histopathology which permits a more biologically significant analysis of the spatial tissue arrangement (Parisot et al., 2018). In addition, generative AI systems, such as generative adversarial networks (GANs) and diffusion models, are altering biomedical imaging with the ability to enhance images, remove noise, reconstruct high-resolution images, and generate synthetic images to be used in training (Karras et al., 2019; Ho et al, 2020). The models are also used in the creation of virtual biopsies, in which AI predicts tissue-level pathology, based on non-invasive imaging. On the whole, these AI tools will greatly boost image capture, interpretation, and one-on-one integration, leading to the framework of more accurate diagnostics and individual approach to medical decisions.

6.2.3 Data Requirements and Challenges

Availability of high-quality imaging datasets in biomedical imaging strongly relates to the successful application of AI in biomedical imaging. To train powerful models that can generalize to a wide variety of patients and imaging conditions, big data is required (Litjens et al., 2017). Nonetheless, it is difficult to get such datasets because of privacy laws, lack of policies regarding data sharing, and sensitivity of medical data. Moreover, medical image annotation also needs a domain knowledge of a radiologist or a clinician, so it is both labor- and time-intensive, and can suffer inter-observer variation (Rajpurkar et al., 2022). The misidentification or non-consistency in annotations may bring bias to AI models and decrease their credibility in the clinical practice. Another important issue is the bias of the dataset. The imbalance in terms of the representation of patient demographics, prevalence of a given disease, the manufacturers of an imaging device, or the particular clinical setting can cause bias and restrict the applicability of AI algorithms to a different population or institution (Obermeyer et al., 2019). Another problem with AI is standardization and interoperability, which enhances the problem of AI implementation. Variability across hospitals and laboratories in imaging protocols, type of scanner, and image format can impair the performance of the model unless these factors are sufficiently accounted for (Erickson et al., 2017). Although the standard known as DICOM (Digital Imaging and Communications in Medicine) offers a widely accepted model of storing and exchanging images, there are still differences in their use and this makes cross-institutional deployment of models a little hard (Bidgood et al., 1997). Attempts to harmonize medical imaging data with the frameworks like HL7-FHIR (Fast Healthcare Interoperability Resource) have played a vital role in enhancing the other clinical and genomic data interoperability and integration of imaging with other clinical and genomic data (Mandel et al., 2016). These issues need thorough measures, such as developing heterogeneous and representative samples, standardized annotation procedures, efficient bias identification, and the embrace of interoperable format of data. This type of measures would make sure that AI models of biomedical imaging are not only accurate but also safe in clinical application to enable the wider implementation and trust in these models in the real-life healthcare setting (Litjens et al., 2017; Mandel et al., 2016).

6.3 AI-Driven Image Acquisition and Enhancement

AI has also improved image acquisition and despite processing biomedical imaging, it has dealt with the problems of radiation exposure, image quality and acquisition efficiency. The first is in low dose imaging, specifically in CT and PET scans, where AI algorithms allow reducing the amount of radiation used without reducing the diagnostic quality. Denoising based on DL can perfectly restore high-quality images out of low-dose acquisitions with reduced exposure to the patient and retain structural and functional information (Wang et al., 2016; Chen et al., 2017).

Besides the minimized noise, AI leads to the improvement of resolutions and removal of artifacts. CNNs and generative models are able to remove motion artifacts, beam-hardening effects, and partial volume distortions in real time, thereby enhancing the quality and diagnostic value of the images (Jin et al., 2017). The methods are particularly in dynamic imaging research, such as cardiac MRI or functional PET, in which patient's motion and physiological changes tend to ruin the image quality. Another such emerging trend is AI-assisted imaging hardware optimization, in which algorithms change scanner parameters (exposure time, field of view, and detector sensitivity) in real time, based on the analysis of the incoming data. It is an adaptive tool that maximizes the quality of the image with minimum scan time and patient pain (Wang et al., 2020). Moreover, intelligent reconstruction algorithms based on compressed sensing and artificial intelligence have shown remarkable potential in accelerating image acquisition. Compressed sensing supports the reconstruction of undersampled sensing, which saves scan time, whereas DL models enhance the reconstruction in high-quality images with fewer artifacts (Hammernik et al., 2018). All these methods are currently being combined in MRI, CT, and newer imaging methods, such as PET/MRI hybrid systems, resulting in shorter scan times, reduced risk, and more accurate imaging processes. Overall, AI is changing medical imaging by improving how we capture and rebuild images, so we get better high-quality images with a reduced radiation dose, reduced scan duration, and increased reliability in the long run to make clinical decisions.

6.4 Automated Image Analysis and Interpretation

6.4.1 Segmentation and Quantification

Automated segmentation and quantification are the key areas of AI-based biomedical imaging. Importantly, DL methods can precisely outline tumor edges, segment organs, and detect single cells of different imaging modalities with the use of U-Net and its different variations (Ronneberger et al., 2015; Milletari et al., 2016). This provides the possibility of a strictly volumetric analysis, radiotherapy planning, and high-throughput cellular investigations in research and pathology (Falk et al., 2019). Compared to manual approaches, automated quantification decreases the inter-observer variation while increasing reproducibility.

6.4.2 Classification and Diagnosis

With the use of AI methods, the classification and diagnosis of clinical imaging are changing. ML and DL algorithms are able to identify cancers, cardiovascular abnormalities, and other neurological conditions at an early stage, with the same accuracy as professional clinicians (Esteva et al., 2017). An example is the high sensitivity and specificity of AI systems used in mammogram, iris scan, and chest X-rays to detect lesions. DL can also detect small functional variations in cardiac imaging; furthermore AI is now used to find the existence of Alzheimer's and Parkinson's diseases at earlier stages of disease progression in neuroimaging, thereby facilitating timely clinical intervention (Ding et al., 2019; Attia et al., 2019).

6.4.3 Functional and Molecular Imaging with AI

Functional and molecular imaging with AI is implemented through radiomics and pathomics, where quantitative imaging phenotypes are farmed in a bid to correlate with clinical outcomes or genetic obtained (Lambin et al., 2012). Radiomics uses MRI, CT, or PET as imaging techniques and computes the features in a high-dimensional fashion, whereas pathomics utilizes histopathology slides to analyze the fine-scale tissue microarchitecture. The integration of imaging and omics data can be presented as the multimodal data fusion in which the AI can give the predictive insights of precision medicine, which can include the prognosis, prediction of therapy responses, and the individual treatment schedule (Aeffner et al., 2019).

6.4.4 Digital Pathology

AI-based analysis of pathology benefits the field of digital pathology. DL-based whole-slide imaging can detect all cancerous areas, proximity of tumor-infiltrating lymphocytes, as well as uncommon pathological patterns otherwise invisible to human observers (Coudray et al., 2018; Campanella et al., 2019). These AI workflows enhance diagnostic throughput, accuracy, and reproducibility, and allow remote consultation and scalable computational pathology research on a large-scale basis to promote scalable and accurate practice of pathology.

6.5 AI in Precision Medicine and Predictive Imaging

This interaction between AI and biomedical imaging has removed new opportunities in the field of precision medicine, which can adjust any diagnostics and therapy to the unique patient characteristics. Radiogenomics is one of the main fields, and it is associated with the connection of quantitative images characteristics or radiomic signatures to the genetic and molecular characterization of the tumor or tissues. Radiogenomic studies enable noninvasive prediction of underlying genetic changes and signal transduction that give a clue of heterogeneity in tumors and informs the decision to use specific therapeutic approaches (Gillies et al., 2015; Lambin et al., 2017). Indicatively, radiomic patterns obtained through MRI or CT may be linked to mutations of genes like EGFR or KRAS, making it possible to predict the response to a therapeutic treatment without necessarily having to repeat a biopsy (Aerts et al., 2014).

AI-based biomarkers are also becoming very useful besides radiogenomics to aid in disease prognosis and the response to treatment monitoring. DL models may use features based on imaging with clinical and laboratory data to forecast outcomes, including survival rates in cancer or the potential development of the disease in cardiovascular and neurological diseases. These AI-driven biomarkers are objective and reproducible measurements that can be used to supplement conventional clinical measurements to ensure earlier intervention and increased quality clinical decision-making. Another radically new use of AI in imaging is personalized treatment planning. The use of AI in radiotherapy, such as creating and segmenting tumors and at-risk organs automatically, applying optimal dose distributions, and its ability to predict possible outcomes can support personalized treatment plans (Bibault and Giraud, 2024). Equally, AI-assisted imaging is applied to plan surgery, direct the execution of minimally invasive surgeries, and tailor patient-specific therapeutic regimens based on surgical and disease anatomy. AI-based longitudinal imaging enables the continuous monitoring and quantification of the progress of the disease over time. AI algorithms can measure minor structural or functional changes that are not easily observed by humans because the data are compared sequentially. This functionality is essential in tracking chronic diseases, monitoring treatment effectiveness, and anticipating relapse or adverse occurrences (Hashimoto et al., 2018). The functioning of AI models combined with the use of time data allows arranging the work with patients as dynamic and adaptive, which is in keeping with the principles of precision medicine. Altogether, AI in precision medicine and predictive imaging alters the conventional principles of diagnostics by providing noninvasive genetic profiling, strong prognostic genes, individual approach to treatment, and constant monitoring, which eventually causes better patient outcomes and increases personalized healthcare.

6.6 Cutting-Edge Trends and Innovations

6.6.1 Foundation Models in Imaging

Lately, developments in AI have borne large pretrained models in biomedical imaging that can generalize across a variety of tasks with only a small quantity of fine-tuning. These models are trained over large datasets and thus are able to trace solid representations of visual and written characteristics concurrently. One of them is vision-language models, including BioMedCLIP and MedCLIP, that combine image and piece of text information to provide intelligent image interpretation, speech-generated report creation, and cross-modal retrieval (Wang et al., 2022; Zhang et al., 2023). These models exploit and capitalize on the mass-trained concept to comprehend the multiple complex interactions of how imaging discoveries relate to clinical scenes to overcome the bridge between radiological images and electronic health record (EHR) knowledge. To provide an example, a vision-language system is capable of drawing attention in one region of an X-ray or MRI scan and produces descriptive text that is agreeable to medical terminology. Segment Anything in Medicine is another well-known innovation; here, general-purpose segmentation models are adapted to biomedical problems to outline anatomical structures, lesions, or tumors with little to no task-specific training (Ma et al., 2024). It has been demonstrated that foundation models in imaging can reduce training large, labeled datasets specifically per task, yielding not only a certain task transferability but also transferability across new imaging modalities or diseases. They can be very promising in clinical deployment, research, and AI-driven discovery due to their flexibility and eventually increase diagnostic accuracy, workflow, and multimodal patient data integration (Wang et al., 2022; Zhang et al., 2023; Ma et al., 2024).

6.6.2 Federated and Distributed Learning

Federated and distributed learning are potentially radical in biomedical imaging AI and allow knowledge sharing between various hospitals and institutions by training models based on the information of multiple hospitals collaboratively. Conventional centralized AI methods demand concentration of the data in a single point, and this brings privacy, security, and regulatory-related issues. Federated learning presents a solution to this issue since it enables the models to be trained locally on the data of every institution, and the only updates about the models are shared and aggregated in one workplace, maintaining patient confidentiality (Li et al., 2020; Sheller et al., 2020).

Federated learning has been utilized in biomedical imaging to accomplish tasks in tumor segmentation, detecting COVID-19 on chest CT scans, and multi-center MRI analysis with success. Through training on heterogeneous datasets across institutions, AI models yield a better ability to generalize on various populations and imaging protocols with minimal bias and a high level of robustness (Li et al., 2020). Furthermore, adopted learning systems facilitate the development of AI that is scalable, i.e. by utilizing the computational power across more than one location in parallel, which speeds up convergence of the model, without requiring invading privacy. Issues have persisted, such as managing data heterogeneity across locations, communication cost, and mitigating model vulnerability to possible adversarial attacks. The downsides are recently being overcome with new devices of secure aggregation protocols, differential privacy, and model personalization, which are enabling federated and distributed learning to be increasingly viable in a practical setting of a healthcare system (Sheller et al., 2020; Bonawitz et al., 2017). Such ways are essential toward supporting cross-institutional AI cooperation and allowing developing more general and clinically viable imaging models.

6.6.3 Explainable AI (XAI)

Explainability has become an important demand in fields of AI systems as they become more common in biomedical imaging to guarantee clinical trust, accountability, and their uptake. Explainable AI (XAI) is a set of approaches that simplify the understanding of AI models to clinicians, radiologists, and other healthcare workers by making the decision-making processes of AI models transparent and interpretable (Samek et al., 2017). In contrast to black-box models, XAI has visual or textual representations of how the prediction was made, and what features or parts of the image were most important to making a diagnosis or classification. XAI in medical imaging, such as saliency map, Grad-CAM (gradient-weighted class activation mapping), SHAP (SHapley Additive exPlanations), and LIME (local interpretable model-agnostic explanations) methods, enable clinicians to confirm the correctness of AI predictions through visual inspection of the localized activities that drove the prediction

(Selvaraju et al., 2017; Lundberg et al., 2017). To provide an example, XAI can be used to detect suspicious lesions on MRI or CT scans in tumor detection or segmentation to make sure that AI predictions and clinical reasoning processes are consistent besides minimizing possible errors in AI predictions.

XAI is also crucial in regulatory approval and clinical integration; explainable models also explain safety, reliability, and human expertise-related evidence, which will help address issues of liability and accountability (Holzinger et al., 2017). Additionally, XAI makes the AI systems and clinicians learn together, with the output being interpretable feedback to refine models and improve the work of diagnostic systems with time. Embracing and adopting AI into real-world clinical care through the incorporation of XAI in imaging pipelines thereby improves the degree of transparency, trust, and conformability to imaging, but does not compromise the degree of diagnostic accuracy and patient safety.

6.6.4 Multimodal Integration

Multimodal integration in biomedical imaging can be described as having the potential that allows both the integration of imaging and other forms of patient data, i.e., EHRs, genomic profiles, and wearable device data to gain more comprehensive and customized insight into the health status of a patient (Miotto et al., 2018). Combining these various data streams in a way allows for the identification of complex correlation between anatomical, functional, molecular, and lifestyle data; this provides AI models with the capacity to predict disease onset, forecast progression, and create a better treatment plan. Combining radiogenomic biomarkers into genomic mutations and gene expression profiles can enable defining of radiogenomic imaging characteristics, which predict tumor aggressiveness or therapy reaction (Shui et al., 2021). To the same effect, the use of continuous heart rate, glucose measurements, or activity tracking data as a longitudinal input form of data retrieved by wearable technology can prove useful in promptly notifying of a cardiovascular event or a metabolic complication alongside imaging and clinical data (Rumsfeld et al., 2016). The multimodal learning AI frameworks, including DL models with cross-modal attention, can successfully include an effective fusion of heterogeneous datasets because the additional information complements each other (Baltrušaitis et al, 2019). Besides improving predictive accuracy, multimodal integration is effective in promoting personalized medicine, since one could inform treatment choices by considering a whole picture of a patient in terms of specific imaging, molecular, and behavioral data. Additionally, such combined methods result into the consistent tracking of disease progression and therapy response for providing actionable information that is stronger than single-modality analysis (Rumsfeld et al., 2016; Miotto et al., 2018; Baltrušaitis et al, 2019; Shui et al., 2021).

6.6.5 Generative AI

Generative AI has also become a revolutionary technology in biomedical imaging, which can produce synthetic images to improve augmentation training and be able to simulate imaging data in order to diagnose patients. GANs or diffusion models are just some examples of techniques capable of producing realistic medical images that replicate complex anatomical structures and pathological variations, thus overcoming the difficulty of limited datasets available with labels in clinical research studies (Yi et al., 2019; Skandarani et al., 2023). Such artificial images can be utilized to increase the size of training data, enhancing the functions and external validity of AI models in segmentation, classification, and diagnosis. Besides, enhancement of datasets, generative AI also allows virtual biopsies, where histopathological properties are noninvasively determined with the help of AI-generated simulation of imaging. The AI models can set the tumor composition, cellular density, or molecular markers without invasive procedures by acquiring the relationships between imaging modalities (e.g., MRI, CT, and PET) and tissue-level features (Zhu et al., 2018). Such a solution can decrease the risk of the patient, decrease the time spent on clinical decision-making, and monitor the progress of the disease over time.

Further, generative AI facilitates domain adaptation, which means that models that are trained to learn a specific imaging modality/institution can be generalized to another through the generation of aligned synthetic images. It will improve cross-center cooperation and make AI implementation in actual clinical practice even stronger (Chartsias et al., 2017). In general, generative AI is transforming the field of biomedical imaging through realistic data generation, less reliance on limited labeled data, and creation of novel uses such as virtual biopsies, and ultimately improving the fields of precision medicine and AI-based diagnostics.

6.7 Clinical Translation and Real-World Applications

Over recent years, there has been a rapid clinical application of AI in the field of biomedical imaging, an area that promises a lot in various specialties of medicine (Table 6.2). The AI models are more frequently used in the oncology imaging field in early cancer detection, lesion segmentation, and therapy response detection. Receptive algorithms inspire networks can detect indivisible tumor patterns in MRI, CT, and PET images in large-scale datasets that are otherwise imperceptible to the eye and provide the opportunity to diagnose cancer earlier and develop treatment plans tailored to individuals. Moreover, with the help of AI-based analyses of radiomics, it is possible to predict both the outcome of therapeutic response and survival, helping to make precision oncology a reality (Esteva et al., 2019).

The field of neurology is an area in which AI is used to dominate neurodegenerative, including the conditions of the stroke, Parkinson's and Alzheimer's diseases, and also neuroacute conditions. Automated MRI and PET scan analysis helps to gain early insight into structural and functional disorders, measure brain atrophy, and imminently decrease in cognition. It is also possible to identify the ischemic areas in acute stroke patients by AI-driven models, which can help clinicians make a quick triage and treatment decision, thus enhancing clinical outcomes (Ding et al., 2019; Baaklini and Hernández, 2025). The field of cardiology that has been fortunate to take advantage of AI-powered imaging tools. The echocardiography, CT angiography, and cardiac MRI images are used in ML and DL algorithms to detect plaques, myocardial segmentation, and functionality assessment of the heart. The tools can accurately measure ventricular performance, ejection-fraction, as well as myocardial perfusion deficits, and they improve diagnosis and risks stratification in cardiovascular disease (Madani et al., 2018).

Other vital roles of AI in infectious disease imaging have also been played especially throughout the COVID-19 pandemic. The use of AI to support chest X-ray and CT images enhanced quick screening, pulmonary involvement detection, and disease progression. Also, screening of tuberculosis has been achieved with AI systems in low-resource areas, which offers automated results in reading chest radiographs and aiding in mass screening initiatives (Lakhani & Sundaram, 2017; Wang et al., 2020). AI is becoming incorporated into surgical guidance and robot-assisted surgery. Intra-operative imaging is processed in real time by AI algorithms and allows them to be accurately localized and find tumor margins and critical vessels. AI supports planning the movements, positioning the instruments, and automated recognition of tissues in robotic surgery, which leads to better accuracy of surgery, decreased time of the operation, and safer treatment of patients (Hashimoto et al., 2018; Yang et al., 2017). Altogether, all of these practical applications indicate the revolutionary effect of AI in clinical processes, presenting improved accuracy in diagnoses, efficiency, and distinct insights regarding the patient in diverse areas of the healthcare system, such as oncology, neurology, cardiology, infectious diseases, and surgery.

6.8 Ethical, Regulatory, and Implementation Challenges

The current rate of AI adoption in biomedical imaging has created several ethical, regulatory, and implementation issues that need to be addressed to ensure secure, equitable, and successful clinical practice. One of the key issues is the confidentiality of data, data security, and patient consent. AI systems involve a lot of sensitive patient information to be trained and validated leading to increased chances of misuse, data breaches, and access by unauthorized personnel. It is essential to comply with the reg-

Table 6.2: Applications of AI in different disease domains.

Disease domain	AI imaging applications	Benefits	Example outputs
Oncology (cancer imaging)	Tumor segmentation, progression tracking, and radiogenomic profiling	Early detection and therapy response prediction	Automated tumor boundary maps and mutation correlation heatmaps
Cardiology	Heart function analysis and plaque detection in vessels	Reduced diagnostic variability and early ischemia detection	Ventricular ejection fraction estimates,= and plaque burden measurements
Neurology	Alzheimer's and Parkinson's diseases and stroke lesion detection	Early-stage disease identification and cognitive decline prediction	Brain atrophy quantification and white matter lesion maps
Infectious diseases	COVID-19 severity grading and tuberculosis detection	Rapid screening and triage in high-burden settings	Lung opacity scoring and infection severity probability
Orthopedics	Bone fracture classification and joint degeneration patterns	Faster emergency diagnosis and surgical planning support	Fracture localization overlays and arthritis progression forecasting
Pathology	Whole-slide histopathology image interpretation	Reduced manual workload and increased precision in grading	Automated cell classification maps and digital pathology heatmaps

ulations like HIPAA (Health Insurance Portability and Accountability Act of 1996) in the United States, GDPR (General Data Protection Regulation) in Europe, and other local privacy regulations; employing mechanisms like anonymization, encryption, and secure federated learning ensures that patient confidentiality is addressed (Jiang et al., 2017; Rieke et al., 2020). Another significant issue is prejudice and equity in AI imaging algorithms. The biases of training datasets can be transferred to AI models and result in differences in their performance based on demographic factors or regions of operation and imaging modalities. For example, a model that was mostly trained on data related to a particular ethnic group will be inefficient at performing its role in patients not in those groups, which could further increase the inequality in healthcare. Current research efforts aim to create fairness-conscious algorithms, strict bias evaluation systems, as well as heterogeneous and representative data to alleviate these risks (Obermeyer et al., 2019; Mehrabi et al., 2021).

Regulatory oversight of AI-based medical imaging remains complex and evolving. Approval pathways and ongoing control require robust evidences of safety, efficiency, and clinical utility to satisfy agencies such as the FDA (United States) and EMA (Europe) (Benjamens et al., 2020). The evolving characteristic of AI, especially adaptive models that constantly update themselves with new information, makes it difficult to set out the usual regulatory routes, validation procedures, and post-market screening measures (Gilbert et al., 2021). Another challenge is in terms of integration into clinical workflows. The AI systems should be complementary to the existing practices and should not interfere with the existing practice, but must be compatible with the existing picture archiving and communication system (PACS), EHRs, and imaging devices. The acceptance of AI in clinicians is related not only to accuracy but also to interpretability, usability, and trust of AI recommendations. Training programs, workflow redesign, and collective decision-making structures are needed to make adoption smooth (Shortliffe and Sepúlveda, 2018). However, the responsibility and accountability in AI-assisted diagnostics is still to be equated. Every case of failure of diagnosis at wrong stage or negative outcomes or technical failures is legally and ethically difficult to determine responsibility. Setting up specifications on the issue of clinician supervision, certification of AI, and the role of the institutions is important to reduce risks and introduce fair use (Price et al., 2019). Addressing these ethical, regulatory, and implementation issues is essential to delivering AI in biomedical imaging that is safe, equitable, and clinically valuable while maintaining public trust and compliance with the law and ethical considerations.

6.9 Future Directions

The next decade of AI in biomedical imaging will emphasize integration, rigor, and responsibility: beyond isolated benchmark, the field must deliver safe, generalizable, and clinically useful systems through tightly coupled advances in technology, science, clinical translation, infrastructure, and society. AI will move from task-specific networks to foundation models that jointly process images, text (reports and notes), and molecular data across scales – from whole-body to organs, tissue, and cell – and adapting to new tasks, diseases, and modalities with zero- or few-shot learning via self-supervised visual pretraining, cross-modal transormers, and structured biological priors for efficient fine-tuning (segmentation, diagnosis, report generation, and outcome prediction).

Credible clinical decision will pair data-driven learning with mechanistic models (biophysics, pharmacokinetics, and anatomy) to respect physical constraints and offer causal reasoning (e.g., physiologically constrained MRI/PET reconstruction, models linking tumor heterogeneity to radiomic patterns, and causal interference frameworks for treatment effects using imaging time series). Deployed systems will learn continually across sites while protecting privacy through secure aggregation, personalized model heads, and verifiable privacy (e.g., differential privacy). The key research goals will include mitigating catastrophic forgetting, optimize communication, and build validation pipelines that prevent performance degradation in any subgroup during updates.

Adoption relies on human-AI teaming: workflows that surface clinically actionable rationales beyond generic saliency, interactive visualization for rapid case review, and interfaces permitting real-time clinician correction and real-time model teaching. The evaluation ought to privilege human-AI team outcomes – accuracy, efficiency, and bias mitigation – rather than model metrics alone. The ecosystem will incorporate regulatory grade validation with reproducible, clinically feasible benchmarks containing longitudinal follow-up, hard clinical endpoints, and demographic variety. The community must have opportunistic, repeatable, and clinically feasible benchmarks containing longitudinal follow-up, hard clinical endpoints, and demographic diversity, standardized validation suites, pre-deployment stress-testing (shifted distributions and rare pathologies), and continuous post-market surveillance, co-developed with regulators for adaptive AI (safe retraining, and transparent versioning).

Imaging will combine with multi-omics (spatial transcriptomics, proteomics, and single-cell sequencing) to reveal localized mechanisms, enabling virtual tissue profiling, and facilitate noninvasive molecular stratification; patient-specific digital twins and predictive simulation will power in silico trials to predict disease progression and optimize personalized dose and radiation plans. Edge and real-time intelligence will deliver low-latency, on-device interface (portable ultrasound, intraoperative microscopes, and bedside CT/portable X-ray) via compression and sensor-algorithm co-design while maintaining calibrated uncertainty and safety.

Equity will be advanced by domain adaptation for low-resource sites, low-shot learning for rare diseases, and efficient dataset creation (active/weak supervision and annotation via cheap crowdsourcing with clinician supervision); furthermore, socio-technical deployment plans must cover reimbursement schemes, local clinician training, and culturally sensitivity interfaces. Progress also requires strong data governance – dynamic consent models, data provenance, and auditable lineages – as well as validated high-fidelity generative models to enable privacy-preserving sharing through federated data catalogs and standardized metadata. MLOps (Machine Learning Operations) will provide end-to-end life cycle management-versioned models, automated retraining, drift monitoring, explainability logs, and open APIs (Application Programming Interfaces) for PACS/EHR integration; so structured reports, provenance metadata and uncertainty interoperate with clinical standards. Ethical and legal frameworks will clarify liability, IP, and permissible application, while new roles (AI stewards, validation officers, and clinical data engineers) ensuring ongoing oversight; finally, sustainability will be addressed via energy efficient algorithms, green data centers, and lifecycle impact assessments, so that safe, equitable AI can scale responsibly. Figure 6.1 represents the pillars and future application areas of AI in biomedical imaging.

Different numbers represent the mutually reinforcing elements of an AI-enabled imaging ecosystem: (1) multimodal integration (fusion of images with genomics and health records for precision medicine), (2) real-time AI diagnostics (on-the-spot decision support during acquisition and procedures), (3) federated and secure learning (privacy-preserving and cross-site model training), (4) AI-driven drug discovery and

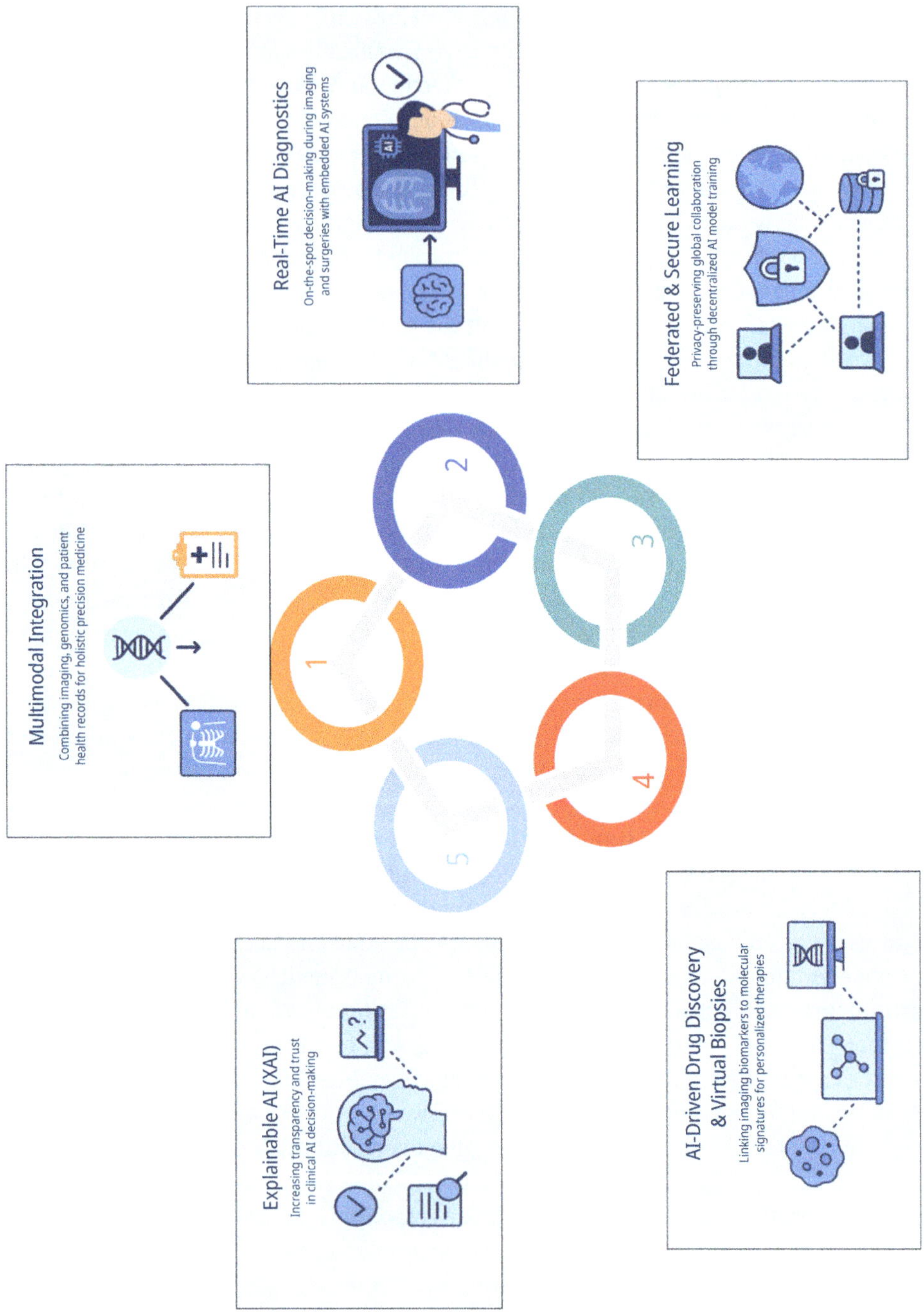

Figure 6.1: AI in biomedical imaging enhances diagnosis, analysis, and predictive insights through advanced algorithms.

virtual biopsies (linking imaging biomarkers to molecular signatures for personalized therapy), and (5) XAI (transparency to build clinician trust). Together, these pillars span current applications and future directions, highlighting how advances in any one area strengthen the overall clinical workflow.

6.10 Conclusions

AI is catalyzing a paradigm shift in biomedical imaging from passive observation to systems that generate mechanistic inferences and promote patient-specific intervention. Advancement in image acquisition, reconstruction, and automated interpretation has improved diagnostic accuracy and efficiency; combined with multimodal integration and principled model design. Real clinical importance, however, will arise not from isolated technical advances, but from deliberate integration into clinical workflows, where human insight and algorithmic strengths are combined to make care safer, faster, and more equitable. Trustworthy deployment demands strong out-of-distribution generalization, calibrated uncertainty, fairness across populations, and robust governance. The models must be proven in realistic, heterogonous clinical environments and sustained throughout their life cycle, with clear acknowledgment of limitations as well as sustained human oversight. Also, workflow design should not impede clinician reasoning; explainability, usability, and explicit accountability must be built in from the outset rather than bolted later. In the future, the most impactful systems will connect scales-linking molecules and cells, to organs and populations, imaging to biology, longitudinal trajectories to patient context. Actionable, regulator-assessable predictions will emerge from hybrids that combine physical modeling and casual reasoning with large-scale learned representations. Achieving this vision requires cross-disciplinary collaboration, new evaluation and implementation standards, and pragmatic approaches for equitable access, so that benefits extend to low-resource settings as well as leading centers. AI is not a replacement for clinical judgment, but an amplifier of it – surfacing subtle patterns earlier, quantifying uncertainty transparently, and freeing clinicians' time for the deeply human work machines cannot do. Realized with rigor, humility, and an unwavering emphasis on patient benefit, AI in biomedical imaging can define twenty-first-century practice by delivering deeper insight, more precise therapy, and more caring care.

References

Aeffner, F., Zarella, M. D., Buchbinder, N., et al. (2019). Introduction to digital pathology: What is digital pathology? *Advances in Anatomic Pathology*, *26*(1), 7–13. https://doi.org/10.1097/PAP.0000000000000196.

Aerts, H. J. W. L., Velazquez, E. R., Leijenaar, R. T., et al. (2014). Decoding tumour phenotype by noninvasive imaging using a quantitative radiomics approach. *Nature Communications*, *5*, 4006. https://doi.org/10.1038/ncomms5006.

Attia, Z. I., Noseworthy, P. A., Lopez-Jimenez, F., et al. (2019). An artificial intelligence-enabled ECG algorithm for the identification of patients with atrial fibrillation during sinus rhythm: A retrospective analysis of outcome prediction. *The Lancet*, *394*(10201), 861–867. https://doi.org/10.1016/S0140-6736(19)31721-0.

Baaklini, M., & Valdés Hernández, M. D. C. (2025). Deep learning for MRI-based acute and subacute ischaemic stroke lesion segmentation – A systematic review, meta-analysis, and pilot evaluation of key results. *Frontiers in Medical Technology*, *7*, 1491197. https://doi.org/10.3389/fmedt.2025.1491197.

Baltrušaitis, T., Ahuja, C., & Morency, L. P. (2019). Multimodal machine learning: A survey and taxonomy. *IEEE Transactions on Pattern Analysis and Machine Intelligence*, *41*(2), 423–443. https://doi.org/10.1109/TPAMI.2018.2798607.

Benjamens, S., Dhunnoo, P., & Meskó, B. (2020). The state of artificial intelligence-based FDA-approved medical devices and algorithms: An online database. *NPJ Digital Medicine*, *3*, 118. https://doi.org/10.1038/s41746-020-00324-0.

Berg, M., Holroyd, N., Walsh, C., West, H., Walker-Samuel, S., & Shipley, R. (2022). Challenges and opportunities of integrating imaging and mathematical modelling to interrogate biological processes. The International Journal of Biochemistry & Cell Biology, 146, 106195. https://doi.org/10.1016/j.biocel.2022.106195.

Bernstein, M. A., King, K. F., & Zhou, X. J. (2004). *Handbook of MRI Pulse Sequences*. Elsevier Inc. https://doi.org/10.1016/B978-0-12-092861-3.X5000-6.

Bidgood, W. D., Horii, S. C., Prior, F. W., & Van Syckle, D. E. (1997). Understanding and using DICOM, the data interchange standard for biomedical imaging. *RadioGraphics*, *17*(3), 739–755. https://doi.org/10.1148/radiographics.17.3.9153684.

Bonawitz, K., Ivanov, V., Kreuter, B., et al. (2017). Practical secure aggregation for privacy-preserving machine learning. *Proceedings of the 2017 ACM SIGSAC Conference on Computer and Communications Security*, 1175–1191. https://doi.org/10.1145/3133956.3133982.

Campanella, G., Hanna, M. G., Geneslaw, L., et al. (2019). Clinical-grade computational pathology using weakly supervised deep learning on whole slide images. *Nature Medicine*, *25*, 1301–1309. https://doi.org/10.1038/s41591-019-0508-1.

Chartsias, A., Joyce, T., Dharmakumar, R., & Tsaftaris, S. A. (2017). Adversarial image synthesis for unpaired multi-modal cardiac data. *International Workshop on Simulation and Synthesis in Medical Imaging*, 3–13. https://doi.org/10.1007/978-3-319-67566-3_1.

Chen, H., Zhang, Y., Kalra, M. K., Lin, F., Chen, Y., Liao, P., . . . & Wang, G. (2017). Low-dose CT with a residual encoder-decoder convolutional neural network. *IEEE Transactions on Medical Imaging*, *36*(12), 2524–2535. https://doi.org/10.1109/TMI.2017.2715284.

Chen, J., Lu, Y., Yu, Q., et al. (2021). TransUNet: Transformers make strong encoders for medical image segmentation. *ArXiv Preprint*. https://doi.org/10.48550/arXiv.2102.04306.

Chen, M., Wang, Y., Wang, Q., Shi, J., Wang, H., Ye, Z., et al. (2024). Impact of human and artificial intelligence collaboration on workload reduction in medical image interpretation: A meta-analysis. *NPJ Digital Medicine*, *7*, 349. https://doi.org/10.1038/s41746-024-01328-w.

Cherry, S. R., Badawi, R. D., Karp, J. S., Moses, W. W., Price, P., & Jones, T. (2018). Total-body PET: Maximizing sensitivity to create new opportunities for clinical research and patient care. *Journal of Nuclear Medicine*, 59(1), 3–12.https://doi.org/10.2967/jnumed.116.184028.

Chudakov, D. M., Matz, M. V., Lukyanov, S., & KA, L. (2010). Fluorescent proteins and their applications in imaging living cells and tissues. *Physiol Rev*, *90*(3), 1103–1163. https://doi.org/10.1152/physrev.00038.2009.

Coudray, N., Ocampo, P. S., Sakellaropoulos, T., et al. (2018). Classification and mutation prediction from non–small cell lung cancer histopathology images using deep learning. *Nature Medicine, 24,* 1559–1567. https://doi.org/10.1038/s41591-018-0177-5.

Ding, Y., Sohn, J. H., Kawczynski, M. G., et al. (2019). A deep learning model to predict a diagnosis of Alzheimer disease by using 18F-FDG PET of the brain. *Radiology, 290*(2), 456–464. https://doi.org/10.1148/radiol.2018180958.

Erickson, B. J., Korfiatis, P., Kline, T. L., Akkus, Z., Philbrick, K., & Weston, A. (2017). Machine learning for medical imaging. *Radiographics, 37*(2), 505–515. https://doi.org/10.1148/rg.2017160130.

Esteva, A., Kuprel, B., Novoa, R. A., et al. (2017). Dermatologist-level classification of skin cancer with deep neural networks. *Nature, 542,* 115–118. https://doi.org/10.1038/nature21056.

Esteva, A., Robicquet, A., Ramsundar, B., et al. (2019). A guide to deep learning in healthcare. *Nature Medicine, 25*(1), 24–29. https://doi.org/10.1038/s41591-018-0316-z.

Falk, T., Mai, D., Bensch, R., et al. (2019). U-Net: Deep learning for cell counting, detection, and morphometry. *Nature Methods, 16,* 67–70. https://doi.org/10.1038/s41592-018-0261-2.

Geroski, T., & Taylor, R. (2023). Big data in multiscale modelling: From medical image data to healthcare decision-making. *Journal of Big Data,* 2023, *10*(1), 14. https://doi.org/10.1186/s40537-023-00763-y.

Gilbert, S., Fenech, M., Hirsch, M., Upadhyay, S., Biasiucci, A., & Starlinger, J. (2021). Algorithm change protocols in the regulation of adaptive machine learning–based medical devices. *Journal of Medical Internet Research, 23*(10), e30545. https://doi.org/10.2196/30545.

Gillies, R. J., Kinahan, P. E., & Hricak, H. (2015). Radiomics: Images are more than pictures, they are data. *Radiology, 278*(2), 563–577. https://doi.org/10.1148/radiol.2015151169.

Ginat, D. T., & Gupta, R. (2014). Advances in computed tomography imaging technology. *Annual Review of Biomedical Engineering, 16,* 431–453. https://doi.org/10.1146/annurev-bioeng-121813-113601.

Giraud, P., & Bibault, J.-E. (2024). Artificial intelligence in radiotherapy: Current applications and future trends. *Diagnostic and Interventional Imaging, 105*(12), 475–480. https://doi.org/10.1016/j.diii.2024.06.001.

Hammernik, K., Klatzer, T., Kobler, E., Recht, M. P., Sodickson, D. K., Pock, T., & Knoll, F. (2018). Learning a variational network for reconstruction of accelerated MRI data. *Magnetic Resonance in Medicine, 79*(6), 3055–3071. https://doi.org/10.1002/mrm.26977.

Hashimoto, D. A., Rosman, G., Rus, D., & Meireles, O. R. (2018). Artificial intelligence in surgery: Promises and perils. *Annals of Surgery, 268*(1), 70–76. https://doi.org/10.1097/SLA.0000000000002693.

Ho, J., Jain, A., & Abbeel, P. (2020). Denoising diffusion probabilistic models. *Advances in Neural Information Processing Systems, 33,* 6840–6851. https://doi.org/10.48550/arXiv.2006.11239.

Holzinger, A., Biemann, C., Pattichis, C. S., & Kell, D. B. (2017). What do we need to build explainable AI systems for the medical domain?. *Reviews in the Medical and Biological Sciences, 49*(1), 1–17. https://doi.org/10.1007/s10916-016-0666-3, https://doi.org/10.1259/bjr.20221031.

James, M. L., & Gambhir, S. S. (2012). A molecular imaging primer: Modalities, imaging agents, and applications. Physiological Reviews, *92*(2), 897–965. https://doi.org/10.1152/physrev.00049.2010.

Jensen, J. H., & Helpern, J. A. (2010). MRI quantification of non-gaussian water diffusion by kurtosis analysis. *NMR in Biomedicine, 23*(7), 698–710. https://doi.org/10.1002/nbm.1518.

Jiang, F., Jiang, Y., Zhi, H., et al. (2017). Artificial intelligence in healthcare: Past, present and future. *Stroke and Vascular Neurology, 2*(4), 230–243. https://doi.org/10.1136/svn-2017-000101.

Jin, K. H., McCann, M. T., Froustey, E., & Unser, M. (2017). Deep convolutional neural network for inverse problems in imaging. *IEEE Transactions on Image Processing, 26*(9), 4509–4522. https://doi.org/10.1109/TIP.2017.2713099.

Jordan, M. I., & Mitchell, T. M. (2015). Machine learning: Trends, perspectives, and prospects. *Science, 349*(6245), 255–260. https://doi.org/10.1126/science.aaa8415.

Karras, T., Laine, S., & Aila, T. (2019). A style-based generator architecture for generative adversarial networks. *IEEE/CVF Conference on Computer Vision and Pattern Recognition*, 4401–4410. https://doi.org/10.1109/CVPR.2019.00453

Lakhani, P., & Sundaram, B. (2017). Deep learning at chest radiography: Automated classification of pulmonary tuberculosis by using convolutional neural networks. *Radiology*, *284*(2), 574–582. https://doi.org/10.1148/radiol.2017162326.

Lambin, P., Leijenaar, R. T., Deist, T. M., et al. (2017). Radiomics: The bridge between medical imaging and personalized medicine. *Nature Reviews Clinical Oncology*, *14*(12), 749–762. https://doi.org/10.1038/nrclinonc.2017.141.

Lambin, P., Rios-Velazquez, E., Leijenaar, R., et al. (2012). Radiomics: Extracting more information from medical images using advanced feature analysis. *European Journal of Cancer*, *48*(4), 441–446. https://doi.org/10.1016/j.ejca.2011.11.036.

LeCun, Y., Bengio, Y., & Hinton, G. (2015). Deep learning. *Nature*, *521*(7553), 436–444. https://doi.org/10.1038/nature14539.

Li, T., Sahu, A. K., Talwalkar, A., & Smith, V. (2020). Federated learning: Challenges, methods, and future directions. *IEEE Signal Processing Magazine*, *37*(3), 50–60. https://doi.org/10.1109/MSP.2020.2975749.

Li, W., Milletari, F., Xu, D., et al. (2020). Privacy-preserving federated brain tumor segmentation. *Medical Image Analysis*, *65*, 101765. https://doi.org/10.1016/j.media.2020.101765.

Litjens, G., Kooi, T., Bejnordi, B. E., et al. (2017). A survey on deep learning in medical image analysis. *Medical Image Analysis*, *42*, 60–88. https://doi.org/10.1016/j.media.2017.07.005.

Lundberg, S. M., & Lee, S. I. (2017). A unified approach to interpreting model predictions. *Advances in Neural Information Processing Systems*, *30*, 4765–4774. https://arxiv.org/abs/1705.07874.

Ma, J., He, Y., Li, F., Han, L., You, C., Wang, B., et al. (2024). Segment anything in medical images. *Nature Communications*, *15*, 654. https://doi.org/10.1038/s41467-024-44824-z.

Madani, A., Arnaout, R., Mofrad, M., & Arnaout, R. (2018). Fast and accurate view classification of echocardiograms using deep learning. *NPJ Digital Medicine*, *1*, 6. https://doi.org/10.1038/s41746-018-0006-1.

Mandel, J. C., Kreda, D. A., Mandl, K. D., Kohane, I. S., & Ramoni, R. B. (2016). SMART on FHIR: A standards-based, interoperable apps platform for electronic health records. *New England Journal of Medicine*, *374*(23), 2247–2250. https://doi.org/10.1056/NEJMp1606159.

Massoud, T. F., & Gambhir, S. S. (2003). Molecular imaging in living subjects: Seeing fundamental biological processes in a new light. *Genes & Development*, *17*(5), 545–580. https://doi.org/10.1101/gad.1047403.

Mehrabi, N., Morstatter, F., Saxena, N., et al. (2021). A survey on bias and fairness in machine learning. *ACM Computing Surveys*, *54*(6), 115. https://doi.org/10.1145/3457607.

Mello-Thoms, C., & Mello, C. A. B. (2023). Clinical applications of artificial intelligence in radiology. *British Journal of Radiology*, 2023 Oct, *96*(1150), 20221031.

Milletari, F., Navab, N., & Ahmadi, S. A. (2016). V-Net: Fully convolutional neural networks for volumetric medical image segmentation. *2016 Fourth International Conference on 3D Vision (3DV)*, 565–571. https://doi.org/10.1109/3DV.2016.79

Miotto, R., Wang, F., Wang, S., Jiang, X., & Dudley, J. T. (2018). Deep learning for healthcare: Review, opportunities and challenges. *Briefings in Bioinformatics*, *19*(6), 1236–1246. https://doi.org/10.1093/bib/bbx044.

Najjar, R. (2023). Redefining radiology: A review of artificial intelligence integration in medical imaging. *Diagnostics (Basel)*, *13*(17), 2760. https://doi.org/10.3390/diagnostics13172760.

Nilsson, N. J. (2014). *Principles of Artificial Intelligence*. Morgan Kaufmann.

Nogales, E., & Scheres, S. H. W. (2015). Cryo-EM: A unique tool for the visualization of macromolecular complexity. *Molecular Cell*, *58*(4), 677–689. https://doi.org/10.1016/j.molcel.2015.02.019.

Ntziachristos, V. (2010). Going deeper than microscopy: The optical imaging frontier in biology. *Nature Methods, 7*(8), 603–614. https://doi.org/10.1038/nmeth.1483.

Obermeyer, Z., Powers, B., Vogeli, C., & Mullainathan, S. (2019). Dissecting racial bias in an algorithm used to manage the health of populations. *Science, 366*(6464), 447–453. https://doi.org/10.1126/science.aax2342.

Parisot, S., Ktena, S. I., Ferrante, E., et al. (2018). Disease prediction using graph convolutional networks. *Medical Image Analysis, 48*, 117–130. https://doi.org/10.1016/j.media.2018.06.001.

Pesapane, F., Codari, M., & Sardanelli, F. (2018). Artificial intelligence in medical imaging: Threat or opportunity?. *European Radiology Experimental, 2*(1), 35. https://doi.org/10.1186/s41747-018-0061-6.

Pinto-Coelho, L. (2023). How artificial intelligence is shaping medical imaging technology: A survey of innovations and applications. *Bioengineering, 10*(12), 1435. https://doi.org/10.3390/bioengineering10121435.

Price, W. N., & Cohen, I. G. (2019). Privacy in the age of medical AI. *Nat Med, 25*, 37–43. https://doi.org/10.1038/s41591-018-0272-7.

Rajpurkar, P., Chen, E., Banerjee, O., & Topol, E. J. (2022). AI in health and medicine. *Nat Med, 28*(1), 31–38. https://doi.org/10.1038/s41591-021-01614-0.

Rieke, N., Hancox, J., Li, W., et al. (2020). The future of digital health with federated learning. *NPJ Digital Medicine, 3*, 119. https://doi.org/10.1038/s41746-020-00323-1.

Ronneberger, O., Fischer, P., & Brox, T. (2015). U-Net: Convolutional networks for biomedical image segmentation. *Medical Image Computing and Computer-Assisted Intervention (MICCAI)*, 234–241. https://doi.org/10.1007/978-3-319-24574-4_28.

Rumsfeld, J. S., Joynt, K. E., & Maddox, T. M. (2016). Big data analytics to improve cardiovascular care: Promise and challenges. *Nature Reviews Cardiology, 13*, 350–359. https://doi.org/10.1038/nrcardio.2016.65.

Russell, S., & Norvig, P. (2021). *Artificial Intelligence: A Modern Approach* (4th ed.). Hoboken: Pearson.

Samek, W., Wiegand, T., & Müller, K. R. (2017). Explainable artificial intelligence: Understanding, visualizing and interpreting deep learning models. *IT – Information Technology, 61*(5–6), 39–46. https://doi.org/10.1515/itit-2017-0003.

Schermelleh, L., Ferrand, A., Huser, T., et al. (2019). Super-resolution microscopy demystified. *Nature Cell Biology, 21*, 72–84. https://doi.org/10.1038/s41556-018-0251-8.

Seibert, J. A. (2004). X-ray imaging physics for nuclear medicine technologists. Part 1: Basic principles of x-ray production. *Journal of Nuclear Medicine Technology*, 32(3), 139–147. https://pubmed.ncbi.nlm.nih.gov/15347692/.

Selvaraju, R. R., Cogswell, M., Das, A., et al. (2017). Grad-CAM: Visual explanations from deep networks via gradient-based localization. *Proceedings of the IEEE International Conference on Computer Vision (ICCV)*, 618–626. https://doi.org/10.1109/ICCV.2017.74.

Sheller, M. J., Reina, G. A., Edwards, B., Martin, J., & Bakas, S. (2020). Federated learning in medicine: Facilitating multi-institutional collaborations without sharing patient data. *Scientific Reports, 10*, 12598. https://doi.org/10.1038/s41598-020-69250-1.

Shortliffe, E. H., & Sepúlveda, M. J. (2018). Clinical decision support in the era of artificial intelligence. *JAMA, 320*(21), 2199–2200. https://doi.org/10.1001/jama.2018.17163.

Shui, L., Ren, H., Yang, X., Li, J., Chen, Z., Yi, C., Zhu, H., & Shui, P. (2021). The era of radiogenomics in precision medicine: An emerging approach to support diagnosis, treatment decisions, and prognostication in oncology. *Frontiers in Oncology, 10*, 570465. https://doi.org/10.3389/fonc.2020.570465.

Skandarani, Y., Jodoin, P.-M., & Lalande, A. (2023). GANs for medical image synthesis: An empirical study. *Journal of Imaging, 9*(3), 69. https://doi.org/10.3390/jimaging9030069.

Suganyadevi, S., Seethalakshmi, V., & Balasamy, K. (2022). A review on deep learning in medical image analysis. *International Journal of Multimedia Information Retrieval*, *11*, 19–38. https://doi.org/10.1007/s13735-021-00218-1.

Szabo, T. L. (2014). *Diagnostic Ultrasound Imaging: Inside Out*. Academic Press. https://doi.org/10.1016/C2009-0-20239-8.

Topol, E. J. (2019). High-performance medicine: The convergence of human and artificial intelligence. *Nat Med*, *25*(1), 44–56. https://doi.org/10.1038/s41591-018-0300-7.

Wang, L., Lin, Z. Q., & Wong, A. (2020). COVID-Net: A tailored deep convolutional neural network design for detection of COVID-19 cases from chest radiography images. *Scientific Reports*, *10*, 19549. https://doi.org/10.1038/s41598-020-76550-z.

Wang, S., Su, Z., Ying, L., Peng, X., Zhu, S., Liang, F., . . . & Liang, D. (2016). Accelerating magnetic resonance imaging via deep learning. *IEEE Transactions on Medical Imaging*, *35*(5), 1225–1238. https://doi.org/10.1109/TMI.2016.2535818.

Wang, Z., Wu, Z., Agarwal, D., & Sun, J. (2022). MedCLIP: Contrastive learning from unpaired medical images and text. In *Proceedings of the 2022 Conference on Empirical Methods in Natural Language Processing (EMNLP)* (pp. 4665–4677). Association for Computational Linguistics. https://doi.org/10.18653/v1/2022.emnlp-main.256.

Weissleder, R., & Pittet, M. J. (2008). Imaging in the era of molecular oncology. *Nature*, *452*(7187), 580–589. https://doi.org/10.1038/nature06917.

Yang, G. Z., Cambias, J., Cleary, K., et al. (2017). Medical robotics – Regulatory, ethical, and legal considerations for increasing levels of autonomy. *Science Robotics*, *2*(4), eaam8638. https://doi.org/10.1126/scirobotics.aam8638.

Yi, X., Walia, E., & Babyn, P. (2019). Generative adversarial network in medical imaging: A review. *Medical Image Analysis*, *58*, 101552. https://doi.org/10.1016/j.media.2019.101552.

Zhang, S., Xu, Y., Usuyama, N., Bagga, J., Tinn, R., Preston, S., et al. (2023). BiomedCLIP: A multimodal biomedical foundation model pretrained from fifteen million scientific image-text pairs. *arXiv Preprint*. https://doi.org/10.48550/arXiv.2303.00915.

Zhu, B., Liu, J. Z., Cauley, S. F., et al. (2018). Image reconstruction by domain-transform manifold learning. *Nature*, *555*, 487–492. https://doi.org/10.1038/nature25988.

Diksha Pathania*, Priyanka Chauhan, Aditya Chauhan, Sonu, Priti Duhan

Chapter 7 Smart Agriculture: AI in Biotechnology for Food Security

Abstract: Artificial intelligence (AI) turned out to be a disruptive technology in the agricultural sector, which can provide a variety of options such as precision farming and crop development, pest control, soil quality tracking, and predicting yields. By integrating biotechnology and high-tech data analytics, AI helps to sustain food production and make this practice less resource-intensive and less harmful to nature. The use of AI in speeding up food security issues in the world is significant, notably in enabling smallholder farmers, enhancing their resistance to climate change, and providing efficient and viable access to technology. The innovations of the future such as AI-robotics, blockchain-based transparency, and climate-sensitive applications, promise immeasurable power to transform the agriculture trend in 2050. Recognizing this vision will require inclusive cooperation, supportive policies, and education to make sure comprehensive adoption. Together, AI and biotechnology pave the way for a resilient, sustainable, and food-secure future.

Keywords: Artificial intelligence, smart agriculture. IoT, crop improvement, machine learning, sustainability

7.1 Introduction

Agriculture is the heart of human existence on earth. Despite this fact, it faces rising pressures in this twenty-first century (Vos & Bellu, 2019) due to climate change, population growth, declining natural resources, and increasing demand of food security with each passing day (Tubiello et al., 2008). Traditional farming methods with old technologies are no longer sufficient to meet the global challenges (Borlaung & Dows-

***Corresponding author**: **Diksha Pathania,** Dairy Technology Division, National Dairy Research Institute (ICAR), Karnal, Haryana, India, e-mail: dikshapathania198@gmail.com
Priyanka Chauhan, SRMIST, Delhi NCR Campus, Ghaziabad, India
Aditya Chauhan, Department of Agriculture, Maharishi Markandeshwar (Deemed to be University), Mullana, Ambala, Haryana, India
Sonu, School of Advanced Chemical Sciences, Shoolini University of Biotechnology and Management Sciences, Solan, Himachal Pradesh, India
Priti Duhan, Department of Biotechnology, K. L. Mehta Dayanand College for Women, Faridabad 121001, Haryana, India

 | https://doi.org/10.1515/9783112214275-007

wel, 2003; Sassenrath et al., 2008). Thus, integration of advanced technology is the requirement of present scenario, rather than an optional choice (Kalbar et al., 2012). Biotechnology contributed significantly in crop resilience as well as crop productivity sector and emergence of artificial intelligence (AI) has opened a new door in smart agriculture field. Collaboration between biological innovation world and data-driven insights of AI leads to the transformation in cultivation, management, and harvesting crops to maintain sustainable agriculture with increased yield (Riaz et al., 2025). AI has emerged as a game changer in agriculture, drastically altering how food is produced, protected, and distributed. Here's how AI technology supports the highest levels of precision, expertise, and adaptability throughout the entire food security and agriculture factor:

- The ever-growing accumulation of massive information provided by satellites, drones, and sensors gives a holistic view of the soil quality, crop conditions, moisture status, and the frequency of insects.
- AI models analyze real-time data, provide profound insights and realistic suggestions to enhance crop productivity and farm management.
- Such disposition of the water, fertilizers, and pesticides in targeted form reduces waste and increases production; this is important to the defensive farmland and helps to prevent damage.
- The cold chain investigating, intelligent logistics platforms, and AI-supported analytics protect food supplies against post-harvest damage, food spoilage, and interruptions.
- AI-based observation and instant alarm systems shield cultivated resources from both environmental and human dangers.

As these creative solutions grow more important to agriculture and food security, industries around the world are entering a new era in which technology is at the heart of resilient, secure, and sustainable food systems. This chapter explores how AI is revolutionizing biotechnology in agriculture – through precision farming, crop improvement, pest management, soil health monitoring, and yield prediction while also addressing its vital role in ensuring sustainable food production and tackling the global food security crisis.

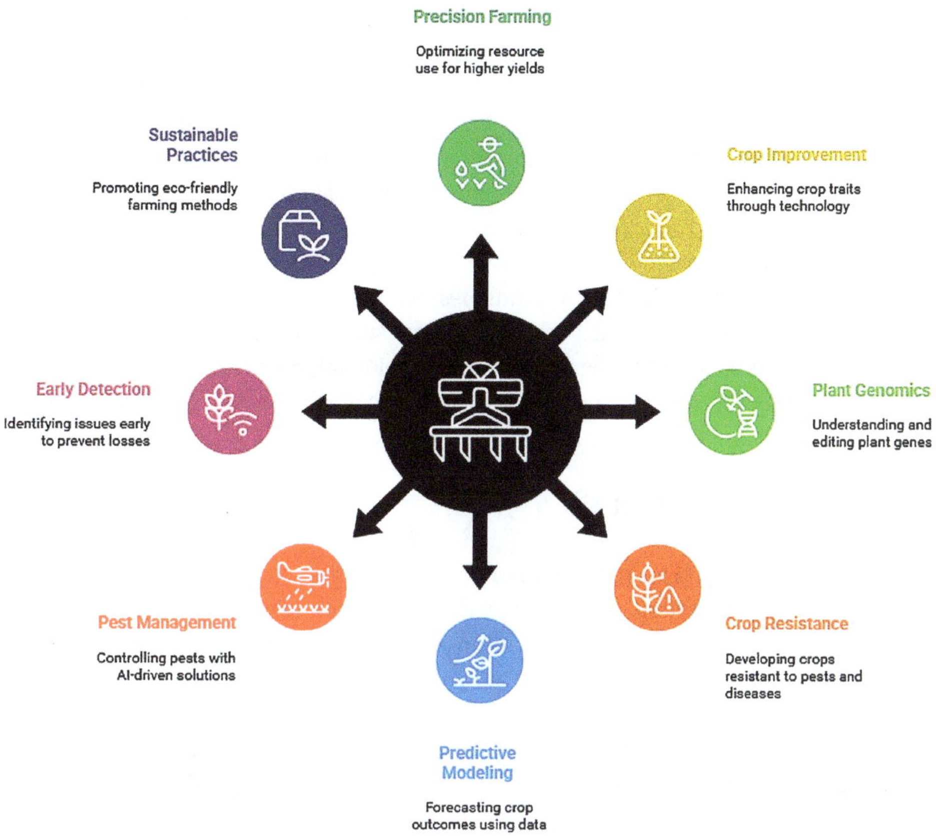

Figure 7.1: AI in agriculture optimizes crop yield, pest control, and resource management.

7.2 AI in Transforming Agriculture

The different applications of AI in modern agriculture are shown in Figure 7.1.

7.2.1 Precision Farming

Agricultural precision farming especially powered by AI is another technology entryway that is altering the industry and saving money (Khan, 2024). In the agricultural field, instead of applying irrigation water, fertilizers, and pesticides uniformly, AI makes it possible to tailor these inputs to specific areas and specific plants (Wei et al., 2024; Ali et al., 2025). This is achieved by the use of smart sensors, drones, satellite imaging, and Internet of Things (IoT) devices (Pearson et al., 2018) that are able to collect information regarding the soil condition, moisture level, nutrients level, crop health, and weather fluctuations

(Mandal et al., 2025). Once the data is collected from specific fields, AI-driven models analyze the data and provide precise information and recommendations such as amount of water (Adinarayana et al., 2024) and site-specific requirement of pest control (Subramanium et al., 2021). In addition to increasing yields, conserving resources, lowering production costs, and minimizing environmental repercussions,, this shift from blanket management to site-specific management makes a significant contribution. Drones that are directed by AI have the potential to identify nutritional deficiencies (Reddy et al., 2025) and early indicators of disease (Basu & Narayan, 2025) before they are visible to the human eye. Furthermore, irrigation techniques that are driven by AI have the capability to estimate crop water constraints based on soil moisture and weather models. Precision farming is driven by AI makes agriculture more ecological, capable, and resistant to satisfy the prospects of consumers for food security by enabling targeted interventions.

7.2.2 Crop Improvement

The development of AI is leading to the new stage of agricultural development as it is the rapid study of the plant genomes and the recognition of their characteristics. The AI systems can find the association between gene traits in large genomes significantly quicker than conventional methods. This will assist the researchers in determining positive qualities, such as the capacity to generate plenty, efficiently use nutrients and endure stress. Machine learning is being used in breeding programs to predict the level of performance of new crop varieties without the need to spend time and money testing these new crops in the field. The application of AI is also being used by researchers to aid them in locating gene targets as well as minimizing off-target effects using CRISPR. This is rendering gene editing more precise and reliable. The utilization of AI and big data analytics in phenotypic analysis offers increased information about the tendencies in the development of plants in a wide range of conditions, which allows crop promotion to be more effective and targeted.

7.2.3 AI in Plant Genomics and Gene Editing

AI transforms the genetics of plants because it accelerates and elevates the ability to identify genetic traits that influence plant production. The patterns and associations between genes and trait can now be discovered through AI and machine learning algorithms that are desired like yield, nutritional value, and stress resilience (Yoosefzadeh Najafabadi et al., 2023). The conventional genomic studies have challenges in dealing with the huge amount and complexity of genetic information. The outcomes of such studies are accelerating mating decisions and minimizing the number of field tests.

AI is also being applied in combination with gene-editing technologies, such as CRISPR, to enhance them in terms of accuracy and efficiency. CRISPR with the help of

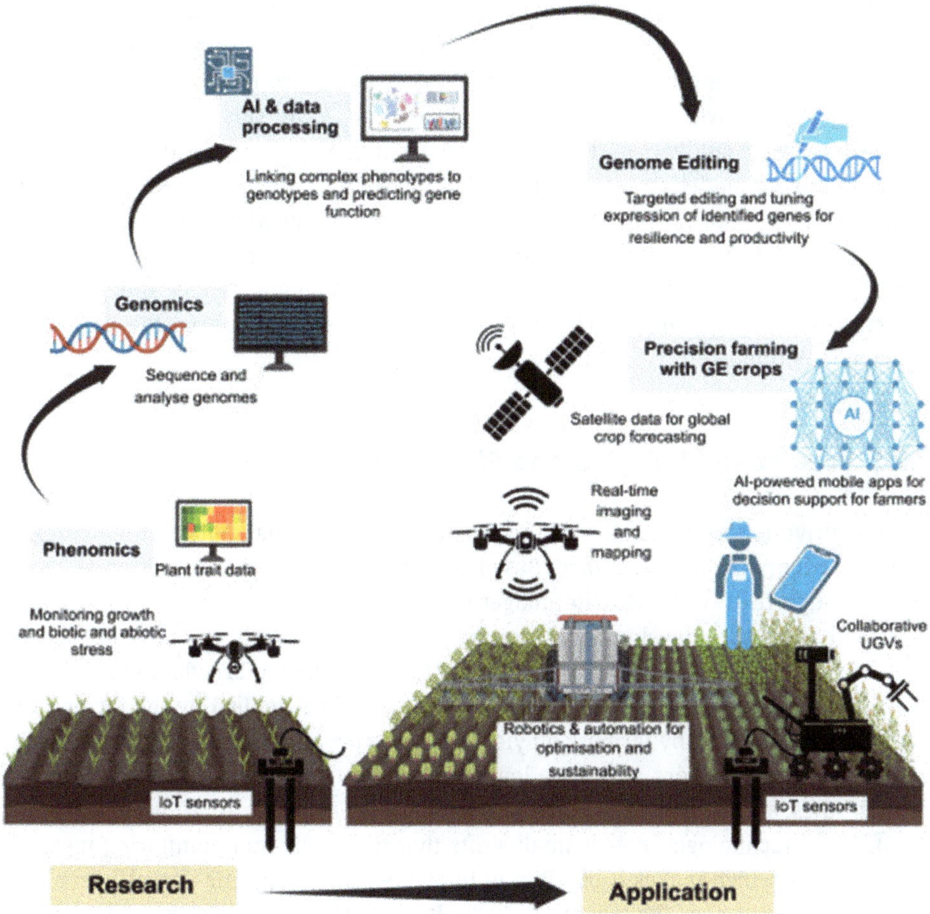

Figure 7.2: AI-driven innovations in crop genomics and precision gene editing.

AI can be used to create new crops that possess some benefits, such as increased production or enhanced resistance to illnesses. This is achieved through forecasting the most desirable target regions and reduction of unnecessary changes (Dara et al., 2024). The interaction between AI and biotechnology leads to more precise crop development (Figure 7.2).

7.2.4 Climate-Resilient Crop Development

Climate change is one of the greatest problems of agriculture in the contemporary world that causes heat waves, droughts, and floods. With the processing of large amounts of data, such as climate forecasts, soil prices (Saxena et al., 2023), and plant stress responses (Sarkar et al., 2023), AI is assisting in the creation of climate-resilient crops (Thingujam et al., 2025). It is possible that the machine learning algorithms may identify some genetic features that can be used to ensure that crops can survive under the adverse weather conditions (Bhuiyan et al., 2023). This will assist scientists to breed or create cultivars that will be in the position to survive in future (Koshariya, 2022). AI during modeling can be used to simulate the development of plants in diverse stress conditions, which can help breeders to find out what plants are resistant to drought, heat, and water consumption (Su et al., 2025). These developments protect the production of food against any unexpected environmental factors and also come up with the environmentally responsible farming systems to serve the increasing global population.

7.2.5 Predictive Modeling and Phenotype Analysis

In the cultivation of crops, AI enhances predictive modeling, enhanced phenotypic analysis, genomics, and resilience. Based on large amounts of information accumulated through field trials, satellite imagery, and laboratory research, AI systems may anticipate the future development of crops under certain conditions and the appearance of pests and diseases (Delfani et al., 2024). By producing breeders that have inbuilt resistance, breeders lessen the quantity of pesticides applied by the farmer and the expenses incurred by farmers on yields. Phenotypic analysis that leverages AI and utilizes pictures, sensors and deep learning can be used to determine plant form, growth rates, and responses to environmental stress (Zhang et al., 2024; Angidi et al., 2025). These technologies detect small shifts that the traditional monitoring methods fail to notice, which makes it possible to implement more specific and efficient crop development processes (Bakala et al., 2020). It is becoming possible to produce crops in a data-driven efficient and needs-focused manner by predictive modeling and phenotypic analysis, which meets the needs of modern agriculture.

7.2.6 Pest and Disease Management

AI is transforming the way we address pests and diseases through identification and diagnosis of agricultural hazards. Computer vision and photo recognition can enable farmers to discover and cure ailments more quickly (Garg et al., 2021). It is possible to prevent pest outbreaks by using predictive analytics, thus changing the reactive approach to the preventive one. As Lawrence et al. (2023) says, satellite drone monitoring helps not only to increase the level of surveillance but also to reduce the quantity of the usage of pesticides. AI makes agricultural protection, food security achievable, stable, and sustainable.

2.7 Early Detection and Predictive Systems

AI has transformed the aspect of pest control in a massive manner because it has simplified and accelerated the process of disease and pest search. Computer vision systems powered by deep learning can look at high-resolution pictures of plants to find early signs of sickness. AI-assisted photo recognition systems that have been trained on huge datasets of ill crops can tell the difference between various diseases, which reduces the frequency of wrong diagnoses. Predictive analytics are now being utilized by people to determine the ways in which illnesses and pests will be propagated. AI prototypes can be based on data on the weather, soils, and insect numbers to demonstrate the mechanism of an infestation and assist farmers. There is an opportunity to monitor large-scale farming processes and monitor the condition of plants in real time across vast regions with the help of drones and satellites (Phang et al., 2023). The proactive strategy would help farmers to transform reactive to preventive strategies and in the process reduce losses without compromising the yield of crops.

7.2.8 Sustainability Agriculture Practices

The application of AI contributes to the enhancement of disease and pest management detection, prediction, and sustainability. AI-powered systems suggest targeted spraying of pesticides to ensure that the pesticides are not sprayed where they are not needed (Balaska et al., 2023). Due to this accuracy, the cost of production by farmers is lowered and also the level of soil contamination and biodiversity loss by pesticides is minimal. Mobile device apps that help farmers diagnose agricultural diseases, drones with AI that spray specific areas, and intelligent decision support systems that give recommendations in relation to when and how much pest control to administrator are all in use. To retain paddy (Panjaitan et al., 2022), rice, maize, and wheat, these processes are currently being experimented in India, China, and the United States. These types of examples show that AI-based pest and disease management are enhancing agricultural output and making it possible to pursue more environmentally responsible farming thus securing food security in the long run.

7.3 AI in Soil Health Monitoring

The soil is an important factor in the amount of food you can produce and the duration of farming time. To grow roots and absorb nutrients, soil needs to have organic matter, minerals, and useful microbes. Crops might be less affected by drought (Sprouls, 2020) and other stressors provided that the soil structure is in good condition. The reason is that the manner of construction of the soil allows it to contain water and air. However, in case the soil is deteriorated, unfertile, compact, or eroded, the production decreases, and more inputs are required. Crop rotation, organic amendment, less tiling, etc., contribute to creating more fertile and diverse soil in terms of microbes. This causes the soil not to be so disturbed and more favorable to the growth of plants (Angon et al., 2023). Farmers can use soil sensors and artificially intelligent predictive algorithms to detect real-time data about the amount of nutrients, pH, and moisture in the soil (Quazi et al., 2022). Combining old and new technologies will allow farmers to increase their productivity and reduce their costs and impact on the environment at the same time (Sharma et al., 2023). Healthy soil is necessary to enhance the productivity of crops, food security, and the continuity of farming to be sustainable to the generations to come (Figure 7.3).

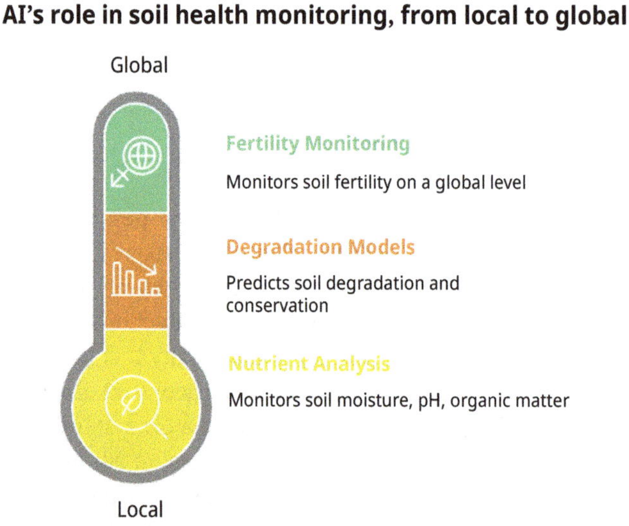

Figure 7.3: AI monitors soil health by analyzing nutrients, moisture, and microbial activity.

7.3.1 AI in Soil Nutrient Analysis and Smart Sensing

The process of soil monitoring and upkeep is being affected by AI these days. Nutrient surveys with the help of AI can possibly create credible maps of fertility with the help of satellite images, soil samples, and remote sensors. It is an improvement in soil testing, which is time-consuming, expensive, and confined in its capabilities. The same maps assist the farmers concentrate their efforts on the use of fertilizers by showing them where the nutrients are deficient and the characteristics of various fields. There are also smart soil sensors that are being installed into the ground to monitor the pH, organic matter, and moisture levels. Using data, AI algorithms can reply in real time to the extent to which the soil is healthy. In addition to saving resources, this helps farmers manage their fertilizer and irrigation.

7.3.2 Predictive Models for Soil Degradation and Conservation

AI would also have the potential to forecast soil runaway and erosion, which are threats to sustainable agriculture. Machine learning techniques may create predictions regarding erosion, salinity, and nutrient depletion based on climate data, land use trends, and topography. Using these prediction models, it is possible to put into action early intervention measures such as cover crops, contour farming, or controlled irrigation. Also, decision support systems powered by AI are being developed to help farmers and policymakers choose the best ways to protect the land. These systems combine historical data and the current state of the field to provide particular suggestions that help keep the soil fertile and the crops growing.

7.3.3 AI for Soil Fertility on Global Level

AI is becoming more and more crucial for keeping an eye on and taking care of soil health on a regional and global scale. Governments and academic organizations are employing AI platforms based on satellites to find damaged land, monitor changes in soil fertility, and plan big restoration initiatives. For example, AI-powered soil mapping projects are being used in nations that are becoming deserts to help them use their resources more effectively. In highly industrialized regions, AI systems are being utilized by precision agriculture companies linking the data of the soil to crop performance. This allows the farmers to adopt land-specific management methods. The use of these applications not only makes farming productive all over the globe but also addresses issues such as climate change, land degradation, and food security. This is why AI has become a crucial instrument of sustainable agriculture.

7.4 AI in Crop Yield Prediction

The machine learning algorithms based on AI, analyzing soil, climate, and past data, are transforming agricultural productivity expectations. It might be less difficult to monitor the real-time progress of crops due to satellite photography and weather forecasts. AI in agricultural risk assessment examines the impact of climate change on agriculture. Once the correct predictions are used, the markets become more stable as they combine the supply and demand into equilibrium and reduce the price fluctuations. Good predictions aid in storage, transportation, and distribution planning, thereby reducing the amounts of losses that occur following harvest.

7.4.1 AI and Data-Driven Yield Prediction

The past methods of estimating crop yields are being substituted by AI that makes use of data and algorithms. The machine learning systems consider the fertility of soil, the development of a crop, and the history of harvest to predict yields. Satellite images are more accurate when compared to ground data in the form of weather prediction and climatic parameters, which makes the model more accurate. Farmers can monitor the performance of their crops in real time and discover problems before they negatively affect the crop yield. The long-term consequences of climate change on crop production are being studied by using AI-driven approaches, as well as assisting agriculture in adapting to drought, heat stress, and unpredictable rainfall (Kumari et al., 2025).

7.4.2 Market Planning and Post-harvest Benefits

AI yield forecasting has influence on markets and supply networks besides estimates of farms. A precise estimation of the yield will stabilize the agricultural markets as this makes the supply and demand match. This reduces price fluctuations and food becomes cheaper. The estimates assist the policymakers and businesses to plan better with regards to imports, exports, and storage. Proper forecasting can be used to reduce the losses incurred after harvesting because of the coordination between harvesting, transportation, and storage. Authorities can also plan additional storage or delivery roads with the help of AI algorithms and in this way, they can calculate how much additional harvest they possess. Due to our inception, we will be able to save the food that will otherwise end up in waste and deliver farm products to our customers with ease and speed. These illustrations demonstrate how the prospect of AI being used to predict the yield of crops could result in more efficient, sustainable, and robust food production. All these are desirable to both the economy and farmers.

7.4.3 AI in Harvesting Index

The most significant measure of resource utilization in the agricultural sector is the harvest index calculated as a ratio of economic output and total biomass. Recent research indicates that artificially intelligent technologies are in the process of exploring and enhancing this index by assembling data of phenotyping, environmental surveillance, and genomics. The application of machine learning algorithms to analyze complex measurements can help identify characteristics that can be used to expand biomass partitioning to grains, fruits, or other items that can be used, which will result in production being more efficient (Liu et al., 2020). Monitoring the stages of crop development and the distribution of dry matter can be performed using computer vision and better imaging with the use of AI, all without any physical contact with the plants. This, in turn, provides real-time estimates of the harvest index. Another use of AI-powered prediction models by researchers is to examine the relationship between soil fertility, irrigation, and climate change, and their impact on biomass distribution. These data can be used to their advantage by plant breeders and agronomists who are interested in developing new types of crops that are characterized by a higher harvest index, better stress tolerance, and climate change resistance. AI is a field-level data and predictive analytics that can be used to maximize the harvest index, implying that agricultural systems can yield more food using fewer resources.

7.4.4 Sustainable Food Production

Sustainable agriculture tries to meet the present demand of food while preserving resources for future generation. This group desires to conserve biodiversity, preserve water, and maintain soil fertility. The studies show that the health of the soil can be improved over the years through application of crop rotation, cover crops, and use of organic amendments. Precision farming also reduces the use of pesticides and fertilizers with the help of AI and the IoT (Sharma et al., 2023). Sustainable agricultural systems ensure that renewable energy is being utilized and that carbon emission is being reduced. The coordinated pest control and intercropping can be used to maintain balance of the ecosystem. Scientists are currently making climate-resilient crop types by using AI-guided genomes and contemporary breeding methods. Drip irrigation and sensor-based irrigation systems can save water. Farmers can make the most of their resources and deal with the consequences of climate change by embracing digital platforms and predictive analytics. Sustainable agriculture is good for rural areas since it raises production while lowering input costs at the same time. It makes sure that food supplies are safe across the globe by combining economics, technology, and the environment.

7.4.5 Sustainable Intensification and Resource Optimization

Agriculture must undergo sustainable intensification in order to produce more food without causing damage to ecosystems to meet the demands of a growing global population and decreasing natural resources. The use of AI allows for more accurate management of water and energy resources. Smart irrigation systems that are driven by AI optimize watering schedules depending on soil moisture, weather, and crop needs, therefore reducing environmental waste and boosting agricultural yields. AI-powered energy management systems make the most of renewable resources, including solar-powered irrigation pumps, to save money and reduce stress on the environment (Al-Khreisat et al., 2025). The application of AI to eliminate activities consuming high resources can be used to reach a compromise between productivity and sustainability.

7.4.6 Climate Resilience and Emission Reduction

In order to ensure the continuation of food production, AI assists in reducing the level of greenhouse gases. With AI-powered models, carbon footprints are also analyzed (the use of fertilizer, the care of livestock, etc.) and recommendations are made on how to reduce emissions. Predictive systems enable you to determine the amount of nitrogen fertilizer to apply to ensure that you are not putting too much of the material and causing too much nitrous oxide to leak out. AI presents the benefits that would accumulate over time to the environment caused by conservation techniques such as non-till farming and managing excess food on farms (Kayusi et al., 2025). These practices decrease emissions and enhance the quantity of carbon held in the soil, which makes farming more resilient to climatic change and more profitable.

7.5 Food Systems Efficiency and Circular Bioeconomy

Sustainable agriculture also covers the management of crops after harvesting and ensuring that there is always efficiency in the use of energy in the food chain. In the case of perishable goods, it is worth having supply chain models that employ AI to determine the demand, locate optimal routes, and reduce the losses in shipping. The sensors of smart storage systems are AI that monitor temperature and humidity to make food last longer. In order to prevent food waste, predictive analytics will be applied in the matching of supply and demand. The waste-to-resource solutions (like converting leftover food into bioenergy or organic fertilizers) are also recommended with the help of AI, which contributes to creating a cyclic bioeconomy (Mehdizadeh et al., 2025). It appears that there is a gradual adoption of AI-based logistics and waste minimization technologies by the less developed countries with emphasis laid on the solutions that are inexpensive and can be adopted by a large population. Developed countries are gaining grounds in most societal aspects. These uses demonstrate that AI is transforming farming to be more productive and establish stronger and more sustainable food systems as shown in Figure 7.4 .

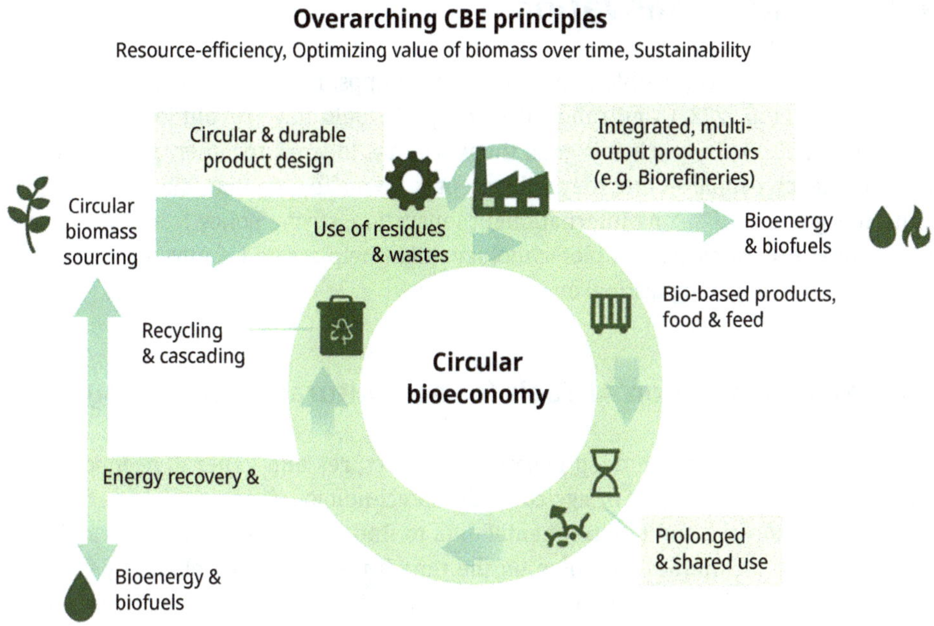

Figure 7.4: Circular bioeconomy promotes sustainable biomass use, recycling, and resource efficiency across industries.

7.6 Provision of Food Security via AI

AI could contribute to the reduction of the world hunger and poverty by developing new technologies in the fields of agriculture, nutrition, and reduction of food waste. Guiding tools, smartphone applications, and precision agricultural solutions that may be powered by AI could assist the smallholder farmers, who produce much of the food that the world produces but lack access to modern technology, to be more productive and resilient. But it is difficult to use among many people due to a lack of infrastructure, bad internet connectivity, and technology divide between rich and poor countries. The inequality in access to agricultural data may further widen the gap between those farmers who have a significant number of resources and those who have less resources. It is very important to deal with ethical and societal issues, such how farmers rely on private tech companies and who owns the data. Strong laws and collaboration across countries make it feasible to use AI in a way that is fair to everyone while also protecting farmers' rights and making sure that agriculture grows in a way that is good for the environment.

7.7 New Age Innovations

By the year 2050, it is possible that climate-smart apps, blockchain-enabled transparency (Anoop et al., 2022), and robots driven by AI would have revolutionized agriculture (Zurek, 2020). To reach this goal, the world has to work together, pass laws that help, and teach people so that they will accept it. To make sure that fair adoption happens, this goal needs strong international cooperation, strong policy frameworks, and educational programs. AI and biotechnology might help create a future for the whole globe that is strong, long-lasting, and food secure.

7.7.1 Next-Generation AI Tools for Agricultural Biotechnology

Advanced AI methods like deep learning architectures and generative models are speeding up breakthroughs in agricultural biotechnology. These methods are using genomic, phenotypic, and environmental data to find trait-gene links that are hidden in noise. These techniques cut down on the time it takes to breed plants by focusing on potential candidate genes and predicting how traits will behave in different conditions. They are currently being used to speed up the process of finding novel crop protection chemicals and modes of action. This change is shown by instances from the industry: AI systems have greatly sped up the process of finding new herbicide targets and other inputs. This shows that computational methods may work with wet-lab experiments to bring forth new ideas more quickly and at a lesser cost.

7.7.2 Integration of AI with Robotics and Blockchain

The combination of AI and robotics is creating autonomous platforms that can plant, weed, spray, and harvest with great accuracy (Meshram et al., 2023). This cuts down on labor shortages and allows farms to run continuously with plenty of data (Munnisunker et al., 2022). Field robots and self-driving tractors now employ AI and computer vision on their devices (Aghi et al., 2020) to find their way, check the health of crops, and make targeted changes. This makes them more efficient and less harmful to the environment. Combining AI with blockchain and other distributed ledgers at the systems level might make supply chains more traceable and trustworthy: AI can check and analyze data from sensors and transactions, while blockchain makes guarantee that provenance records can't be changed. This helps cut down on fraud, speed up recalls, and establish trust in claims about where products come from and how long they last (Lopes et al., 2018). Real-world deployments and pilots, such as autonomous herders and tractors and blockchain traceability pilots, illustrate that this technology convergence is quickly advancing from the experimental stage to the operational stage.

7.7.3 Climate-Smart AI, the 2050 Vision

Advanced AI models are increasingly being applied to climate-smart agriculture: they combine high-resolution climate projections, remote sensing, and crop models to optimize planting windows, resource allocation, and varietal choices under future climates. Looking toward 2050, a plausible vision is a resilient (Zhao et al., 2023), AI-enabled global agriculture that couples predictive analytics, autonomous systems, and circular bioeconomy practices to produce more food with fewer inputs while buffering climate shocks. It is necessary to make targeted investments in multidisciplinary research, open and ethical data infrastructures, the development of farmer and extension service capacity, and legislative frameworks that enable equitable access, data governance, and public-private partnerships in order to accomplish the goal of global organizations promoting inclusive AI deployment through collaborations and standards while monitoring the impacts on society and the environment.

7.8 Conclusion

AI has changed farming in a big way by making the entire food supply chain more accurate, productive, and sustainable. AI assists farmers to surpass the capacity of the human mind in several fields such as crop production, insect control, soil wellness, and harvesting timing. To get biotechnology and AI to collaborate, predictive algorithms that require gene-editing practices may be applied to develop crops resistant to climate change, rich in nutrients, and capable of yielding high volumes of food. This relationship will ensure that the further agricultural systems are not only productive but also environmentally friendly. To fill the digital divide that still prevails among regions, the future of the world food security should focus on equitable access to AI-powered technology, research, enabling policy frameworks, and educating farmers. Not only can AI in agriculture be used to make technology better, but also to make the food systems strong, open to all people, and long-lived. Biotechnology and AI could be used to assist people in feeding the growing population and at the same time be environmentally responsible by integrating new concepts and responsibility.

References

Adinarayana, S., Raju, M. G., Srirangam, D. P., Prasad, D. S., Kumar, M. R., & Veesam, S. B. (2024 Jun 17). Enhancing resource management in precision farming through AI-based irrigation optimization. In *How Machine Learning is Innovating Today's World: A Concise Technical Guide* (pp. 221–251).

Aghi, D., Mazzia, V., & Chiaberge, M. (2020 May 25). Local motion planner for autonomous navigation in vineyards with a RGB-D camera-based algorithm and deep learning synergy, *Machines*, *8*(2), 27.

Ali, A., Hussain, T., & Zahid, A. (2025 Apr 4). Smart irrigation technologies and prospects for enhancing water use efficiency for sustainable agriculture, *AgriEngineering*, *7*(4), 106.

Al-Khreisat, A., Abdelal, Q., Al-Kilani, M., Al-Bakri, J., Damra, I., & Al-Kilani, O. (2025). Water-energy-carbon cost of compensating atmospheric water losses from open irrigation ponds in arid regions. *Global Journal of Environmental Science and Management*, *11*(4), e728181.

Angidi, S., Madankar, K., Tehseen, M. M., & Bhatla, A. (2025 Feb 28). Advanced high-throughput phenotyping techniques for managing abiotic stress in agricultural crops – A comprehensive review, *Crops*, *5*(2), 8.

Angon, P. B., Anjum, N., Akter, M. M., KC, S., Suma, R. P., & Jannat, S. (2023). An overview of the impact of tillage and cropping systems on soil health in agricultural practices. *Advances in Agriculture*, *2023*(1), 8861216.

Anoop, V. S., Adarsh, S., & Asharaf, S. (2022 Jun 27). Artificial intelligence and the blockchain in smart agriculture: Emergence, opportunities, and challenges. In *Artificial Intelligence and Smart Agriculture Technology* (pp. 43–55).

Bakala, H. S., Singh, G., & Srivastava, P. (2020 Dec 30). *Smart Breeding for Climate Resilient Agriculture. InPlant Breeding-current and Future Views*. IntechOpen.

Balaska, V., Adamidou, Z., Vryzas, Z., & Gasteratos, A. (2023 Jul 25). Sustainable crop protection via robotics and artificial intelligence solutions, *Machines*, *11*(8), 774.

Basu, A., & Narayan, A. (2025 Jun 4). The role of machine learning in transforming agricultural practices: Insights into crop yield optimization and disease detection. *Iran Journal of Computer Science, 8*(4): 1199–217

Bhuiyan, M. M., Rahaman, M. M., Aziz, M. M., Islam, M. R., & Das, K. (2023 Dec 15). Predictive analytics in plant biotechnology: Using data science to drive crop resilience and productivity, *Journal of Environmental and Agricultural Studies*, *4*(3), 77–83.

Borlaug, N. E., & Dowswell, C. R. (2003 May). Feeding a world of ten billion people: A 21st century challenge. *In Proceedings of the International Congress in the Wake of the Double Helix: From the Green Revolution to the Gene Revolution* (. 27–31). Ames: Iowa State Press.

Delfani, P., Thuraga, V., Banerjee, B., & Chawade, A. (2024 Oct). Integrative approaches in modern agriculture: IoT, ML and AI for disease forecasting amidst climate change, *Precision Agriculture*, *25*(5), 2589–2613.

Garg, G., Gupta, S., Mishra, P., Vidyarthi, A., Singh, A., & Ali, A. (2021 Sep 1). CROPCARE: An intelligent real-time sustainable IoT system for crop disease detection using mobile vision, *IEEE Internet of Things Journal*, *10*(4), 2840–2851.

Kalbar, P. P., Karmakar, S., & Asolekar, S. R. (2012 Dec 30). Selection of an appropriate wastewater treatment technology: A scenario-based multiple-attribute decision-making approach. *Journal of Environmental Management*, *113*, 158–169.

Kayusi, F., Wasike, J., & Chavula, P. (2025). The role of mulching in reducing greenhouse gas emissions and enhancing soil health among smallholder farmers in Zambia, Malawi, Kenya, and Tanzania: An AI-Driven approach. *LatIA*, (3), 75.

Khan, N. (2024 Oct 29). Unlocking innovation in crop resilience and productivity: Breakthroughs in biotechnology and sustainable farming. *Innovation Discovery*, *1*(4), 3046–403X.

Koshariya, A. K. (2022). Climate-resilient crops: Breeding strategies for extreme weather conditions. *Plant Science Archives*, *1*(03), 2518–6965.

Kumari, K., Mirzakhani Nafchi, A., Mirzaee, S., & Abdalla, A. (2025 Mar 20). AI-driven future farming: Achieving climate-smart and sustainable agriculture, *AgriEngineering*, *7*(3), 89.

Lawrence, I. D., Agnishwar, J., & Vijayakumar, R. (2023). Drone technology in agriculture for surveillance and inspection. *European Chemical Bulletin*, *12*(Special Issue 12), 1253–1263.

Liu, W., Hou, P., Liu, G., Yang, Y., Guo, X., Ming, B., Xie, R., Wang, K., Liu, Y., & Li, S. (2020 Nov). Contribution of total dry matter and harvest index to maize grain yield – A multisource data analysis, *Food and Energy Security*, *9*(4), e256.

Lopes, V., & Alexandre, L. A. (2018 Sep 30). An overview of blockchain integration with robotics and artificial intelligence. *arXiv Preprint arXiv:1810.00329*.

Mandal, S., Yadav, A., Panwar, R., Kumar, S. M., Karthick, A., Priya, A., Vijayakumar, R., & Ganesh, S. S. (2025 Aug). Smart water management for SDG 6: A review of AI and IoT-enabled solutions, *Water Conservation Science and Engineering*, *10*(2), 1–5.

Mehdizadeh, M., Omidi, A., Matindike, R., Nigussie, Z. G., Ikegwu, T. M., Agu, H. O., Onen, P., Gumula, I., Ezzat, S. M., & Merghany, R. M. (2025 Aug 8). Agri-waste valorization: Pathways to sustainable bioenergy and biochemical innovation. *Circular Economy and Sustainability*, 1–31, 5.

Meshram, A. T., Vanalkar, A. V., Kalambe, K. B., & Badar, A. M. (2022 Mar). Pesticide spraying robot for precision agriculture: A categorical literature review and future trends, *Journal of Field Robotics*, *39*(2), 153–171.

Munnisunker, S., Nel, L., & Diederichs, D. (2022 Mar 18). The impact of artificial intelligence on agricultural labour in Europe. *Journal of Agricultural Informatics*, *13*(1), 46–54.

Panjaitan, S. D., Dewi, Y. S., Hendri, M. I., Wicaksono, R. A., & Priyatman, H. (2022 Nov 9). A drone technology implementation approach to conventional paddy fields application. *Ieee Access*, *10*, 120650–120658.

Pearson WA. *Drones in precision agriculture: Corn field testing and sampling through drone mapping* (Master's thesis, California State University, Fresno). 2018

Phang, S. K., Chiang, T. H., Happonen, A., & Chang, M. M. (2023 Nov 7). From satellite to UAV-based remote sensing: A review on precision agriculture. *Ieee Access*, *11*, 127057–127076.

Qazi, S., Khawaja, B. A., & Farooq, Q. U. (2022 Feb 17). IoT-equipped and AI-enabled next generation smart agriculture: A critical review, current challenges and future trends. *Ieee Access*, *10*, 21219–21235.

Reddy, C. H., Bhat, M. R., Kankanawadi, N., & Gowda, N. P. (2025 Aug 16). Artificial intelligence in the new era of Sericulture, *Journal of Scientific Research and Reports*, *31*(8), 788–803.

Riaz, M., Yasmeen, E., Saleem, B., Hameed, M. K., Saeed Almheiri, M. T., Saeed Al Mir, R. O., Alameri, G., Khamis Alghafri, J. S., & Gururani, M. A. (2025 Jun 19). Evolution of agricultural biotechnology is the paradigm shift in crop resilience and development: A review. *Frontiers in Plant Science*, *16*, 1585826.

Sarkar, N. C., Mondal, K., Das, A., Mukherjee, A., Mandal, S., Ghosh, S., Bhattacharya, B., Lawes, R., & Huda, S. (2023 Feb 17). Enhancing livelihoods in farming communities through super-resolution agromet advisories using advanced digital agriculture technologies, *Journal of Agrometeorology*, *25*(1), 68–78.

Sassenrath, G. F., Heilman, P., Luschei, E., Bennett, G. L., Fitzgerald, G., Klesius, P., Tracy, W., Williford, J. R., & Zimba, P. V. (2008 Dec;). Technology, complexity and change in agricultural production systems, *Renewable Agriculture and Food Systems*, *23*(4), 285–295.

Saxena, R., Joshi, A., Joshi, S., Borkotoky, S., Singh, K., Rai, P. K., Mueed, Z., & Sharma, R. (2023 Jan 1). The role of artificial intelligence strategies to mitigate abiotic stress and climate change in crop production. In *Visualization Techniques for Climate Change with Machine Learning and Artificial Intelligence* (pp. 273–293). Elsevier.

Sharma, A., Sharma, A., Tselykh, A., Bozhenyuk, A., Choudhury, T., Alomar, M. A., & Sánchez-Chero, M. (2023 Oct 14;). Artificial intelligence and internet of things oriented sustainable precision farming: Towards modern agriculture, *Open Life Sciences*, *18*(1), 20220713.

Sprouls, J. M. (2020). *Growing Trees in a Gravel Bed Stormwater Retention System as a Novel Approach to Stormwater Management in Urban Sites*.

Su, X., Xu, J., Li, C., & Wei, M. (2025 Jul 18). The research progress on crop stress resistance and environmental adaptability improvement based on artificial intelligence, *Advances in Resources Research*, *5*(3), 1589–1610.

Subramanium, *et al.*, 2021 site specific pest control by use of drones.

Thingujam, D., Gouli, S., Cooray, S. P., Chandran, K. B., Givens, S. B., Gandhimeyyan, R. V., Tan, Z., Wang, Y., Patam, K., Greer, S. A., & Acharya, R. (2025 Aug 29). Climate-resilient crops: Integrating ai, multi-omics, and advanced phenotyping to address global agricultural and societal challenges, *Plants*, *14*(17), 2699.

Tubiello, F., Schmidhuber, J., Howden, M., Neofotis, P. G., Park, S., Fernandes, E., & Thapa, D. Climate change response strategies for agriculture: Challenges and opportunities for the 21st century. Agriculture and rural development discussion paper. 2008 Jan 1; 42.

Vos, R., & Bellù, L. G. (2019 Jan 1). Global trends and challenges to food and agriculture into the 21st century. *Sustainable Food and Agriculture*, 11–30. https://doi.org/10.1016/B978-0-12-812134-4.00002-9.

Wei, H., Xu, W., Kang, B., Eisner, R., Muleke, A., Rodriguez, D., deVoil, P., Sadras, V., Monjardino, M., & Harrison, M. T. (2024 Jun). Irrigation with artificial intelligence: Problems, premises, promises, *Human-Centric Intelligent Systems*, *4*(2), 187–205.

Yoosefzadeh Najafabadi, M., Hesami, M., & Eskandari, M. (2023 Mar 23). Machine learning-assisted approaches in modernized plant breeding programs, *Genes*, *14*(4), 777.

Zhang, X., Ibrahim, Z., Khaskheli, M. B., Raza, H., Zhou, F., & Shamsi, I. H. (2024 Sep 3). Integrative approaches to abiotic stress management in crops: Combining bioinformatics educational tools and artificial intelligence applications, *Sustainability*, *16*(17), 7651.

Zhao, J., Liu, D., & Huang, R. (2023 Feb 13). A review of climate-smart agriculture: Recent advancements, challenges, and future directions, *Sustainability*, *15*(4), 3404.

Zurek, M., Hebinck, A., & Selomane, O. (2020). *Food and Agriculture Systems Foresight Study: Implications for Climate Change and the Environment-synthesis.*

Imran Sheikh, Pooja Saini, Aaliya Ali, Syed Irfan Yaqoob, Varruchi Sharma*, Anil Kumar Sharma*, Anupam Sharma, Poonam Bansal, Vandana Sharma, J.K. Sharma, Anil Panwar

Chapter 8 AI in Biomanufacturing: Unlocking Efficiency and Innovation

Abstract: Artificial intelligence (AI) is disrupting biomanufacturing by transforming age-old, labor-intensive processes into a system of data-rich, highly adaptive, and efficient tools that enhance efficiency, reproducibility, and sustainability. This chapter discusses how the technologies of machine learning, deep learning, reinforcement learning, natural language processing, and digital twins are transforming the biomanufacturing workflow in the upstream, monitoring, downstream, innovation, and sustainability arenas. In upstream processing, AI-based genome design, media optimization, and predictive modeling enable the development of strains and cultures to be achieved much faster and more resource-efficiently. Adaptive control is enabled by smart sensors, process analytical technology, and real-time analytics, which minimize variability and batch failures because they allow control by design (quality by design) and predictive maintenance. At the down-end, AI is used to improve purification and separation, yield improvement, predictive modeling, and robotics to provide GMP-compliant automation. Case studies focus on vaccine development, biologics, biosimilars, precision fermentation, cultured meat, food proteins, enzymes, and secondary metabolites, highlighting how AI can be utilized to accelerate discovery, reduce costs,

***Corresponding authors: Varruchi Sharma,** Department of Biotechnology and Bioinformatics, Sri Guru Gobind Singh College, Sector 26, Chandigarh 160019, India, e-mail: Varruchi.sharma2013@gmail.com

***Corresponding authors: Anil Kumar Sharma,** School of Biological Sciences, Amity University, Mohali 140306, Punjab, India, e-mail: anibiotech18@gmail.com

Imran Sheikh, Aaliya Ali, UCRD, Chandigarh University, Mohali, India

Pooja Saini, Department of Biotechnology, Eternal University, Baru Sahib, 173101, Himachal Pradesh, India

Syed Irfan Yaqoob, AIT-CSE, Chandigarh University, Ghauran, Mohali 140413, Punjab, India

Anupam Sharma, Department of Physics, Guru Kashi University, Talwandi Sabo, Bathinda 151302, Punjab, India

Poonam Bansal, Department of Bio-sciences & Technology, MMEC, Maharishi Markandeshwar (Deemed to be University), Mullana, Ambala 133207, India

Vandana Sharma, J.K. Sharma, Department of Physics, MMEC, Maharishi Markandeshwar (Deemed to be University), Mullana, Ambala 133207, India

Anil Panwar, Department of Bioinformatics and Computational Biology, CCS Haryana Agricultural University, Hisar 125004, India

 | https://doi.org/10.1515/9783112214275-008

and scale up production. The aspects of sustainability are addressed through AI-based waste minimization, energy-saving bioprocessing, model bioproducts based on circular biomarketing, and innovative environmental sensing. In addition to these opportunities, there are still challenges, including fragmented and noisy data, low interpretability, regulatory certification, cybersecurity concerns, and skill gaps in the workforce, which hinder widespread adoption. Ethical and regulatory viewpoints emphasize transparency, accountability, and harmonious global systems to facilitate the responsible adoption of practices. The AI is poised to become a transformative agent of change and a foundational pillar of accuracy, stability, and green innovation in biomanufacturing, the potential of which hinges on its effective governance, cross-disciplinary partnerships, and scalable and reliable deployment.

Keywords: Artificial intelligence in biomanufacturing, process analytical technology (PAT), predictive modeling in bioprocessing, sustainable and green biomanufacturing, AI, multi-omics integration with AI

8.1 Introduction

The biomanufacturing of the modern era of biotechnology has gone beyond the previous boundaries of fermentation vessels and antibody-production pipelines into an integrated, data-rich continuum that extends to encompass organism and strain design, high-throughput screening, upstream cultivation (fermentation and cell culture), downstream purification and formulation, quality control and assurance, facility operation, supply-chain coordination, and regulatory compliance. Beyond that, the space is increasingly moving toward flexible biofoundries, continuous and modular manufacturing platforms, cellular and gene-therapy manufacturing, personalized biologics, and vaccines. New syntheses in the field describe this shift as one of automation, process analytical technologies (PAT), digitalization, and more sophisticated modeling, thus transforming biomanufacturing into a coordinated cyber-physical ecosystem instead of a sequence of disconnected unit operations (Isoko et al., 2024). It is against this background that artificial intelligence (AI), including machine learning (ML), deep neural networks, reinforcement learning (RL), probabilistic modeling, and hybrid physics-informed methods have emerged as a transformative driver enabling the conversion of the enormous, heterogeneous data streams that are currently present in modern facilities into actionable insights. ML can use multi-omics and phenotype data to predict beneficial genetic edits and media formulations. ML can predict unmeasured critical quality attributes (CQAs) using online process measurements. ML can create predictive, soft-sensor models. Nonetheless, conventional biomanufacturing still entails deep-seated challenges, limiting productivity gains and inflating costs. The capital and operational cost of producing a biologic remains high due to the high cost of media, consumables, single-use or stainless-steel infrastructure, and energy-intensive downstream purification operations. Scalability is not easily achieved due to nonlinear responses to scale in many biological processes, which produce reduced yields, changed impurity profiles, or process instabilities. Reproducibility and lot-to-lot consistency due to biological heterogeneity, variability in raw materials, and, furthermore, data governance and standardization issues, lack of sensor coverage in most legacy plants, and conservative regulatory processes that appropriately prioritize safety over speed all work against AI adoption. Industry reports and reviews have, over and over again, pointed to gaps in the theory-to-practice disconnects of AI, such as the lack of curated, high-quality training datasets, interpretability requirements of models in regulated settings, or the combination of physics-based first-principles and purely data-driven models to guarantee extrapolation reliability with scale-up (Schmidt et al., 2022). To surmount these established and maintained barriers and reap the benefits, this chapter aims to achieve four major objectives. The first is to define rigidly and cartographically the modern biomanufacturing landscape, including the upstream, downstream, quality, operational, and sustainability subdomains, and establish the location of new modalities, including biofoundries, continuous manufacturing, and custom, tailored biologics, within that structure to allow readers to grasp the scale and interdependencies of the field. A second goal is the synthesis and critical assessment of the implementation of particular AI tools, such as multivariate statistical process control, ensemble ML models, deep learning (DL), reinforcement learning, and hybrid digital-twin models to the design and production cycle to optimize yield, variability, and time-to-market; supplemented with substantive collective examples and performance metrics in recent scholarly literature and business case studies. The third goal reveals and evaluates the technical, organizational, regulatory, and ethical obstacles that stop extensive use of AI, focusing on data quality and interoperability, sensor and PAT adoption, explainable AI toolkits, and PAT expansion, and presents measures that can be readily implemented to address this issue, including federated learning, synthetic data creation, explainable AI toolkits, and expand PAT, all contribute to trust and utility. The fourth goal is a future-oriented research and implementation package that reflects high-value opportunities of AI-based digital twins of virtual scale-up testing, autonomous laboratories of strain optimization, reinforcement-based, use-of-real-time-control, and life cycle optimized sustainability, and outlines the measurements and validation studies, which are needed to apply laboratory gains to a regulatory law-compliant production enterprise. To justify these aims, this chapter builds upon the modern syntheses of digital-twin and PAT integration, positing a loop of quantitative data through the use of PAT and the need to simulate scale-dependent phenomena without being forced to challenge them through physical experiments. This chapter provides a virtual simulation environment for such virtual experiments (Schmidt et al., 2022). It also reflects current industry thinking, which emphasizes the need for human-machine harmonization, demonstrating how to upskill operators, introduce a human-in-the-loop approach, and establish oversight of critical decision-making and governance arrangements that balance responsibility with automation. Sustainability and economics are the two terms that are directly addressed in this chapter; quantitative analyses have been performed regarding how AI-enabled solutions, i.e., process optimization, predictive maintenance, yield improvement, and continuous versus batch trade-offs may reduce material and energy consumption, decrease the green-house-gas intensity per unit product, and reduce water and waste footprint: all of which are improved as a biomanufacturing becomes more circular and aligned with the broader circular-bioeconomy objectives and a target policy agenda. This chapter ends with a bibliography of selected recent additional sources and tabular, comparative views that empathize distinct AI methodologies to biomanufacturing subtasks, such as strain design, media optimization, online CQA prediction, fault detection, scale-up transfer functions, downstream chromatographic conditions, and supply-chain coordination, which proposes a set of uniform performance measures and knowledge-sharing practices, with which, were institutive, would put progress here on an escalating curve. By offering an integrated, evidence-based, and practical treatment of where AI has already delivered value, where substantive gaps remain, and how practitioners can responsibly move from pilot to production, this chapter seeks to be both a primer for newcomers and an operational playbook for researchers and manufacturing leaders committed to unlocking efficiency, reproducibility, and sustainable innovation in the next generation of biomanufacturing (Butean et al., 2025).

8.2 Overview of AI Technologies in Biomanufacturing

Within an accelerating biomanufacturing environment, groups of essential AI systems, such as ML, DL, natural language processing (NLP), RL, and the paradigms of digital twins, AI-driven automation, robotization and system integration, and Edge and Cloud computing, are emerging and are foundational underpinnings of smart, resilient, and adaptive bioprocessing. ML gives this the structure to enable predictive modeling, strain selection, media optimization, scale-up predictions, and the monitoring of key quality attributes (CQAs) and fiduciary data streams without the size of explicit structure similar to multivariate statistical modeling, instead of with state-of-the-art supervised learning pipelines (Helleckes et al., 2023). A subtype of ML, DL, which supports multiple hidden layers, expands on this ability, particularly in the case of unstructured data, i.e., images, video, high-frequency sensor data, sensitive feature extraction, anomaly detection, and pattern recognition in cell-growth imaging, bioreactor camera feeds, or microbial morphology, which classical ML would overlook. DL also supports surrogate modeling when mechanistic models are complex or not available. NLP has not yet found extensive application in the biomanufacturing sector. However, its ability to extract knowledge from unstructured data, such as equipment manuals, batch test data, regulatory texts, and literature review texts, is of useful application in process troubleshooting, automated report writing, and even the literature-based correlation of reaction parameters, which is actually coming to fruition.

Adaptive decision making in uncertain, dynamic, or poorly understood aspects of the process can be achieved with RL: more recent experiments demonstrate that the RL agents of multi-setpoint and multi-trajectory tracking manage microbial culture systems under uncertainty, and are more stable and competent in complex control. Experimental usage in photobioreactors demonstrates that RL agents capable of cloning a behavior with reduced error and control effort make their control decisions. Digital twins, or virtual representations of physical components within manufacturing systems e.g., cells, bioreactors, and downstream purification units, serve as a critical integrative framework. By leveraging real-time sensor data through IoT, they utilize ML/DL models to forecast system states and fault modes, simulate "what-if" scenarios such as scale-up, parameter variations, and equipment malfunctions, enable predictive maintenance, and support continuous improvement loops. Apart from that, recent overviews of the Bioprocessing 4.0 point to the growing list of uses to design, optimize, and operate data miners with digital twins, such as the effectiveness of the equipment or in-goods, identification of out-of-spec batches early thereof, occurrence of out-of-context supply-chain context, or process anomalies prediction (Isoko et al., 2024). These capabilities are further applied in AI-driven automation and robotics, automated sampling, high-throughput, and automation of cell-culture, liquid-handling, and downstream purification, and the ability to combine lab automation and AI models to schedule and correct errors; human error is reduced, resulting in improved reproducibility and fine temporal resolution to collect and control data. Distributed sensors, actuators, and environmental monitors help in obtaining the continuous measurement of the key process parameters, such as temperature, pH, dissolved oxygen, cell density, and nutrient concentrations. The operated data streams can be used to feed ML and DL models, or to run digital twins representations, thus allowing inference and control in real time. Being located close to sensing or processing devices, edge computing is essential to the needs of low-latency, real-time decision-making processes, including anomaly detection and control domain operation. Cloud computing, on the other hand, offers distributed storage, powerful model training, such as DL and RL, cross-site model distribution, and remote monitoring dashboards. The need to balance the distribution of resources between the edge and cloud infrastructure is therefore critical, and thus trade-offs between latency, bandwidth, data privacy, and regulatory compliance are negotiated. In addition, modern architecture paradigms combine federated learning, allowing distributed model training without exchanging raw samples, with hybrid physics-informed ML methods to overcome the problems related to data sparsity, understanding, and extrapolation of models outside of their training space. Together, these technologies form a multilayered stack of sensing and IoT, edge/cloud infrastructure, core AI, including ML, DL, NLP, RL, digital twins, automation, and robotics, and thus form the framework that is sometimes called smart biomanufacturing. Several important recent case studies have demonstrated the multiplicative benefits of the combination, including greater packaged yield consistency, minimized downtime, faster process development, lower cost of goods, increased product quality, and higher sustainability of activity, e.g., reduced waste and improved resource utilization. Nevertheless, there are difficulties in completely obviating all levels, delivering model interpretability, particularly under regulatory oversight, sustaining healthy data chains, understanding legacy equipment limitations, securing cybersecurity and information integrity, balancing computational finance with the collected benefits, and meeting good manufacturing practice (GMP) and regulatory validation examinations of AI-pinpointed controls. This chapter will first map out these core AI technologies, summarizing their methodological principles, strengths, and limitations; then will survey recent applications in biomanufacturing that illustrate their integration. Next, we will examine system architectures that combine IoT, automation, digital twins, edge and cloud infrastructure; and finally, we will look forward to trends such as federated learning, hybrid models combining mechanistic and data-driven approaches, more autonomous RL-based control, and more precise NLP-assisted knowledge extraction, all to illustrate how AI technologies can deliver smart, robust, efficient, and innovative biomanufacturing.

8.3 AI Across the Biomanufacturing Workflow

8.3.1 Upstream Processing

Upstream processing, which describes the system of laboratory and pilot-scale processes that convert engineered cells into high-throughput biological product streams of consistently high quality, is undergoing rapid change driven by AI. Data-driven techniques are opening up new bounds of speed, accuracy, and resource optimization in strain design, media and input control, and process regulation. Traditionally, strain engineering and synthetic biology were synonymous with manual approaches that were characterized by wet-chip cycles and the use of intuition; however, more modern models focused on learning sequence-functionally relationships, suggesting candidate changes, and ranking experimental campaigns, repositioning genome design as a model to be actively optimized in real time, not a slothful handicraft. Recent examples of autonomous and ML-driven platforms incorporating large models have demonstrated that closed-loop design-build-test-learn models can rapidly enhance enzymes and pathways with minimal human supervision (Singh et al., 2025). Concrete, reproducible improvements in activity and stability have been demonstrated at the molecular level for ML-based frameworks that solve protein and pathway engineering, utilizing a combination of representation learning, directed evolution datasets, and interpretable surrogate models. These gains translate into increased yields of the upstream processes and robustness of the processes (Kouba et al., 2023). The optimal use of AI, which should be integrated with genetic optimization, is in media optimization and the choice of raw inputs. Rather than experimenting with one factor at a time, supervised and active-learning methodologies quickly explore multidimensional nutrient spaces, revealing previously nonintuitive interactions between carbon, nitrogen, trace elements, and supplements that enable the most productive response and the least amount of wasteful extravagance. Research on the application of ML to media formulation reveals that data-efficient algorithms can help reduce experimental workload and also distinguish between native and heterologous product formation strategies. These insights reduce development life cycles and minimize the environmental impact of upstream campaigns (Aida et al., 2023). Systematic reviews of algorithmic methods for cell-culture media also provide further simplification of how we can learn to cluster, Bayesian optimization, and ensemble methods have evolved into general, transferable toolkits available to the bioprocess scientist to find faster-to-scale alternatives (Zhou et al., 2023). Examples aside, the formulation, raw material selection, and control over resources, as well as efficient input management, have become key areas to which AI can be applied to convert typically noisy supplier and batch data into meaningful risk signals that can be acted upon. Most predictive models can accurately predict the effects of lot-to-lot variation, based on historical batch genealogy, supplier certificates, and spectral or chemical fingerprint data, allowing the implementation of on-the-fly rules to replace lots or adjust set-points to optimize inventory, thereby upholding product quality and controlling unreasonable hold cases or rework. Combined with digital procurement reporting and near real-time analytics, these models help a company to shift from traditionally conservative, source-specification sourcing strategies to dynamic, data-driven procurement strategies that save money and reduce supply risk, as well as regulatory compliance failures. Predictive modeling of fermentation and cell culture conditions is tailored to both reactor and process scale, and combines classical statistical algorithms, deep recurrent neural networks, physics-informed ML hybrids, and provides highly accurate short-term predictions of cellular proliferation dynamics, metabolite accumulation, and impurity profiles. Such projections, in turn, feed model-predictive controllers and Bayesian experimental planners, ensuring that productivity remains intact in the presence of perturbations. According to the latest reviews and methodological articles, the schedules of ML learned fermentation designs exhibit sensor fusion, soft sensor inference into otherwise independent variables, and transfer training to produce predictions that can scale and strain diversity, a feat that is crucial in accomplishing a replication of lab findings on manufacturing in the enterprise (Wang et al., 2025). A combination of more advanced sensor networks, such as optical, spectroscopic, and bioreactor telemetry, in conjunction with richer omics and analytics readouts and more robust learning algorithms, also makes it possible to apply digital twin concepts to upstream processing. These virtual replicas operate concurrently with the physical process, enabling the testing of alternative feeding strategies, ramp rates, and induction conditions in silico, and recommending low-risk experimental or control adjustments. Digital twins are decision amplifiers, and instead of the numerous physical experiments that must be conducted to scale up a system, digital twins can offer interpretable scenario analysis for scale-up, helping engineers synthesize yield, impurity profiles, and resource consumption. In addition to predictive modeling, AI has been useful in prioritizing and risk managing the upstream processing. By measuring uncertainty in deep-learning models and explaining the phenomena underlying performance with state-of-the-art methods, teams can invest in experimentally-intensive inquiries, producing the best payoff, with a backup strategy that is conservative in the absence of extrapolation out of the samples where the model was trained. It is a risk-sensitive automation that reduces development schedules without missing regulatory traceability, which is an essential part of the manufacturing of therapeutic products. These benefits must be attained by a very strict focus on data hygiene, interoperability, and governance. All the most successful deployments will be characterized by a sound metadata capturing piece, as well as a training set that exhibits the regimes of operation of interest, and an integration layer that translates machine-learned results into available process-control systems and quality-management systems. When used wisely, AI-enhanced upstream processing can speed up strain and media evolution and advance a more sustainable paradigm of biomanufacturing, thus minimizing the loss of water and nutrients, reducing energy spent on optimizing feed, and preventing batch failures through predictive intervention. As a result, the measures of productivity become bound with environmental stewardship. With the maturation of the field, the competition between biotechnology manufacturers involved in AI-guided genome design, ML-based media optimization, intelligent selection of raw materials, and predictive fermentation control will continue to define competitive advantage, hence turning the traditional artisanal construction of processes into a reproducible, automated, and resource-saving pipeline that is both scientifically innovative and operationally resilient.

8.3.2 Bioprocess Monitoring and Control

Going beyond ad hoc, offline measurement, modern bioprocess measurement and control has increasingly adopted a data-focused, comprehensive paradigm, which forms the basis of modern biomanufacturing. This paradigm shift can be triggered by the application of high-resolution, finely resolved sensor networks and real-time analytics, as well as by AI use, in which process analytical technology (PAT) and quality by design (QbD) become a central objective of operation and not a standard to meet. This transition is fueled by variations of fine, high-resolution sensors, real-time analytics, and AI. In this scheme of things, PAT and QbD become primary operational principles, instead of being an obligatory requirement. State-of-the-art biosensors, such as electrochemical, optical, and miniaturized spectroscopic probes, are now able to detect high-frequency signals that are related to classical parameters of critical processes such as pH, dissolved oxygen, temperature, and optical density, and to new metabolite/product parameters that are associated with quality. Such instruments contribute to in situ testing of cellular physicochemical conditions, and therefore most of the tedious laboratory procedures that used to be predominant in the sphere can be replaced (Vo et al., 2024). Over this instrumentation layer is a running feed of real-time measurements and edge KPIs, which, in combination with efforts to minimize the time delay between measurement capture and process intervention. This way, signal preprocessing, feature extraction, and end sensor closed loop adaptations may reduce end-to-end variability and define design spaces that downstream controls must protect (Isoko et al., 2024). AI PAT is additionally enhanced by known streams of noisy, multimodal sensors and spectral fingerprints, such as NIR and Raman and mass spectrometry images, which are used as scalable solid estimates of CQAs and critical process parameters. Through this, the knowledge gained via large-scale historical batch data can anticipate drift and possibly suggest innovative operating regimes that adhere to QbD principles and thus make QbD a more vibrant and responsive manufacturing philosophy (Vora et al., 2023). Beyond the parameter estimation task, the AI-based methodologies that build on the classical multivariate statistical process control task have evolved into modern DL, ensemble models, and hybrid physics-informed networks, which have shown effective performance on fault detection and diagnosis. Primordial signs of contaminant, cell-culture stress, sensor contamination, or stiction of control valves become evident in multistream time-series data, despite traditional alarms being triggered much in advance, allowing automated intervention to maintain yields and product quality. As fault detection and diagnosis, anomaly prediction, and prognostics expand, anomaly prediction systems with IoT sensor networks build health index models and tools that project the available future useful life, enabling facilities to transition to condition-based maintenance strategies, instead of calendar-based ones, in order to reduce the use of system opportunities and decrease operating costs without sacrificing the ability to regulate (Rana 2025). Importantly, such AI systems are used to the greatest effect when they are integrated into Bioprocessing 4.0 designs (digital twins, standardized data fabrics, and interoperable PAT toolboxes) such that simulated process replicas can complement blue oceans of sparse empirical data and facilitate virtual (or real) experiments whose outcomes pose neither material cost nor regulatory risk (Isoko et al., 2024). Issues over model interpretability, model data integrity, and transferability have stimulated solution hybrid methodologies involving mechanistic, first-principles models, plus data-driven residual learners; meanwhile, these paradigms have increased the importance of rigorous model life cycle management and continuous revalidation protocols, and explainability methods that are robust enough to satisfy process engineers and regulatory agencies (Auclair & Rathore 2025). Regulatory and QbD viewed through the lens of AI-enhanced PAT support risk-based control strategies that are based on probabilistic predictions of CQA excursions: risk matrices and control action libraries are explained using risk-based predictions, and automated corrections are well documented and effected within the established preapproved design spaces, helping to accelerate batch release and reduce lot-to-lot variability (Yang et al., 2025). However, a critique of these theoretical applications meet significant challenges in real life – among them are the heterogeneity of legacy instrumentation, costs borne by data labeling, and cyber-physical security, as well as the need to invest in common standards (data schemas, model validation protocols), human-in-the-loop defenses, and pilot deployments that produce size and cell-line scale-varying exemplars of information that can be generalized by models (Isoko et al., 2024). In the future, with the integration of advanced sensors, robust edge analytics,

AI-based PAT, and predictive maintenance, an even more robust, efficient, and quality-focused biomanufacturing model will be realized, in which in-process intelligence, along with continuous control by predicting operations, does not just observe the operations, but proactively codesign operating strategies, truncates its development cycles, and integrates QbD into everyday practice, not only preserving traceability and regulatory compliance but also actively monitoring its shape and controls (Vo et al., 2024).

8.3.3 Downstream Processing

The downstream processing, which consists of sequential steps of purification, separation, and formulation that follow cell culture or fermentation, is gradually becoming a center of innovation in AI and ML due to two areas of necessity: cost reduction and yield maximization. One of the most significant developments in this field is the implementation of AI-driven optimization of purification and separation processes. Hybrid mechanistic/data-driven and data-driven only surrogate models have been applied to chromatography platforms in various modes, including ion-exchange, multicolumn, continuous, and periodic counter-current modes. The models enable real-time or cycle-to-cycle optimization of process variables, including flow rates, buffer composition, column switching programs, and gradient elution schedules, thereby minimizing recovery and purity losses to the lowest levels and reducing the need for human trial and error. An example of this is a recent case study of the ion-exchange chromatography purification of monoclonal antibodies, where hybrid surrogate models achieved a recovery of almost 82% instead of the approximately 78% obtained by purely mechanistic models, without violating purity constraints; the all-data-based models also enabled within-cycle updates at the minute scale to reduce process drift or perturbation. Constant purification, such as periodic counter-current chromatography with low-titer enzymes, has also been useful in achieving the purification advantages of design optimization and operational optimization, resulting in reduced buffer usage, increased productivity, and purity within industrially relevant conditions (Ko et al., 2024). To complement such advances in purification, ML-based predictive models help achieve cost savings and yield improvements by analyzing past data to predict important outcomes, including impurity profiles, product loss, chromatographic breakthrough, and optimal harvest timing. In the particular application of optimizing fermentation downstream harvests, stochastic dynamic models that model biological kinetics and impurity build-up have been proposed to make decisions in the presence of uncertainty. Such devices are advantageous in practice due to the high increase in yields and limiting purity rates (when used in the facility) to a higher extent (Wang et al., 2024). These broad reviews also highlight the assumption that a combination of multivariate statistics-based ML and hybrid modeling can be utilized in QbD frameworks in downstream operations that are capable of explaining multifactor nonlinear combinations among process factors and quality properties that conventional models cannot forecast (Walsh et al., 2022). Robotics and

AI combined with automation, particularly at larger scales that meet the requirements of GMP, are another benefit. Conditional control algorithms that use real-time sensor data responses have emerged as a necessity to control automated chromatography and filtration systems, inline buffer conditioning, multicolumn mode switching, and continuous downstream modalities. Back to the downstream purification stage, automated batch-comparison approaches paired with multivariate root-cause analysis software have allowed one of the largest biopharmaceutical firms to generate right-first-time batches. As a result, the approach has saved the required amount of recirculation in purification by more than 6%; therefore, restricting the generation of waste and operational expenses. At the same time, in a manufacturing process that requires a high level of strict adherence to GMP, robotics are being willing to perform tasks like the alignment of the stopper, pick-and-place tasks, and some sterile inline tasks. These robotic processes are controlled by AI systems that apply computer-vision algorithms and use digital twins to verify those processes, thus ensuring sterility standards, first-air HEPA circulation, clean room space layout, and traceability distribution. A good example is a robotic station at Optima Pharma, where a system has the wrong stopper, whereby both robots accidentally replaced the stoppers in a continuous high-speed filling and sealing process. The faulty stopper was subsequently replaced, with no additional intervention performed on the component. Its equipment is powered by an inline camera to deliver real-time feedback and control, relying on AI. Such a situation is an example of the synergy between automation and AI in terms of maximizing throughput and maintaining the application of GMP even in the presence of unfavorable downstream conditions. Together, these advancements signify that downstream processing has ceased to be only a cost center, but it has equally transformed into a platform that is based on empirically optimized, dithering, batch systems with wide safety margins. Instead, it is an automatic improvement in the system rather than a reactive one. Yield-degrading events, including protein breakthrough or aggregation, are predicted and prevented with the help of ML algorithms. It is also possible to design purification schemes in silico with the help of AI devices and optimize them over time. Robotics ensures the precision and reproducibility of physical operations that are regulated by GMP, and digital twins that have real-time analytics provide traceable and auditable control that regulators can examine. In order to realize the potential of such technologies fully, a number of challenges need to be addressed: ensuring large-scale, high-quality datasets; enabling regulator comprehension and adoption of ML models; establishing an interface between robotics and the complex predictive fluidic systems; and ensuring a balance between automated operations and necessary human intervention. Finally, there is a promise of lower costs, improved outputs, and high regulatory confidence in site-level processing controlled by AI that would all be instrumental goals in the domain of precision medicine (Table 8.1).

Table 8.1: Key AI applications across the biomanufacturing workflow.

Biomanufacturing area	AI applications	Key outcomes	References
Upstream processing	AI-guided strain engineering, ML-based media optimization, and predictive modeling of fermentation	Faster design-build-test cycles; improved yields; reduced variability; and optimized resource use	Aida et al. (2023) and Wang et al. (2025)
Bioprocess monitoring and control (PAT/QbD)	Smart sensors, soft sensors, digital twins, anomaly detection, and predictive maintenance	Real-time monitoring; early fault detection; enhanced consistency; and reduced batch failures	Isoko et al. (2024), Rana (2025), and Auclair and Rathore (2025)
Downstream processing	ML for chromatography optimization, hybrid mechanistic-data models, robotics, and automation	Improved purity and recovery; reduced buffer consumption; higher throughput; and GMP-compliant automation	Ko et al. (2024) and Wang et al. (2024)
Innovation (generative AI, RL, multi-omics)	Generative protein design, RL-based bioprocess control, metabolic modeling, and multimodal omics integration	Accelerated enzyme/pathway engineering; optimized control strategies; and improved phenotype predictions	Loeffler et al. (2024) and Cheng et al. (2023)
Vaccines, biologics, food proteins	mRNA sequence optimization, IVT optimization, biosimilar modeling, and precision fermentation	Improved antigen expression; higher mRNA yields; optimized biologics profiles; and enhanced protein production	Rosa et al. (2022), Bas and Duarte (2024), and Chakraborty et al. (2025)
Sustainability and green biomanufacturing	AI-based waste reduction, energy optimization, circular bioeconomy modeling, and environmental sensing	Lower energy use; waste minimization; efficient biomass utilization; and enhanced environmental compliance	Sutton (2010) and Kaya (2025)
Regulatory and ethical considerations	AI governance, model life cycle management, explainability, data privacy, and biosecurity	Safer deployment; regulatory compliance; transparency; and reduced ethical risks	Weiner et al. (2025) and Niazi (2025)

8.4 AI for Biomanufacturing Innovation

AI is producing revolutionary advancements in the field of biomanufacturing by providing generative, as well as prescriptive, abilities that span the spectrum of molecular design, small-scale, and large-scale manufacturing. AI is poised to improve mechanistic reasoning, and it is combined with high-dimensional data to shorten the development cycle and save money, as well as increase the effectiveness of biologically derived therapeutics. Generative AI at the molecular frontier, such as language-model analogues, diffusion models, and conditional generative models, has begun to show the ability to sample protein and small-molecule sequence/structure space in directed, controllable ways, improving new biomolecules and biocatalysts with desired properties (stability, activity, and binding specificity) in ways that brute-force screening would fail to hit. Models that can directly produce conditioned protein sequences and structures, including the chroma and diffusion-based protein design programs, as well as useful toolkits for quickly generating leads and engineering enzymes, have been documented in landmark studies and surveys (Ingraham et al., 2023). These generative systems are not new; however, when integrated with high-throughput functional screening and active-learning loops, they significantly accelerate the design -build -test cycle, enable high-resolution exploration of high-fitness variants in sequence space, and facilitate the engineering of custom biocatalysts optimized for downstream processing conditions (e.g., sterilization). Such designs, including heat-tolerant and solvent-resistant enzymes, reduce downstream purification burdens and enhance overall process economics (Loeffler et al., 2024). In addition to its generative characteristics, RL offers an effective control framework for scaling up complex, stochastic, and multi-objective production process design. RL agents can suggest dynamic feeding, aeration, temperature, and induction plans in both fed-batch and continuous cultures that are higher-performing in sequence interaction with simulated or real process environments by concluding policies that provide tradeoffs that integrate productivity, resource use, and quality measures over time. Early studies of coculture and batch bioprocess control show that RL can be used to address nonlinearity, as well as uncertainty, with more recent approaches to methodology involving the addition of model-based RL, behavior cloning, and domain adaptation, indicating that it may be applied practically in pilot scale and continuous process environments in which safe exploration and sample efficiency are key factors (Treloar et al., 2020). More importantly, the potential of RL can be further increased when implemented in digital twins and hybrid simulators, where mechanistic models are supplemented with trained surrogates. Such a combination, with learning, allows for extensive virtual experimentation and direct policy optimization before proceeding to GMP (Isoko et al., 2024). At the systems and cellular level, AI-based metabolic modeling has developed beyond the analysis of systems in terms of static flux-balance into hybrid neural-based modeling frameworks, which combine sparse experimental constraints, enzyme kinetics, and regulatory signals, thereby enhancing phenotype predictions by genotype, and in addition, enabling the suggestion of pathway edits that maximize titer, productivity, and carbon performance. Recent reviews and primary studies show that ML can dramatically accelerate the construction of genome-scale metabolic models, facilitate the identification of rate-limiting enzymes, propose strategies for combinatorial gene knockouts or overexpressions, and quantify uncertainty, thereby proposing minimal-intervention design plans that maintain host viability. These studies have facilitated the redesign of pathways and enzyme replacements that bring significant welfare in product titers in either a microbial or a mammalian system by integrating ML: with constraint-based modeling and multi-objective optimization (Cheng et al., 2023). However, the most challenging progress lies in integrating multi-omics data, which encompasses genomics, transcriptomics, proteomics, metabolomics, and, in an ever-growing scope, single-cell and single-spat data, to drive the action calculus of biomanufacturing by linking multi-omics services with AI. Aligning and harmonizing different omics layers through multimodal AI pipelines enables the identification of regulation bottlenecks that are not apparent in individual layers, supporting the detection of changes in enzyme activity and off-target metabolic flux that impact the quality or titer of a product. Together with interpretability algorithms and mechanistic models, these pipelines amplify the causal hypotheses that can be subject to experimental verification, translating the correlative patterns into practical engineering instructions. Ongoing reviews and empirical inquiries in 2023–2025 highlight novel approaches to methodological innovations in data integration, such as representation learning, graph-based learning networks, and multimodal transformer networks, and emphasizes practical case studies in which integrated omics and AI have far surpassed conventional applications of omics and AI by improving host selection, media design, and stress-response engineering, in a requirement of determining reproducible scale-up (Nam et al., 2024). Their practical implementation as a reliable source of manufacturing returns depends on a pragmatic combination of technologies: the generative and predictive models need to be combined with automated construction and screening platforms; the RL agents need verifiable simulators and safe deployment channels; the metabolite engineering solutions need verification of fluids and quantifiable measurement predicates, as well as multi-omics should be supported by a strict set of data measures and rich experimental metadata in order to ensure reproducibility. Data from bioprocessing 4.0 architectures, including digital twins, standardized data fabrics, and model life cycle governance, can provide the required infrastructure to complete the loop between in silico design and production-scale. These models support ensemble designs to carry out candidate design, which generative models and ML predictors propose to screen high-performing designs, RL algorithms to advise optimal cultivation policies, and metabolic models to refine pathway changes based on omics-produced feedback (Isoko et al., 2024). However, though practical progress has been made, there are still serious problems. To reach predictive accuracies similar to those reached millions of years ago, generative models would have to tackle both long-range biochemical constraints and manufacturability priors, and otherwise, these features would be either coded or harshly screened. Reinforcement learning faces issues of sample efficiency and safety in the case of moving the simulation settings to actual production. It requires high-quality transport and kinetic information to use genome-scale models. Furthermore, the combination of multi-omics data is hindered by the issue of batch effect, discrepancies in the preprocessing protocols, and the absence of a more general approach to circumvent these issues. These problems will require significant funding to define standard benchmarks, create data consortia to produce publicly and privately owned digital twins to allow transfer learning across strains and scales, and transparent model-validation frameworks to abide by regulatory expectations (Mondal et al., 2023). The biomanufacturing innovation implemented by AI, according to the existing academic discussion, cannot be defined as a single groundbreaking algorithm implementation; instead, it is a multifactorial collection of converged skills. These include: expanding an extended repertoire of biomolecules and biocatalysts, enhancing dynamic and high-performance operation regimes by the aid of reinforcing learning strategies, transforming genetic data into high-yield phenotypes through in silico genome-scale models, and integrating multimodal makeup of omics data, which provides mechanistic data accrediting design goals to actual results. Scalable, digitized, and auditable pipelines provide the background to all of these developments and enable precision medicine and next-generation biologics. It is based on two forces interdependent in learning: the width of algorithmic ingenuity and the complexity of data engineering, cross-disciplinary relations, and active involvement of player regulators. This kind of coordinated activity is necessary in the high event that the transformational potential of AI is delivered to the full potential in secure, healthy, and sustainable systems of production (Table 8.1; Figure 8.1).

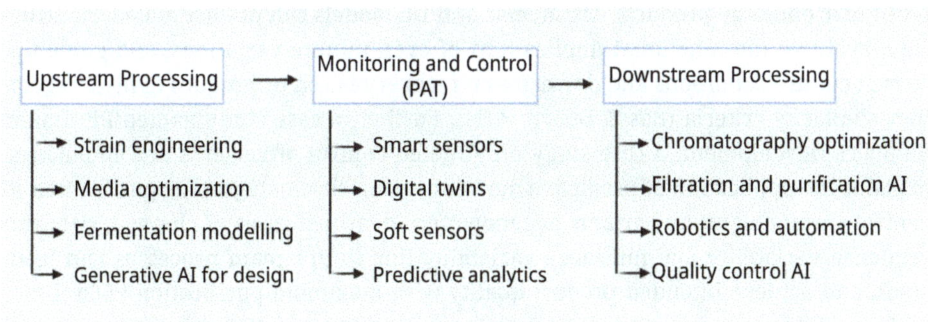

Figure 8.1: Incorporating AI into biomanufacturing streamlines processes, optimizes production, ensures quality control, and enhances efficiency in the development of biologics and bio-based products.

8.5 Case Studies and Applications

AI is having a significant impact in various biomanufacturing fields, including but not limited to vaccine development, food protein production, biologics, enzymes, and the increasing production of metabolic products. These developments emphasize the potential changeability of AI-oriented innovation, as well as the challenges that come with the process of scaling AI-related implementations. One clear example of AI use is in vaccine production, specifically in the COVID-19 mRNA programs. Linear design algorithm, which was designed by Baidu Research, used generative algorithms, based on natural-language processing, to optimize the secondary structure and codon usage of the SARS-CoV-2 spike antigen mRNA. The designed resultant designs possessed significantly improved in vitro stability, protein production, and in vivo antigen reactions, such as up to only about 23 times better than a codon-only optimization benchmark, and have been computed in minutes instead of weeks (Bagabir et al., 2022). Meanwhile, Bayesian optimization was used to screen millions of reaction-parameter combinations to optimize the in vitro transcription reaction for RNA vaccines, including temperature, enzyme concentration, NTPs, and buffer composition. This method increased mRNA production twofold and reduced assay time by about half, with industry-grade yields of approximately 12a-L reverted in about 2 h and only required a limited number of experimental cycles (Rosa et al., 2022). In addition to the mRNA domain, AI has been utilized to identify antigenic peptides and epitopes, optimize lipid nanoparticle formulations to achieve superior stability and delivery, and predict cold-chain logistics and supply-chain bottlenecks in the event of a pandemic (Ghosh et al., 2023). In the context of bio-logic-biomolecule production, AI is being used to navigate the complexity of protein therapeutics, particularly in cases involving posttranslational modifications, glycosylation variants, charge, and size heterogeneity, all of which pose a threat to efficacy and immunogenicity. According to a recent study of biosimilar oncology products, ML models and DL models can predict molecular structure, evaluate the functional implications of even minute variations, and guide upstream culture conditions and downstream recovery-based techniques to meet regulatory similarity criteria (Bas & Duarte 2024). Further, a case study named Biosimilar Biologics Development, a case study on targeted control of variants and impurities, explains the application of predictive modeling using chromatographic purification to control chromatography variants dependent on identified cases of charged, size, and sequence variability and minimize variability due to upstream processes and feedstock, and achieve intended product quality with maximum purification yield. Users of AI in biologics production, including those at Pfizer and Novartis, have reported utilizing it to optimize manufacturing and supply chain logistics, resulting in higher yields, lower losses, and reduced waste. The data may be used to conform to or surpass regulatory standards (Saini et al., 2025), which is only part of the 25% mentioned, including food proteins, biofuels, bioplastics, and microbial and cell-based biomanufacturing. The microbial and cell-based biomanufacturing of food proteins, biofuels, and bioplastics is a significant area of application. Integrated AI and CRISPR fusion, as well as DL approaches, are gaining momentum in effectiveness, particularly in precision fermentation. To provide an example, a review carried out recently (2025) suggests that, by using transcriptomic datasets in transcriptomics-guided cell response models, a superset of ethyl-based models, DL models were capable of discerning promoter-gene affinities to produce alternative proteins by approximately 300, not only increasing the quantity but also decreasing the metabolic load simultaneously. Similarly, to overcome the primer-binding problem, instruments like AutoCRISpr can be used to predict any off-target effects, thereby enhancing strain-engineering cycles, safety, and efficiency (Priyadharshini et al., 2025). The microscale growth systems used in high-throughput instances during the production of microbial proteins from lignocellulosic hydrolysates, such as those of the Venus flytrap, have generated time-series datasets. The datasets are useful in generalizing kinetic models to mixed-substrate conditions, thereby streamlining both upstream growth and downstream recovery processes (Banks et al., 2024). AI is also utilized in selecting algal, microbial strains, and mutagenesis plans or strategies to improve the production of bioactive compounds, biofuels, lipid precursors, and bioplastics precursors. According to a recent survey article on AI-assisted enzyme discovery and pathway engineering for nutraceuticals, pigment synthesis, aromatic compound synthesis, and fuel synthesis are presented (Chakraborty et al., 2025). In prescription fermentation and cultured meat, AI-based scaling frameworks are sequentially eliminating primary blockers, such as strain optimization, along with media formulation, process scale-up, feedstock cost, and downstream processing. Priyadharshini et al. (2025) demonstrate that AI can accelerate CRISPR target optimization and implement reinforcement-learning-based control in bioreactors, achieving up to ~60% improvement in reactor performance. This approach also markedly enhanced the production of proteins, metabolites, and other value-added compounds, and in some cases, increased supply-chain efficiency several-fold. The case study on "Three Examples of Precision Fermentation Using a Microbial Platform" by BRAIN Biotech also utilized fermentation on a microbial basis to produce scalable amounts of representative proteins from plant, animal, and gut probiotic sources. Additionally, it implemented AI-assisted selection of microbial host systems, expression systems, and fermentation conditions to achieve food-grade yields. Still, there are obstacles, particularly related to media costs and legislative support for cultured meat, including concerns about safety, reliability, and capacity. However, as these case studies presuppose, AI can bring more than a significant change, but it is essential to achieving large-scale viability. In the field of industrial enzymes and secondary metabolites, AI has enabled progress in both enzyme discovery and the optimization of their synthesis in industry. Metagenomic mining, combined with structural prediction algorithms such as AlphaFold, has enabled the analysis of large datasets of enzyme sequences to predict thermostability, substrate specificity, and catalytic potential, informing mutant libraries or rational design interventions, thereby reducing the number of required lab cycles (Amore & Philip 2023). Regarding secondary metabolites, such as antibiotics, pigments, and nutraceuticals, recent metabolic engineering projects spanning 2024–2025 include AI-assisted pathway redesign, the integration of omics data (genomics, transcriptomics, metabolomics) to maximize yield, and coupled enzyme engineering with regulatory network modeling. These measures aim to minimize the generation of by-products, optimize the selected precursor levels, and enhance product concentration in fermentation at the industrial level (Chakraborty et al., 2025). Also, the use of mass AI in the production process has already started. Software-on-a-service AI developed by Aizon and linked with Amazon Web Services to support pharmaceutical manufacturing has enabled the construction of digital twins of bioreactors, installation of Internet-of-Things sensors, and process optimization using ML, which is said to have increased yield by over 10% and drastically cut computation or operating overhead, and maintained Good Practice (G x P) at the same time. Taken as a whole, these case studies indicate that AI is used not only in discovery or affinity rules upstream but also in completely integrated workflows in which antigen or protein design, strain engineering, reaction optimization, process scaling, purification, regulatory compliance, and cost considerations are co-optimized. At the same time, a series of ongoing challenges are mentioned: access to high-quality, labeled data at industrial levels; the reproducibility and translatability of these solutions to other strains, hosts, and scales; regulatory enabling novel modalities of production and manufacturing such as cultured meat, precision-fermented proteins, and biosimilars, reducing the costs of supply chains and media, as well as obligation dollar separation in predictions; and increasing the predictive model interpretability and safety. However, it is now more likely to believe that biomanufacturing is set to reach an inflection point: AI will cease being a peripheral utility, a scaler, a cost controller, a quality assurer, and an enabler of innovation in vaccines and biologics, as well as in food proteins, enzymes, and secondary metabolites.

8.6 AI for Sustainability and Green Biomanufacturing

AI has become a key facilitator to progress toward sustainability in the biomanufacturing industry, introducing groundbreaking solutions to waste reduction, resource optimization, the circular bioeconomy, energy efficiency, and environmental regulation. Such benefits enable implementing significant improvements in operational efficiency and mitigating environmental effects, especially when AI technologies are used, contributing to the sustainability of more sustainable bioeconomic trends. Waste management in biomanufacturing is being redefined by the introduction of AI-based approaches. ML algorithms can also accurately predict the needs of raw materials during a particular production run, preventing excessive use and reducing the amount of waste. Moreover, AI can help to monitor biological processes in real-time, identify inefficiencies more quickly, and take corrective measures. Such a dynamic method does not just lower expenses, it also lowers the environmental footprint of biomanufacturing processes. This notion of the circular bioeconomy is currently becoming very popular, and AI is essential in its implementation. ML-based predictive analytics can predict changes in biomass supply, optimize harvesting, and reduce waste throughout the entire supply chain, i.e., the farm, the factory, and the consumer. AI is environmentally friendly, as it is used to simplify the transformation of biowaste into useful energy and materials, and to create new opportunities via economic processes supported by the bioeconomy. Energy use is a key concern in biomanufacturing, and AI solutions are currently being used to streamline energy use in bioprocessing. Due to the example, AI algorithms may analyze the data obtained in bioreactors and adjust the temperature, pH, and aeration levels so that the energy is used efficiently without affecting the product quality. The optimization results in a lower carbon footprint, low costs of operations, and general low energy use by biomanufacturers. Such biomanufacturing activities should also be directed toward compliance with environmental laws. The application of AI to environmental surveillance enables automated sampling and analysis of environmental data, providing accurate and actionable data that is used to make quick decisions (Sutton 2010). Additionally, AI agents revolutionize environmental control and surveillance in manufacturing plants, providing efficiency, sustainability, and operational optimization like never before (Kaya 2025). With the help of AI's ability to process and analyze extensive ecological data, manufacturers will be assured that they achieve sustainability goals, optimizing production quality and minimizing waste. AI is having a revolutionary impact on ensuring sustainability in biomanufacturing. AI enables biomanufacturers to operate more sustainably through waste minimization, resource utilization, circular bioeconomy platforms, energy-efficient bioprocessing, and improved environmental monitoring and compliance. With further advancement of AI technologies, their sequential funding in biomanufacturing operations will be considered as a decisive step in creating a more sustainable and circular bioeconomy.

8.7 Regulatory, Ethical, and Policy Perspectives

Regulatory, ethical, and policy frameworks constitute the foundational framework that will guide the development of AI in biomanufacturing as a reliable, safe, and socially responsible technology, or as a systemic risk and source of social skepticism. This field has to be managed through the synthesis of agile algorithmic competencies that have long-term principles of security, transparency, justice, and accountability. To regulators, the transition of theoretical position statements to operational guidance is fast. The United States Food and Drug Administration (FDA) has published risk-based frameworks and guidance tools regarding the application of AI in regulatory decision-making, along with software-as-a-medical-device, AI/ML-enabled software-as-a-medical-device, context-of-use, model-credibility, continuous monitoring, and documented change control. This tendency means that future regulatory requirements (including model life cycle management and auditable provenance) will become compulsory prerequisites to regulatory acceptance (Opderbeck 2019). Similarly, regional regimes such as the AI Act within the European Union and the European Medicines Regulatory Network incorporate direct commitments concerning high-risk AI systems, including the need to provide high-quality training data, human supervision, risk reduction, and technical documentation. The future state of AI guidance through the medicinal product life cycle, as defined by the European Medicines Agency in its work plans, serves as an illustration of the need to ensure consistency of standards across jurisdictions (Jiang et al., 2017). The policies and ethical discussions reveal some recurring themes that have direct, practical implications for biomanufacturing. To begin with, data acceptance and protection must ensure that datasets are representative, compiled where applicable, and anonymized to prevent reidentification, particularly with artificially generated models. Second, discrimination and tolerance will need to be reduced, engaging models not to propagate inequities, often due to distorted datasets or unrestrained proxy procedures that may compromise access to, the quality of, or the safety of biologics. Third, transparency and explainability require the specific production of interpretable evidence to the regulators and clinicians on why the models are making specific predictions or control decisions. Fourth, responsibility and liability need to include clarifying the responsibility in cases where an AI-driven decision leads to batch failure, safety indicators, or supply-chain disruptions. Fifth, dual-use and biosecurity requirements necessitate a system that prevents the use of the capabilities inherent in this type of model for creating dangerous agents, such as base generative sequence design. Global health authorities and ethicists have reiterated these moral priorities, saying that (AI) design should focus on the human aspects and uphold rights-respecting principles that would put the abstract principles in use (Weiner et al., 2025). Some practical leverage can be used to hasten responsible adoption on a policy and implementation basis. To start with, unified model validation procedures and benchmark datasets for standard biomanufacturing processes can help reduce regulatory uncertainty, for example, through PAT analytics and impurity prediction. Secondly, auditability and post-deployment surveillance could be provided with the help of mandated model cards, data lineage records, and versioned digital twins. Third, an infrastructure of public-privacy organizations must develop common, de-anonymized data in industrial form and pretest scenarios where regulators, industry, and academia could test solutions in the real-world environment. Fourth, the labor force will also be AI-capable to collaborate with operators, quality functions, and regulators to assess the output of the models and put in place actions to intervene safely. Fifth, they are able to buy fair benefits in terms of incentive alignment, i.e., buying calculable environmental footprint reductions or culls which can be assigned to AI. The risks of the international supply chain being hampered will result in international harmonization being required to prevent fragmentation. The terms and risk categories and need requirements within the evidence in the FDA, EMA, WHO, and other national working agencies will have their correlations to guarantee that redundancies are minimized and compliant applications are fastened. A balance that satisfies both society in terms of innovation and stability should be in the mind of the regulators and policymakers as they talk openly to the people about the benefits and the limitations, and subsequent sensible post-market follow-up of the unwanted harms, as well as clear redress mechanisms. In this way, the potential of AI to facilitate the generation of better yields, minimize waste, and accelerate the development of therapeutics can be achieved without compromising safety, equity, or ethical principles (Bouderhem 2024).

8.8 Challenges and Limitations

Although the application of AI to biotechnology and biomanufacturing is potentially transformative, it remains fraught with numerous challenges and limitations, which manifest in technical, regulatory, organizational, and ethical aspects, directly affecting the scalability and reliability of its application. One limiting technical problem is the availability and quality of data; biomanufacturing results in heterogeneous data on multi-omics platforms, sensors, and batch records, which are generally fragmented, noisy, proprietary, and sparsely labeled, thus conflicting with high-quality, high-performance model training (Zampieri et al., 2023). Moreover, there is no standard data format among the various information management systems used by laboratories, and there is limited interoperability between them, which limits the harmonious implementation of AI pipelines into existing workflows. Another significant obstacle is model interpretability; even DL and generative AI can predict molecular interactions or optimize a pathway, or control a process, but their black box aspect often conflicts with regulatory requirements of transparency, explainability, and reproducibility in the context of GMP (Niazi 2025). In addition to the problem of algorithmic opaqueness, scaling generalization gives rise to even further problems; bioinformatics ML models, being typically trained on the anaerobic laboratory size and pilot-scale bioreactors, perform worse when deployed onto the anaerobic industrial scale because of scalar interactions and mass-transfer and metabolic change nonlinearities, leading to the so-called challenge of reliable scaling. The issue of cybersecurity and data management is a growing concern, as sensitive genomic, process, and patient-related information becomes increasingly digitalized, raising questions about unauthorized access, data theft, and the dual purposes of this technology (WHO, 2024). In the organizational aspect, the skills gap only expands: a successful adoption of AI would require the ability of biomanufacturers to combine knowledge in the fields of computational science, bioengineering, and regulatory affairs, but employee training in cross-disciplinary communication and skill development, and the implementation is minimal. In addition, implementing more sophisticated AI tools, including the infrastructure upgrades required for real-time sensor networks or compliance-mandated model validation, can be a barrier to small and medium-sized enterprises, which may also exacerbate inequities in access to innovation. There are also ethical and societal restraint issues, such as the risk that the algorithm may be biased against specific strains, production conditions, or types of patients, which could perpetuate disparities in access to biologics and biosimilars. These problems are further complicated by policy uncertainty, as the FDA, EMA, and health agencies worldwide prepare AI guidelines.

Conflicting principles and shifting regulations pose a threat to compliance, hindering the emerging adoption of AI across the industry (Fakhouri et al., 2024). Altogether, the challenges mentioned above highlight that the application of AI in biomanufacturing must be done with care. However, ongoing investment in data infrastructure, explainability studies, and interdisciplinary education, as well as harmonized regulatory frameworks worldwide, is essential to deliver impactful and sustainable results that merit trust and equality.

8.9 Conclusion

AI is rapidly transforming biomanufacturing from a traditional assembly of identifiable unit operations to a data-driven ecosystem, making systems more efficient, replicable, and competitive. Throughout the workflow, including strain design and optimization of media upstream, real-time monitoring, downstream purification, and supply chain, the growth of ML, RL, and digital twins will rapidly reduce costs, minimize waste, and accelerate time-to-market. Case studies of developing vaccines, biologics, biosimilars, food proteins, cultured meat, and industrial enzyme development not only offer performance benefits but also the possibility to transform the current health and food systems worldwide. Nevertheless, the ongoing problems related to data quality, interpretability, regulatory validation, and workforce preparedness underscore the need for responsible adoption to be accompanied by strong governance, ethical scrutiny, and collaboration across disciplines. As AI continues to develop, AI-biotechnology collaboration presents a new era of accuracy, sustainability, and environmentally friendly manufacturing in the biomanufacturing industry, creating technological advances and solutions aligned with social and environmental demands.

References

Aida, H., Uchida, K., Nagai, M., Hashizume, T., Masuo, S., Takaya, N., et al. (2023). Machine learning-assisted medium optimization revealed the discriminated strategies for improved production of the foreign and native metabolites. *Computational and Structural Biotechnology Journal*, *21*, 2654–2663.

Amore, A., & Philip, S. (2023). Artificial intelligence in food biotechnology: Trends and perspectives. *Frontiers in Industrial Microbiology*, *1*, 1255505.

Auclair, J., & Rathore, A. (2025). *Biopharmaceutical Characterization in the Age of Artificial Intelligence*.

Bagabir, S. A., Ibrahim, N. K., Bagabir, H. A., & Ateeq, R. H. (2022). COVID-19 and artificial intelligence: Genome sequencing, drug development and vaccine discovery. *Journal of Infection and Public Health*, *15*, 289–296.

Banks, M., Taylor, M., & Guo, M. (2024). High throughput parameter estimation and uncertainty analysis applied to the production of mycoprotein from synthetic lignocellulosic hydrolysates. *Current Research in Food Science*, *9*, 100908.

Bas, T. G., & Duarte, V. (2024). Biosimilars in the era of artificial Intelligence – International regulations and the use in oncological treatments. *Pharmaceuticals, 17*, 925.

Bouderhem, R. (2024). Shaping the future of AI in healthcare through ethics and governance. *Humanities and Social Sciences Communications, 11*, 1–12.

Butean, A., Cutean, I., Barbero, R., Enriquez, J., & Matei, A. (2025). A review of artificial intelligence applications for biorefineries and bioprocessing: From data-driven processes to optimization strategies and real-time control. *Processes, 13*, 2544.

Chakraborty, P., Kumar, R., Karn, S., Patel, P., & Gosai, H. (2025). Recent trends in metabolic engineering for microbial production of value-added natural products. *Biochemical Engineering Journal, 213*, 109537.

Cheng, Y., Bi, X., Xu, Y., Liu, Y., Li, J., Du, G., et al. (2023). Machine learning for metabolic pathway optimization: A review. *Computational and Structural Biotechnology Journal, 21*, 2381–2393.

Fakhouri, T. H., Liu, Q., & ElZarrad, M. K. (2024). Artificial intelligence in regulatory decision-making for drug and biological products. In *Artificial Intelligence in Clinical Practice* (pp. 373–375). Elsevier.

Ghosh, A., Larrondo-Petrie, M. M., & Pavlovic, M. (2023). Revolutionizing vaccine development for COVID-19: A review of AI-based approaches. *Information, 14*, 665.

Helleckes, L. M., Hemmerich, J., Wiechert, W., von Lieres, E., & Grünberger, A. (2023). Machine learning in bioprocess development: From promise to practice. *Trends in Biotechnology, 41*, 817–835.

Ingraham, J. B., Baranov, M., Costello, Z., Barber, K. W., Wang, W., Ismail, A., et al. (2023). Illuminating protein space with a programmable generative model. *Nature, 623*, 1070–1078.

Isoko, K., Cordiner, J. L., Kis, Z., & Moghadam, P. Z. (2024). Bioprocessing 4.0: A pragmatic review and future perspectives. *Digital Discovery, 3*, 1662–1681.

Jiang, F., Jiang, Y., Zhi, H., Dong, Y., Li, H., Ma, S., et al. (2017). Artificial intelligence in healthcare: Past, present and future. *Stroke and Vascular Neurology, 2*, 230–243.

Kaya, C. (2025). Intelligent environmental control in plant factories: Integrating sensors, automation, and AI for optimal crop production. *Food and Energy Security, 14*, e70026.

Ko, K., Kim, M-J., Kim, D., Seo, K., & Lee, S. (2024). Design and optimization of a continuous purification process using ion-exchange periodic counter-current chromatography for a low-titer enzyme. *Biotechnology and Bioprocess Engineering, 29*, 721–732.

Kouba, P., Kohout, P., Haddadi, F., Bushuiev, A., Samusevich, R., Sedlar, J., et al. (2023). Machine learning-guided protein engineering. *ACS Catalysis, 13*, 13863–13895.

Loeffler, H. H., He, J., Tibo, A., Janet, J. P., Voronov, A., Mervin, L. H., et al. (2024). Reinvent 4: Modern AI–driven generative molecule design. *Journal of Cheminformatics, 16*, 20.

Mondal, P. P., Galodha, A., Verma, V. K., Singh, V., Show, P. L., Awasthi, M. K., et al. (2023). Review on machine learning-based bioprocess optimization, monitoring, and control systems. *Bioresource Technology, 370*, 128523.

Nam, Y., Kim, J., Jung, S.-H., Woerner, J., Suh, E. H., D-g, L., et al. (2024). Harnessing artificial intelligence in multimodal omics data integration: Paving the path for the next frontier in precision medicine. *Annual Review of Biomedical Data Science, 7*, 225–250.

Niazi, S. K. (2025). Regulatory perspectives for AI/ML implementation in pharmaceutical GMP environments. *Pharmaceuticals, 18*, 901.

Opderbeck, D. W. (2019). Artificial intelligence in pharmaceuticals, biologics, and medical devices: Present and future regulatory models. *Fordham Law Review, 88*, 553.

Priyadharshini, D., Muthuvel, I., Saraswathy, S., Kavitha, P., & Jegadeeswari, V. (2025). Precision to plate: AI-driven innovations in fermentation and hyper-personalized diets. *Frontiers in Nutrition, 12*, 1659511.

Rana, S. (2025). AI-driven fault detection and predictive maintenance in electrical power systems: A systematic review of data-driven approaches, digital twins, and self-healing grids. *American Journal of Advanced Technology and Engineering Solutions, 1*, 258–289.

Rosa, S. S., Nunes, D., Antunes, L., Prazeres, D. M., Marques, M. P., & Azevedo, A. M. (2022). Maximizing mRNA vaccine production with Bayesian optimization. *Biotechnology and Bioengineering, 119*, 3127–3139.

Saini, J. P. S., Thakur, A., & Yadav, D. (2025). AI driven innovations in pharmaceuticals: Optimizing drug discovery and industry operations. *RSC Pharmaceutics, 2*, 437–454.

Schmidt, A., Helgers, H., Lohmann, L. J., Vetter, F., Juckers, A., Mouellef, M., et al. (2022). Process analytical technology as key-enabler for digital twins in continuous biomanufacturing. *Journal of Chemical Technology & Biotechnology, 97*, 2336–2346.

Singh, N., Lane, S., Yu, T., Lu, J., Ramos, A., Cui, H., et al. (2025). A generalized platform for artificial intelligence-powered autonomous enzyme engineering. *Nature Communications, 16*, 5648.

Sutton, S. (2010). The environmental monitoring program in a GMP environment. *Journal of GXP Compliance, 14*, 22.

Treloar, N. J., Fedorec, A. J., Ingalls, B., & Barnes, C. P. (2020). Deep reinforcement learning for the control of microbial co-cultures in bioreactors. *PLoS Computational Biology, 16*, e1007783.

Vo, T. S., Hoang, T., Vo, TTBC., Jeon, B., Nguyen, V. H., & Kim, K. (2024). Recent trends of bioanalytical sensors with smart health monitoring systems: From materials to applications. *Advanced Healthcare Materials, 13*, 2303923.

Vora, L. K., Gholap, A. D., Jetha, K., Thakur, R. R. S., Solanki, H. K., & Chavda, V. P. (2023). Artificial intelligence in pharmaceutical technology and drug delivery design. *Pharmaceutics, 15*, 1916.

Walsh, I., Myint, M., Nguyen-Khuong, T., Ho, Y. S., & Ng, S. K. (2022). Lakshmanan M Harnessing the potential of machine learning for advancing “quality by design” in biomanufacturing. In *MAbs* (Vol. 1, pp. 2013593). Taylor & Francis.

Wang, B., Xie, W., Martagan, T., Akcay, A., & Ravenstein, B. (2024). Biomanufacturing harvest optimization with small data. *Production and Operations Management, 33*, 2381–2400.

Wang, -Z.-Z., Zeng, D.-W., Zhu, Y.-F., Zhou, M.-H., Kondo, A., Hasunuma, T., et al. (2025). Fermentation design and process optimization strategy based on machine learning. *BioDesign Research, 7*, 100002.

Weiner, E. B., Dankwa-Mullan, I., Nelson, W. A., & Hassanpour, S. (2025). Ethical challenges and evolving strategies in the integration of artificial intelligence into clinical practice. *PLOS Digital Health, 4*, e0000810.

World Health Organization. (2024). WHO principles for ethical human genomic data collection, access, use and sharing. Geneva: World Health Organization.

Yang, S., Hu, X., Zhu, J., Zheng, B., Bi, W., Wang, X., et al. (2025). Aspects and implementation of pharmaceutical quality by design from conceptual frameworks to industrial applications. *Pharmaceutics, 17*, 623.

Zampieri, G., Campanaro, S., Angione, C., & Treu, L. (2023). Metatranscriptomics-guided genome-scale metabolic modeling of microbial communities. *Cell Reports Methods, 3*(1).

Zhou, T., Reji, R., Kairon, R. S., & Chiam, K. H. (2023). A review of algorithmic approaches for cell culture media optimization. *Frontiers in Bioengineering and Biotechnology, 11*, 1195294.

Kumkum Verma*, Priti Duhan, Anil Kumar Sharma, Poonam Bansal, Pooja Verma, Amit

Chapter 9 Transformative AI in Healthcare Biotechnology

Abstract: Artificial intelligence (AI) is transforming healthcare and biotechnology by improving illness diagnosis, treatment, and prevention. AI uses machine learning, deep learning, and natural language processing to quickly analyze large biological datasets, identify complicated patterns, and speed up decision-making. AI algorithms in drug development speed up target discovery and compound screening, saving time and money. AI in diagnostics improves image interpretation, disease identification, and real-time monitoring. AI is transforming personalized healthcare by creating predictive models that tailor therapies to an individual's genetic and clinical profile. This review examines the application of AI technologies in biotechnology, including drug discovery, genetic engineering, diagnostics, and customized medicine. This review examines current AI uses in healthcare biotechnology, identifies pioneering developments, and analyzes future options for enhancing AI's effect while guaranteeing ethical and fair deployment. Combining AI and biotechnology can significantly improve global health outcomes.

Keywords: Artificial intelligence, healthcare biotechnology, drug discovery, personalized medicine, medical diagnostics

9.1 Introduction

The combination of artificial intelligence (AI) with biotechnology has triggered a paradigm change in the healthcare business, altering diagnostics, treatments, and personalized medicine. AI, which includes machine learning (ML), deep learning (DL), and

*__Corresponding author: Kumkum Verma__, ICAR-IARI, Regional Station, Karnal, Haryana, India, e-mail: Kumkum.verma89@gmail.com

Priti Duhan, Department of Biotechnology, K. L. Mehta Dayanand College for Women, Faridabad, Haryana, India

Anil Kumar Sharma, Department of Biotechnology, School of Biological Sciences, Amity University Punjab, Mohali 140306, India

Poonam Bansal, Department of Biosciences and Technology, MMEC, Maharishi Markandeshwar (Deemed to be University), Mullana, Ambala 133207, Haryana, India

Pooja Verma, ICAR-IIWBR, Karnal-Haryana, India

Amit, Division of Genetics and Plant Breeding, ICAR-IIWBR, Karnal, Haryana, India

 | https://doi.org/10.1515/9783112214275-009

natural language processing (NLP), has demonstrated enormous potential for evaluating complicated biological data, expediting drug discovery, and enabling precision medicine (Esteva et al., 2019; Topol, 2019a). AI is developing as a potent tool in healthcare biotechnology, which combines biological processes with engineering to generate medical goods, for uncovering unique insights from genomic, proteomic, and metabolomic data, thereby improving illness prediction, prevention, and treatment. AI plays a transformational role in biotechnology, particularly in drug discovery and development, where it drastically reduces the time and expense associated with traditional methods. AI-powered platforms can find promising therapeutic candidates, forecast efficacy and toxicity, and optimize lead molecules at unprecedented speed and accuracy (Zhavoronkov et al., 2019). Furthermore, AI algorithms make it easier to comprehend large biological datasets, which helps with targeted therapy design and biomarker identification (Ching et al., 2018). In clinical practice, AI-powered solutions are transforming medical imaging, robotic surgery, and decision-support systems. These developments not only improve diagnostic accuracy but also allow for real-time, tailored therapies (Jiang et al., 2017). Furthermore, AI improves biotechnological research by automating laboratory operations, AI-guided CRISPR editing, and synthetic biology design, pushing the boundaries of what is feasible in modern medicine. This chapter will provide an overview of how AI is being incorporated into biotechnology, which has the potential to provide healthcare in a more inventive, efficient, and egalitarian manner as the healthcare industry continues to embrace digital transformation. To guarantee safe and efficient implementation, this quick development also calls for strong frameworks for data privacy, ethical governance, and clinical validation (He et al., 2019).

9.2 Foundations of Artificial Intelligence in Biotechnology

AI in biotechnology is built on the ability to model and comprehend high-dimensional biological data in order to address complicated challenges in healthcare and life sciences (Figure 9.1). AI makes genetic sequence analysis, illness biomarker identification, laboratory process automation, and intelligent treatment development easier by applying algorithmic learning. This section delves into the core AI technologies – ML and DL, NLP, computer vision, and reinforcement learning (RL) – that form the foundation of AI-driven biotechnology.

9.2.1 Machine Learning and Deep Learning

ML methods allow systems to understand patterns from data and make predictions or judgments without requiring explicit programming. In biotechnology, ML has transformed applications that range from gene expression analysis and drug discovery to disease classification, as well as synthetic biology modeling (Libbrecht & Noble, 2015). DL, a subfield of ML, uses artificial neural networks with numerous layers to extract complicated features from big datasets. It has shown excellent performance in modeling nonlinear interactions in omics data, such as genomics, proteomics, and metabolomics (Min et al., 2017). Deep neural networks are also used to predict protein structures, as seen in DeepMind's AlphaFold, which was a breakthrough in computational biology (Jumper et al., 2021).

9.2.2 Natural Language Processing (NLP)

NLP allows machines to comprehend, interpret, and produce human language. In biotechnology, NLP is essential for extracting knowledge from unstructured biomedical literature, clinical notes, patents, and research papers. Tools such as BioBERT and PubMedBERT are trained specifically on biomedical corpora to enhance information retrieval and biomedical named entity recognition (Lee et al., 2020). NLP also supports hypothesis generation, drug repurposing, and semantic search in biomedical databases, accelerating scientific discovery by automating literature curation and synthesis of evidence.

9.2.3 Computer Vision and Image Analysis

Computer vision applies AI techniques to interpret visual data, playing a crucial role in medical imaging, microscopy, and high-throughput screening. In biotechnology, AI-powered image analysis is used to detect cellular morphology changes, analyze histopathological slides, quantify fluorescence microscopy data, and classify tumor subtypes or microbial colonies. Deep convolutional neural networks have demonstrated radiologist-level performance in identifying disorders from X-rays, MRIs, and CT scans (Litjens et al., 2017), and they are currently being incorporated into automated diagnostic processes in biotechnology research.

9.2.4 Reinforcement Learning and Autonomous Systems

RL is a sort of ML in which an agent develops optimal behavior from interactions with its surroundings using rewards or punishments. In biotechnology, RL is emerging in applications such as (a) autonomous lab systems for experiment planning and robotic control, (b) synthetic pathway optimization in metabolic engineering, (c) adaptive clinical trial design, and (d) protein folding simulations. Autonomous platforms like robotic chemists or biofoundries utilize RL to efficiently explore complex chemical and biological design spaces with minimal human intervention (MacLeod et al., 2020).

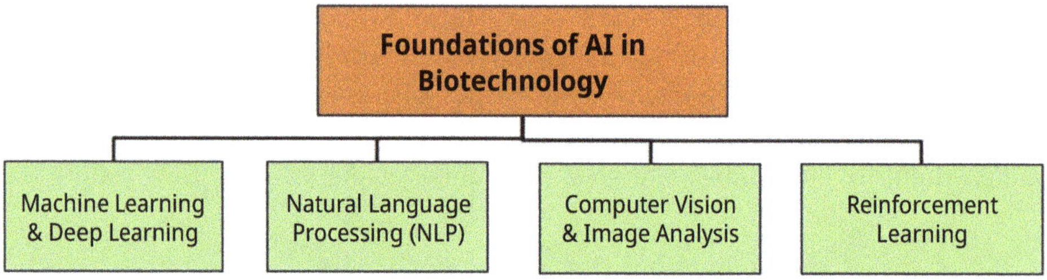

Figure 9.1: AI foundations in biotechnology enable intelligent analysis, prediction, and biological innovation.

9.3 AI Applications in Biomedical Research

AI has emerged as a critical tool in biomedical research, driving advances in fields as diverse as drug discovery, genomics, proteomics, and systems biology. AI's ability to detect hidden patterns in big biological datasets allows researchers to build more effective medicines, interpret genetic information, and simulate intricate biological systems. The following subsections explore how AI is revolutionizing critical areas of biomedical science.

9.3.1 Drug Discovery and Development

Traditional drug development involves a lengthy and expensive procedure with a high failure rate. AI speeds this workflow by allowing for data-driven target selection, virtual screening, and lead optimization, as well as toxicity prediction (Ekins et al., 2019). ML models analyze high-throughput screening data, chemical structures, and pharmacokinetics to identify promising drug candidates efficiently. AI platforms such as AtomNet, DeepChem, and Insilico Medicine use DL for structure-based drug design and generative modeling. Recent breakthroughs, such as AI-generated drug molecules entering clinical trials, demonstrate the practical viability of these technologies (Zhavoronkov et al., 2019). Moreover, RL and Bayesian optimization are used to design compounds with desirable properties, reducing the number of wet-lab experiments needed to find viable therapeutics (Segler et al., 2018).

9.3.2 Genomics and Precision Medicine

AI is central to genomic data interpretation, helping decode vast amounts of sequencing data to uncover disease-associated variants and regulatory elements. DL tools such as DeepSEA and SpliceAI predict the functional consequences of noncoding mutations and splicing events (Zhou & Troyanskaya, 2015; Jaganathan et al., 2019). In precision medicine, AI enables the stratification of patient populations, the prediction of disease risk, and personalized treatment recommendations based on genomic profiles. AI-powered platforms combine genomic, transcriptomic, and clinical data to create predictive models for cancer therapy selection, uncommon illness detection, and pharmacogenomics (Kourou et al., 2015).

9.3.3 Protein Structure Prediction and Engineering

AI has made fundamental contributions to protein science, particularly in structure prediction and protein design. The most significant success is AlphaFold, which predicted 3D protein structures with near-experimental accuracy from amino acid sequences (Jumper et al., 2021). This development has far-reaching implications for drug target identification, enzyme engineering, and synthetic biology, allowing the creation of novel proteins with specific activities. Other AI technologies, such as RosettaFold, ProGen, and ProteinMPNN, enhance protein engineering capabilities by maximizing binding affinity, stability, and specificity.

9.3.4 Systems Biology and Computational Modeling

The goal of systems biology is to gain a comprehensive understanding of the intricate connections between genes, proteins, metabolites, and cells. AI enhances systems biology through network inference and pathway modeling, revealing regulatory and metabolic interactions; dynamic simulations of cellular processes using ML and hybrid mechanistic-AI models; and predictive modeling of disease progression and treatment response. Tools such as CellNet, GeneMANIA, and Biocellion utilize AI to reconstruct gene regulatory networks and simulate tissue-scale behavior (Libbrecht & Noble, 2015; Ma et al., 2018). AI also supports digital twin models for personalized simulations of physiology and therapy outcomes, a promising frontier in systems medicine.

9.4 Clinical Applications and Decision Support

AI is transforming clinical practice by allowing for faster, more accurate, and tailored healthcare delivery. From early disease identification to real-time therapeutic recommendations, AI helps doctors all the way through the care process. This section delves into AI's key clinical applications, with a focus on medical imaging, predictive analytics, and decision support, as well as personalized treatment strategies.

9.4.1 AI in Medical Imaging and Diagnostics

AI has transformed medical imaging, with skills that rival or outperform human picture interpretation. DL, specifically convolutional neural networks, has enabled the speedy and exact interpretation of complex imaging modalities such as MRI, CT, X-rays, and histopathological slides (Litjens et al., 2017).

Applications include (a) early cancer detection (e.g., lung, breast, and skin cancers), (b) tumor segmentation and classification, (c) retinal image analysis for diabetic retinopathy and glaucoma, and (d) radiomics, which extracts high-dimensional features from images to aid prognostic models. Food and Drug Administration (FDA)-approved AI products such as IDx-DR (for diabetic retinopathy identification) and Aidoc (for stroke triage) demonstrate the increasing clinical trust in AI-assisted diagnostics.

9.4.2 Predictive Analytics and Risk Stratification

AI-powered predictive analytics estimate patient outcomes and identify high-risk people by leveraging large-scale electronic health records (EHRs), wearable sensor data, and even lab results. These models help forecast hospital readmissions, disease start (e.g., sepsis, heart failure), and adverse drug reactions; monitor patient deterioration in ICU settings; and stratify patients based on genetic risk factors or lifestyle variables (Rajkomar et al., 2018). Random forests, gradient boosting, and recurrent neural networks are popular ML techniques for predicting models that are constantly updated as new data becomes available.

9.4.3 AI-Assisted Clinical Decision Support Systems

Clinical decision support systems use AI to make real-time, evidence-based recommendations to healthcare providers. These systems can assess clinical data, detect anomalies, and suggest diagnostic or treatment approaches (Sendak et al., 2020). Its main features include alerts for drug interactions, dosing errors, and lab abnormalities; guidelines integration for chronic disease management, and dynamic learning from institutional data to improve over time. AI-enhanced clinical decision support systems has demonstrated benefits in reducing diagnostic errors, improving workflow efficiency, and standardizing care, especially in complex or high-stakes environments like oncology and critical care.

9.4.4 Personalized Treatment Planning

AI provides personalized treatment planning by combining several data sources (genomic profiles, imaging, EHRs, and patient preferences) to create tailored therapy plans. AI models predict tumor response to various chemotherapy or immunotherapy regimens based on molecular signatures. Personalized risk models guide interventions for arrhythmias, myocardial infarction, and heart failure. AI supports adaptive treatment plans in depression or anxiety based on behavioral and biological feedback. AI-powered tools such as IBM Watson for Oncology and Tempus synthesis organized and unstructured data to help oncologists personalize evidence-based treatments to specific patients (Gulshan et al., 2016).

9.5 Biomanufacturing and Process Optimization

The use of AI in biomanufacturing is transforming the way biological products such as vaccines and monoclonal antibodies, as well as cell-based therapies, are manufactured, monitored, and refined. By leveraging real-time data analytics, ML, and automation, AI enhances process efficiency, product consistency, and regulatory compliance, all while reducing operational costs and downtime. This section explores how AI is transforming key components of modern bioprocessing and manufacturing workflows.

9.5.1 AI in Bioprocess Monitoring and Control

Modern bioprocesses are complex, involving multiple variables that must be tightly regulated to maintain product quality. AI models enable real-time monitoring and control by predicting process behavior, detecting anomalies, and recommending adjustments to maintain optimal conditions (Parthiban et al., 2023). The applications include (a) soft sensors for estimating unmeasured process variables (e.g., cell viability, metabolite levels) using surrogate data and ML; (b) model predictive control powered by AI for dynamically adjusting pH, temperature, oxygen, and nutrient feed rates; and (c) digital twins that simulate and optimize entire bioreactor systems based on historical and live process data (Canzoneri et al., 2024). These technologies facilitate a shift from traditional open-loop systems to adaptive, closed-loop bioprocessing that responds intelligently to fluctuations.

9.5.2 Quality Assurance and Predictive Maintenance

AI improves quality by design and real-time release testing by constantly evaluating product quality parameters during the production process. ML algorithms trained on historical batch data may predict critical quality attributes along with critical process parameters, identify batches as likely to pass or fail quality inspections, and perform early problem identification and root cause investigation (Rathore et al., 2023). In predictive maintenance, AI uses sensor data from equipment (e.g., vibration, temperature, and pressure) to foresee failures and schedule timely interventions, minimizing unplanned downtime and maintaining compliance with good manufacturing practice.

9.5.3 Automation in Biotech Manufacturing

The use of AI-driven automation in biotech manufacturing leads to enhanced throughput, reproducibility, and cost efficiency. Fully integrated biomanufacturing systems can automate upstream processes such as media preparation, inoculation, and cell cultivation; control downstream workflows including chromatography, filtration, and purification; and coordinate with robotic arms and lab-scale liquid handlers for precision and consistency. AI is also crucial in adaptive scheduling and inventory management, ensuring optimal resource allocation and minimizing waste (Lee et al., 2015). Furthermore, cloud-based platforms and Industrial Internet of Things devices facilitate centralized control and remote process analytics, paving the way toward “smart factories” in biotechnology.

9.6 AI-Driven Innovation in Healthcare Delivery

AI is creating a paradigm shift in healthcare delivery by improving accessibility, efficiency, and patient involvement with digital tools and real-time analytics. With growing demands on healthcare systems worldwide, AI enables scalable, patient-centric innovations such as virtual assistants, remote care, and population-level disease monitoring. These advances help reduce clinician workload, increase diagnostic accuracy, and deliver timely interventions even in resource-constrained settings.

9.6.1 Virtual Health Assistants and Chatbots

AI-powered virtual health assistants and chatbots are increasingly employed to help with patient education, symptom triage, appointment scheduling, and medication adherence. These systems evaluate user queries using NLP and ML to provide personalized, context-aware responses. Tools such as Babylon Health, Ada, and Woebot show how AI chatbots may provide preliminary medical advice and behavioral support, as well as mental health counseling. Clinical chatbots trained on huge medical datasets can triage symptoms with similar accuracy to human practitioners (Miner et al., 2016). These solutions are available around the clock and can help to decrease administrative burdens, particularly in primary care as well as mental health services. These assistants are also linked to EHR systems and hospital portals to improve care continuity and patient involvement.

9.6.2 Remote Monitoring and Telemedicine

AI improves remote patient monitoring and telemedicine by allowing for continuous, real-time analysis of health data via wearables, cellphones, and home sensors. AI algorithms identify variations from baseline values and forecast undesirable outcomes, allowing for timely action. The main applications are chronic disease management (e.g., diabetes, hypertension, COPD) via wearable biosensors that monitor vitals and transmit alerts; postoperative monitoring, in which AI detects early signs of complications such as infection or cardiac instability; and AI-augmented teleconsultations, where pre-visit data is summarized to assist clinicians with decision-making and documentation (Topol, 2019b). This hybrid model of care delivery is especially valuable in rural or underserved regions, bridging access gaps and ensuring continuity of care.

9.6.3 AI in Public Health Surveillance

AI is becoming increasingly important for real-time public health surveillance because it enables automated data mining, trend recognition, and outbreak prediction. AI can detect early warning indications of public health hazards by evaluating data from a variety of sources, such as social media, search engines, hospital records, and mobile devices. Examples include:

BlueDot and HealthMap utilize NLP and ML to monitor worldwide disease epidemics like Zika and COVID-19 (Kraemer et al., 2020). Predictive models that estimate flu transmission, vaccination impact, and geographic hotspots for vector-borne diseases. AI-driven syndromic surveillance using EHRs and emergency department data to monitor emerging patterns in respiratory, gastrointestinal, and neurological symptoms. By enabling faster, data-driven responses, AI strengthens the resilience of public health infrastructure and supports effective resource allocation and risk communication.

9.7 Ethical, Regulatory, and Societal Considerations

As AI becomes more integrated into healthcare, as well as biotechnology, its development and deployment have complicated ethical, legal, and social implications. Ensuring responsible innovation necessitates addressing issues about data privacy, algorithmic bias, accountability, and the broader implications on health inequalities and the medical workforce. This section discusses the fundamental ethical and regulatory considerations necessary for the safe and equitable integration of AI in healthcare systems, as represented in Figure 9.2.

9.7.1 Data Privacy and Security

The effectiveness of AI in healthcare is strongly reliant on access to enormous amounts of sensitive patient data, such as genomic, clinical, and behavioral information. This raises fundamental questions about data privacy, consent, and security. AI systems must comply with data protection requirements such as the Health Insurance Portability and Accountability Act in the United States and the General Data Protection Regulation in the European Union (Shabani & Marelli, 2019). Techniques such as differential privacy, federated learning, and homomorphic encryption are increasingly being used to train AI models while maintaining patient confidentiality (Rieke et al., 2020). Robust cybersecurity processes and audit methods are required to mitigate the risks of cyberattacks, data abuse, and reidentification of anonymized data. Transparent data governance and patient-centered consent procedures are critical to sustaining public trust.

9.7.2 Bias and Fairness in AI Algorithms

AI systems educated on biased datasets can increase existing health inequities, particularly among underrepresented groups. Common sources of bias include (a) historical disparities in data collecting (e.g., underrepresentation of minorities, women, or low-income groups), (b) sampling bias in training datasets, and (c) algorithmic bias resulting from model architecture or feature selection (Obermeyer et al., 2019). Such biases can lead to inaccurate diagnoses, inappropriate treatment recommendations, and inequities in resource allocation. To mitigate bias, it is essential to use diverse and representative datasets, apply model explainability and transparency tools (e.g., SHAP,

LIME), and conduct independent audits and ethical assessments during AI deployment. Fair AI requires continuous performance monitoring across demographic subgroups to ensure equitable outcomes.

9.7.3 Regulatory Frameworks and Approvals

AI solutions in healthcare frequently operate as Software as a Medical Device, needing compliance with regulatory requirements. Regulatory bodies, such as the United States FDA, European Medicines Agency, and MHRA (UK), are developing frameworks for AI oversight (Topol, 2019b). The FDA's AI/ML Software as a Medical Device Action Plan outlines a risk-based approach to approval, focusing on transparency, real-world performance, and change control (Brown et al., 2021). Challenges include regulating "black-box" models, updating oversight as AI systems evolve, and harmonizing international standards. Global regulatory harmonization is needed to accelerate innovation while safeguarding patient safety and efficacy.

9.7.4 Societal Impact and Workforce Transformation

AI-driven automation will significantly alter the healthcare and biotech workforce, raising both opportunities and challenges, such as routine tasks (e.g., image analysis, data entry, monitoring) may be increasingly automated, allowing clinicians and researchers to focus on complex, human-centered activities; new roles will emerge in AI system oversight, data science, model validation, and digital ethics, necessitating interdisciplinary training; and concerns regarding job displacement, de-skilling, and digital disparities emphasize the importance of inclusive transition initiatives (Floridi et al., 2018). In the broader societal context, AI must be integrated with health equality objectives, ensuring that technological gains are dispersed equally among populations.

9.8 Challenges and Future Directions

Despite AI's immense potential to revolutionize healthcare and biotechnology, a number of technological, governmental, and societal barriers continue to impede its widespread use. Addressing these issues is critical for supporting responsible innovation and maximizing AI's benefits for healthcare and biomanufacturing systems. AI models rely heavily on the quality and diversity, as well as the volume of training data. In healthcare, challenges such as missing data, variability in EHRs, and data silos among institutions might limit model performance and generalizability. In biotechnology,

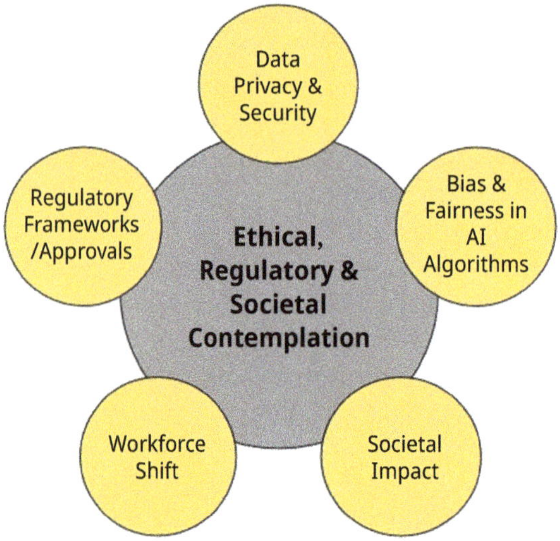

Figure 9.2: Ethical, regulatory, and societal considerations guide responsible AI and biotechnology applications.

batch heterogeneity and inadequate process data in early development impede strong AI modeling. The "black-box" aspect of many AI models, particularly DL, raises questions about explainability. Clinicians and regulators frequently require transparent, auditable reasoning from AI systems, particularly in high-stakes applications such as diagnosis and therapy recommendation (Samek et al., 2017). AI systems frequently suffer from adoption barriers or delays due to workflow misalignment, insufficient digital infrastructure, and a lack of integration with existing tools. Bridging the gap between AI research prototypes, along with deployable systems, remains a significant challenge. Regulatory frameworks are still evolving to meet the specific issues provided by adaptive and self-learning AI systems. Ethical problems, including bias, autonomy, accountability, and data ownership, add to the complexity (Morley et al., 2020). AI deployment demands a multidisciplinary workforce with expertise in computer science, biomedical engineering, ethics, and domain knowledge. Upskilling clinicians, regulators, and bioengineers is critical for responsible adoption.

Techniques such as federated learning, synthetic data generation, and differential privacy will be critical in overcoming data-sharing barriers while maintaining confidentiality and facilitating collaboration across institutions and countries. Future systems will prioritize explainability, accessibility, and clinician-AI codesign, ensuring that AI complements rather than replaces human competence. Explainable AI tools will be critical for establishing confidence and facilitating regulatory acceptance. The integration of AI with multi-omics, digital twins, and quantum computing may allow for the real-time design of highly tailored treatments, improved enzymes, or synthetic pathways, increasing the pace of discovery and innovation. Moreover, collaborative efforts toward standardized datasets, benchmarking protocols, and harmonized regulatory pathways will streamline the validation and deployment of AI tools globally, promoting innovation with accountability. Emphasis on inclusive design, global equity, and environmental sustainability will shape the next phase of AI innovation. Ensuring access to AI technologies for low-resource settings will be critical in closing global health disparities.

9.9 Conclusion

AI is clearly changing the face of healthcare and science. AI systems enable previously unattainable precision, efficiency, and scalability in fields ranging from customized medicine and virtual health aides to smart biomanufacturing and drug discovery. However, in order to realize AI's full potential, obstacles such as data governance and bias, regulation, as well as human-machine collaboration, must be addressed. As the area advances, a multidisciplinary and morally sound approach will be required. AI has the potential to not only improve clinical outcomes and treatment pipelines but also to rethink the whole architecture of modern healthcare and life sciences.

References

Brown, N. A., Carey, C. H., & Gerry, E. I. (2021). FDA releases action plan for artificial intelligence/Machine learning-enabled software as a medical device. *The Journal of Robotics, Artificial Intelligence & Law, 4.*

Canzoneri, M., Horner, M., & Banerjee, P. (2024). Digital twins in biomanufacturing. In *Bioconvergence Revolution: Challenges and Opportunities in Research Productivity* (pp. 59–81). Cham: Springer Nature Switzerland.

Ching, T., Himmelstein, D. S., Beaulieu-Jones, B. K., Kalinin, A. A., Do, B. T., Way, G. P., . . . & Greene, C. S. (2018). Opportunities and obstacles for deep learning in biology and medicine. *Journal of the Royal Society Interface, 15*(141), 20170387.

Ekins, S., Puhl, A. C., Zorn, K. M., Lane, T. R., Russo, D. P., Klein, J. J., . . . & Clark, A. M. (2019). Exploiting machine learning for end-to-end drug discovery and development. *Nature Materials, 18*(5), 435–441.

Esteva, A., Robicquet, A., Ramsundar, B., Kuleshov, V., DePristo, M., Chou, K., . . . & Dean, J. (2019). A guide to deep learning in healthcare. *Nature Medicine, 25*(1), 24–29.

Floridi, L., Cowls, J., Beltrametti, M., Chatila, R., Chazerand, P., Dignum, V., . . . & Vayena, E. (2018). AI4People – An ethical framework for a good AI society: Opportunities, risks, principles, and recommendations. *Minds and Machines, 28*, 689–707.

Gulshan, V., Peng, L., Coram, M., Stumpe, M. C., Wu, D., Narayanaswamy, A., . . . & Webster, D. R. (2016). Development and validation of a deep learning algorithm for detection of diabetic retinopathy in retinal fundus photographs. *jama, 316*(22), 2402–2410.

He, J., Baxter, S. L., Xu, J., Xu, J., Zhou, X., & Zhang, K. (2019). The practical implementation of artificial intelligence technologies in medicine. *Nature Medicine, 25*(1), 30–36.

Jaganathan, K., Panagiotopoulou, S. K., McRae, J. F., Darbandi, S. F., Knowles, D., Li, Y. I., . . . & Farh, K. K. H. (2019). Predicting splicing from primary sequence with deep learning. *Cell, 176*(3), 535–548.

Jiang, F., Jiang, Y., Zhi, H., Dong, Y., Li, H., Ma, S., . . . & Wang, Y. (2017). Artificial intelligence in healthcare: Past, present and future. *Stroke and Vascular Neurology*, *2*(4), 230–243.

Jumper, J., Evans, R., Pritzel, A., Green, T., Figurnov, M., Ronneberger, O., . . . & Hassabis, D. (2021). Highly accurate protein structure prediction with AlphaFold. *nature*, *596*(7873), 583–589.

Kourou, K., Exarchos, T. P., Exarchos, K. P., Karamouzis, M. V., & Fotiadis, D. I. (2015). Machine learning applications in cancer prognosis and prediction. *Computational and Structural Biotechnology Journal*, *13*, 8–17.

Kraemer, M. U., Yang, C. H., Gutierrez, B., Wu, C. H., Klein, B., Pigott, D. M., . . . & Scarpino, S. V. (2020). The effect of human mobility and control measures on the COVID-19 epidemic in China. *Science*, *368*(6490), 493–497.

Lee, J., Yoon, W., Kim, S., Kim, D., Kim, S., So, C. H., & Kang, J. (2020). BioBERT: A pre-trained biomedical language representation model for biomedical text mining. *Bioinformatics*, *36*(4), 1234–1240.

Lee, S. L., O'Connor, T. F., Yang, X., Cruz, C. N., Chatterjee, S., Madurawe, R. D., . . . & Woodcock, J. (2015). Modernizing pharmaceutical manufacturing: From batch to continuous production. *Journal of Pharmaceutical Innovation*, *10*, 191–199.

Libbrecht, M. W., & Noble, W. S. (2015). Machine learning applications in genetics and genomics. *Nature Reviews Genetics*, *16*(6), 321–332.

Litjens, G., Kooi, T., Bejnordi, B. E., Setio, A. A. A., Ciompi, F., Ghafoorian, M., . . . & Sánchez, C. I. (2017). A survey on deep learning in medical image analysis. *Medical Image Analysis*, *42*, 60–88.

Ma, J., Yu, M. K., Fong, S., Ono, K., Sage, E., Demchak, B., . . . & Ideker, T. (2018). Using deep learning to model the hierarchical structure and function of a cell. *Nature Methods*, *15*(4), 290–298.

MacLeod, B. P., Parlane, F. G., Morrissey, T. D., Häse, F., Roch, L. M., Dettelbach, K. E., . . . & Berlinguette, C. P. (2020). Self-driving laboratory for accelerated discovery of thin-film materials. *Science Advances*, *6*(20), eaaz8867.

Min, S., Lee, B., & Yoon, S. (2017). Deep learning in bioinformatics. *Briefings in Bioinformatics*, *18*(5), 851–869.

Miner, A. S., Milstein, A., Schueller, S., Hegde, R., Mangurian, C., & Linos, E. (2016). Smartphone-based conversational agents and responses to questions about mental health, interpersonal violence, and physical health. *JAMA Internal Medicine*, *176*(5), 619–625.

Morley, J., Floridi, L., Kinsey, L., & Elhalal, A. (2020). From what to how: An initial review of publicly available AI ethics tools, methods and research to translate principles into practices. *Science and Engineering Ethics*, *26*(4), 2141–2168.

Obermeyer, Z., Powers, B., Vogeli, C., & Mullainathan, S. (2019). Dissecting racial bias in an algorithm used to manage the health of populations. *Science*, *366*(6464), 447–453.

Parthiban, S., Vijeesh, T., Gayathri, T., Shanmugaraj, B., Sharma, A., & Sathishkumar, R. (2023). Artificial intelligence-driven systems engineering for next-generation plant-derived biopharmaceuticals. *Frontiers in Plant Science*, *14*, 1252166.

Rajkomar, A., Oren, E., Chen, K., Dai, A. M., Hajaj, N., Hardt, M., . . . & Dean, J. (2018). Scalable and accurate deep learning with electronic health records. *NPJ Digital Medicine*, *1*(1), 18.

Rathore, A. S., Nikita, S., Thakur, G., & Mishra, S. (2023). Artificial intelligence and machine learning applications in biopharmaceutical manufacturing. *Trends in Biotechnology*, *41*(4), 497–510.

Rieke, N., Hancox, J., Li, W., Milletari, F., Roth, H. R., Albarqouni, S., . . . & Cardoso, M. J. (2020). The future of digital health with federated learning. *NPJ Digital Medicine*, *3*(1), 119.

Samek, W., Wiegand, T., & Müller, K. R. (2017). Explainable artificial intelligence: Understanding, visualizing and interpreting deep learning models. *arXiv Preprint arXiv:1708.08296*.

Segler, M. H., Preuss, M., & Waller, M. P. (2018). Planning chemical syntheses with deep neural networks and symbolic AI. *Nature*, *555*(7698), 604–610.

Sendak, M. P., D'Arcy, J., Kashyap, S., Gao, M., Nichols, M., Corey, K., . . . & Balu, S. (2020). A path for translation of machine learning products into healthcare delivery. *EMJ Innovations*, *10*, 19–00172.

Shabani, M., & Marelli, L. (2019). Re-identifiability of genomic data and the GDPR: Assessing the re-identifiability of genomic data in light of the EU general data protection regulation. *EMBO Reports, 20*(6), e48316.

Topol, E. (2019a). *Deep Medicine: How Artificial Intelligence Can Make Healthcare Human Again*. Hachette UK.

Topol, E. J. (2019b). High-performance medicine: The convergence of human and artificial intelligence. *Nature Medicine, 25*(1), 44–56.

Zhavoronkov, A., Ivanenkov, Y. A., Aliper, A., Veselov, M. S., Aladinskiy, V. A., Aladinskaya, A. V., . . . & Aspuru-Guzik, A. (2019). Deep learning enables rapid identification of potent DDR1 kinase inhibitors. *Nature Biotechnology, 37*(9), 1038–1040.

Zhou, J., & Troyanskaya, O. G. (2015). Predicting effects of noncoding variants with deep learning-based sequence model. *Nature Methods, 12*(10), 931–934.

Priti Duhan*, Kumkum Verma, Poonam Bansal, Anil Kumar Sharma, Anupam Sharma, Pooja Verma

Chapter 10
Artificial Intelligence: Transforming the Landscape of Computational Biology

Abstract: Artificial intelligence (AI) has emerged as a transformative force in computational biology, enabling the analysis of complex biological systems with unprecedented speed and accuracy. This review explores the integration of AI techniques, including machine learning, deep learning, and data-driven modeling, into key areas such as genomics, proteomics, systems biology, and drug discovery. By leveraging large-scale biological datasets, AI-driven approaches have enhanced pattern recognition, functional annotation, and predictive modeling, overcoming limitations of traditional computational methods. This chapter also discusses recent advancements in algorithm development, data integration, and interpretability, highlighting their impact on biological research and clinical applications. Additionally, current challenges such as data quality, model transparency, ethical considerations, and reproducibility are examined. By synthesizing recent progress and future prospects, this review underscores the pivotal role of AI in reshaping computational biology and accelerating biological insights and biomedical innovation.

Keywords: Artificial intelligence, computational biology, machine learning, bioinformatics, systems biology

10.1 Introduction

In recent years, artificial intelligence (AI) has appeared as a transformative power across scientific disciplines, notably reshaping the domain of computational biology. The proficiencies of machine learning (ML), deep learning (DL), natural language proc-

*Corresponding author: **Priti Duhan**, Department of Biotechnology, K. L. Mehta Dayanand College for Women, Faridabad, Haryana, India, e-mail: pduhanbiotech@kuk.ac.in
Kumkum Verma, ICAR-IARI, Regional Station, Karnal, Haryana, India
Poonam Bansal, Department of Biosciences and Technology, MMEC, Maharishi Markandeshwar (Deemed to be University), Mullana, Ambala, Haryana, India
Anil Kumar Sharma, Department of Biotechnology, Amity University, Sector 82A, IT City Road, Sahibzada Ajit Singh Nagar 140306, Punjab, India.
Anupam Sharma, Department of Physics, Guru Kashi University, Talwandi Sabo, Bathinda 151302, Punjab, India
Pooja Verma, ICAR-IIWBR, Karnal, Haryana, India

 | https://doi.org/10.1515/9783112214275-010

essing, and other AI paradigms have enabled researchers to analyze complex biological data with unprecedented precision and scale. The incorporation of AI into computational biology is fast-tracking the discoveries in genomics, proteomics, drug discovery, personalized medicine, and systems biology, offering novel solutions to previously intractable biological challenges. Computational biology traditionally relies on mathematical modeling, algorithmic approaches, and high-performance computing to interpret biological systems. However, the arrival of high-proficiency technologies, such as next-generation sequencing, mass spectrometry, and single-cell omics, has resulted in massive, multidimensional datasets that challenge conventional computational methods. AI, particularly ML and DL, offers advanced tools to mine these multifaceted datasets, expose hidden patterns, calculate biological outcomes, and generate new hypotheses. AI algorithms can model nonlinear, high-dimensional biological relationships that are often inaccessible to traditional statistical approaches (Libbrecht & Noble, 2015). Furthermore, improvements in AI hardware, such as graphical processing units and cloud computing, have made it feasible to apply sophisticated models to large-scale biological problems (Ching et al., 2018). AI's impact on life sciences spans multiple domains. AI-driven models have greatly enhanced genome annotation, variant calling, and functional genomics. DL architectures such as convolutional neural networks (CNNs) and recurrent neural networks (RNNs) have been used to predict regulatory elements, splicing patterns, and chromatin accessibility (Zhou & Troyanskaya, 2015; Kelley et al., 2016). Tools like DeepVariant (Poplin et al., 2018) outperform traditional pipelines in variant detection accuracy. AI algorithms are revolutionizing protein structure prediction, as exemplified by DeepMind's AlphaFold, which forecasts protein 3D structures with near-experimental accuracy (Jumper et al., 2021). In proteomics, ML techniques are used for peptide identification, posttranslational modification prediction, and mass spectrometry data analysis (Tran et al., 2019). AI hastens drug discovery by identifying novel drug targets, predicting drug-target interactions, and optimizing molecular properties (Vamathevan et al., 2019). Companies like In-silico Medicine and Benevolent AI leverage AI to shorten the drug development timeline significantly. In precision medicine, AI mixes genomic, clinical, and environmental data to tailor patient-specific remedies (Esteva et al., 2019). AI models help in decrypting multifaceted biological networks, including gene regulatory networks, protein-protein interaction networks, and metabolic paths (Greene et al., 2015). Reinforcement learning and graph neural networks (GNNs) are increasingly applied to model dynamic cellular systems. Computer vision, a subfield of AI, has dramatically improved medical image analysis. Algorithms now surpass human experts in diagnosing certain cancers, retinal diseases, and neurological disorders based on imaging data (Ardila et al., 2019; Gulshan et al., 2016).

The following review aims to explore the existing applications of AI in various subfields of computational biology. It will highlight the transformative influence of AI-driven methodologies on biological data analysis, interpretation, and discovery. Recent breakthroughs and case studies illustrating the role of AI in genomics, proteomics, metabolomics, the discovery of drugs, and systems biology will also be discussed. Moreover, the limitations, challenges, and ethical considerations associated with the deployment of AI in biological research will also be examined alongwith its emerging trends and future directions where AI is likely to further revolutionize computational biology.

10.2 Foundations of AI in Computational Biology

AI in computational biology primarily draws upon two major subfields: ML and DL. Both fields let computers discover patterns from data and make decisions or predictions without having to be told how to do each task.

10.2.1 Machine Learning and Deep Learning Basics

10.2.1.1 Machine Learning (ML)

ML algorithms acquire input data to build predictive models. In computational biology, ML is used for tasks like classification (e.g., disease vs. healthy), regression (e.g., gene expression level prediction), and clustering (e.g., grouping cells by phenotype). Commonly used ML models comprise decision trees, random forests, and support vector machines, in addition to gradient boosting algorithms (Libbrecht & Noble, 2015).

10.2.1.2 Deep Learning (DL)

DL is a subsection of ML that hires artificial neural linkages with manifold layers to model complex, hierarchical interactions between data. DL excels at tasks involving large, high-dimensional, and unstructured datasets such as sequences of genomes, protein structures, and data imaging (Ching et al., 2018). DL models contain CNNs, RNNs, long short-term memory networks, and transformers. The flexibility and scalability of DL models make them applicable for addressing the challenges that are being posed by modern high-throughput biological data.

10.2.2 Common AI Algorithms Used

Several AI algorithms have become staples in computational biology due to their robustness, interpretability, and performance on diverse biological tasks, as discussed in Table 10.1.

Table 10.1: Common AI algorithms used in computational biology.

Algorithm	Applications in computational biology	References
Neural networks (CNNs)	Genomic sequence annotation, regulatory element prediction, and protein structure prediction	Zhou and Troyanskaya (2015) and Jumper et al. (2021)
Recurrent neural networks (RNNs), LSTM	Modeling biological sequences, transcriptomics, and time-series omics data	Singh et al. (2016)
Transformers (e.g., BERT, AlphaFold's Evoformer)	Protein folding and language modeling of DNA/RNA/protein sequences	Rives et al. (2021) and Jumper et al. (2021)
Autoencoders	Dimensionality reduction and noise reduction in omics data	Tan et al. (2015)
Generative adversarial networks (GANs)	Data augmentation, synthetic biology design, and image synthesis	Lan et al. (2020)

10.2.3 Integration with Bioinformatics Pipelines

AI models are increasingly integrated into bioinformatics pipelines, enhancing both the efficiency and accuracy of biological data analysis. AI assists in cleaning, transforming, and normalizing high-throughput biological data to reduce noise and bias (Lan et al., 2018). Techniques like autoencoders, PCA, and recursive feature elimination identify key biological features while reducing data complexity (Hu et al., 2022). AI-driven predictive models classify diseases, predict drug responses, and infer gene functions directly within existing bioinformatics pipelines (Libbrecht & Noble, 2015). AI enables the incorporation of genomics, transcriptomics, proteomics, metabolomics, and clinical data for a holistic systems biology view (Huang et al., 2017). AI-powered tools facilitate the visualization of complex biological networks, pathways, and molecular interactions (Greene et al., 2015). AI is increasingly embedded into end-to-end platforms that automate sample processing, quality control, data analysis, and result reporting, enabling scalable and reproducible research (Pal et al., 2020).

10.3 Applications of AI in Computational Biology

AI has rapidly advanced the ability to analyze complex biological data, allowing researchers to explore systems and mechanisms that were previously intractable. Given below are the key areas where AI is making transformative contributions.

10.3.1 Genomics and Transcriptomics

The beginning of better sequencing technologies has instigated massive volumes of genomic and transcriptomic data. AI represents a critical role in deciphering this information through:

a) **Variant calling and genome annotation:** AI-based tools such as Google DeepVariant use DL to categorize genetic variants with higher precision than traditional approaches (Poplin et al., 2018).
b) **Functional genomics:** Models like DeepSEA (Zhou & Troyanskaya, 2015) and Basset (Kelley et al., 2016) predict regulatory element activity and the impact of noncoding variants.
c) **Splicing and isoform prediction:** RNN-based models predict alternative splicing patterns from genomic sequences (Jaganathan et al., 2019).
d) **Gene expression prediction:** DeepChrome uses histone modification data to calculate gene expression levels using CNNs (Singh et al., 2016).
e) **Single-cell omics:** AI models cluster and classify cell populations on the basis of single-cell RNA-seq data (Aibar et al., 2017).

10.3.2 Proteomics and Structural Biology

Proteins are central to all cellular processes, and AI is revolutionizing our understanding of their structure and function. AlphaFold2 by DeepMind predicts protein 3D arrangements with atomic-level accuracy, solving a decades-old problem in structural biology (Jumper et al., 2021). ML algorithms analyze mass spectrometry data to improve peptide-spectrum matching and quantify proteins (Tran et al., 2019). AI models predict sites of phosphorylation, methylation, and ubiquitination (Li et al., 2020). DL frameworks predict PPI networks that inform drug discovery and pathway modeling (Hashemifar et al., 2018).

10.3.3 Systems Biology and Network Analysis

AI enables the incorporation of diverse datasets to model multifarious biological systems holistically. ML algorithms reconstruct regulatory networks from multi-omics data (Marbach et al., 2016). GNNs model complex PPI networks for pathway analysis and disease mechanism discovery (Zitnik & Leskovec, 2017). Reinforcement learning and DL are used to model metabolic flux and pathway optimization (King et al., 2016). AI helps simulate cell signaling pathways and model cellular dynamics over time (Eduati et al., 2018).

10.3.4 AI in Drug Discovery and Development

Drug discovery is amongst the most encouraging fields for AI applications, offering the potential to reduce costs and accelerate timelines. AI integrates genomic, proteomic, and clinical data to classify new drug targets (Vamathevan et al., 2019). Generative models such as generative adversarial networks and variational autoencoders design unique molecular organizations with desired biological properties (Zhavoronkov et al., 2019). ML models forecast binding affinity among molecules and protein targets (Öztürk et al., 2018). AI optimizes patient recruitment, stratification, and trial design, enhancing trial success rates (Wong et al., 2019). ML algorithms predict adverse drug reactions based on molecular properties and patient data (Xu et al., 2015).

10.3.5 Personalized and Precision Medicine

AI facilitates the evolution from one-size-fits-all dealings to individualized therapies. AI analyzes patient genomes to inform targeted therapies (Kourou et al., 2015). ML models predict subtypes of tumorous cells and treatment responses based on multi-omics data (Chaudhary et al., 2018). AI predicts how individual genetic variations affect drug metabolism and efficacy (Whirl-Carrillo et al., 2012). AI integrates data from wearable devices with clinical and genomic data for continuous health monitoring. AI assists clinicians by providing evidence-based references based on patient-specific data (Esteva et al., 2019).

10.4 AI in Biomedical Image Analysis

Biomedical image analysis is amongst the most rapidly advancing fields of AI application in life sciences. By leveraging DL and computer vision, AI systems can process complex imaging data with high precision, improving both research and clinical decision-making, as discussed in Figure 10.1.

10.4.1 Microscopy Image Interpretation

Microscopy is essential for understanding cell and tissue morphology, subcellular structures, and biological processes. The integration of AI allows for automated, scalable, and more objective image analysis. CNNs like U-Net and Mask R-CNN accurately segment cells and organelles in complex microscopic images (Ronneberger et al., 2015; He et al., 2017). Tools such as Cellpose and DeepCell have become widely used for robust segmentation across diverse imaging modalities. AI models classify cell types, states, and subcellular phenotypes in high-content screening assays (Kraus et al., 2017). This facilitates drug screening and functional genomics. DL models enhance resolution in low-quality or noisy images, such as fluorescence microscopy data (Weigert et al., 2018), improving visualization without hardware upgrades. ML models extract quantitative features (size, shape, texture) from cells and tissues, enabling statistical analysis of morphology (Christiansen et al., 2018).

10.4.2 Medical Imaging and Diagnostics

AI is transforming diagnostic radiology, pathology, and other medical imaging disciplines by providing highly accurate, rapid, and reproducible analyses. AI algorithms detect and classify abnormalities in CT, MRI, ultrasound, and X-ray images. For example, Google's DL model has achieved an expert-level performance in the detection of lung cancer from low-dose CT scans (Ardila et al., 2019). AI-powered image analysis systems can automatically classify tumor subtypes, grade cancer severity, and predict patient prognosis from digitized pathology slides (Coudray et al., 2018). AI models developed by DeepMind have achieved human-level accuracy in diagnosing retinopathy in diabetic patients and age-related macular degeneration from retinal scans (Gulshan et al., 2016; De Fauw et al., 2018). DL models interpret echocardiograms, segment cardiac structures, and detect arrhythmias from electrocardiogram data (Hannun et al., 2019). CNNs accurately classify skin lesions and predict malignancy, often matching dermatologist-level performance (Esteva et al., 2017).

10.4.3 Spatial Transcriptomics

Spatial transcriptomics integrates gene expression profiling with spatial localization, enabling researchers to map transcriptional activity within intact tissues. AI models deconvolute gene expression at each spatial capture spot to estimate the composition of different cell types (Cable et al., 2022). DL models identify spatial domains and tissue architectures based on both expression and histology data (Andersson et al., 2020). AI frameworks integrate 3D transcriptomics with single-cell RNA-sequences, histology, and imaging mass cytometry to generate comprehensive tissue atlases (Dai et al., 2022). In cancer, spatial AI models predict how gene expression patterns correlate with tumor microenvironments, aiding in prognosis and therapy design (Moncada et al., 2020).

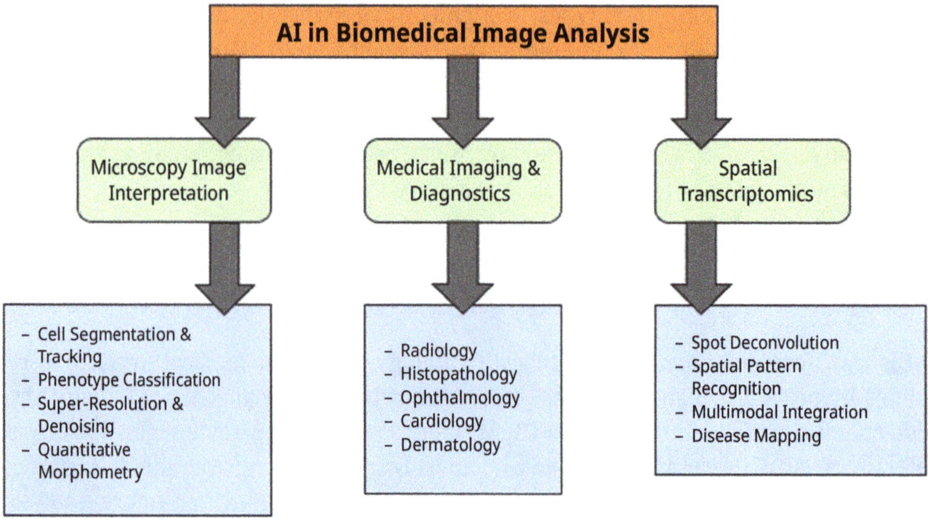

Figure 10.1: AI enhances biomedical image analysis by enabling accurate, automated detection and diagnosis.

10.5 AI for Multi-omics Data Integration

The complication of biological systems ascends from the interchange of multiple molecular layers, including genomics, transcriptomics, proteomics, metabolomics, and epigenomics. Multi-omics amalgamation provides a holistic understanding of biological functions, but poses many significant analytical challenges. AI methods, especially ML and DL, have emerged as prevailing tools to address these complexities.

10.5.1 Challenges of High-Dimensional Data

Multi-omics datasets are inherently high-dimensional, heterogeneous, and sparse, introducing several obstacles for traditional computational approaches. The number of features (genes, proteins, metabolites) frequently exceeds the number of samples, increasing the threat of over-fitting and reducing model generalizability (Li et al., 2019). Different omics layers have distinct data structures, scales, and noise levels (Huang et al., 2017). For example, transcriptomic data is count-based, while proteomic data may be continuous and noisy. Multi-omics studies often suffer from incomplete data across modalities for certain samples, requiring robust imputation or data fusion methods (Liu et al., 2023). Capturing meaningful interactions across omics layers (e.g., gene-protein-metabolite interactions) is nontrivial due to nonlinear and context-specific dependencies (Huang et al., 2017). AI-based approaches, particularly DL, propose promising solutions to overcome these challenges through effective feature learning, dimensionality reduction, and data fusion.

10.5.2 Fusion Models and Representation Learning

10.5.2.1 Fusion Models

AI models for multi-omics integration generally adopt one of three strategies: early fusion, intermediate fusion, or late fusion (Menyhárt & Győrffy, 2021):

Early fusion: This concatenates raw or preprocessed features from unrelated omics layers into a single input vector. This approach is simple but may struggle with heterogeneity (Zhang et al., 2018).

Intermediate fusion: Learns modality-specific representations through separate neural network branches before integrating them via shared layers, capturing inter-omics relationships.

Late fusion: Independently trains models on each omics layer and combines their outputs through ensemble methods or meta-learning (Zitnik et al., 2019).

10.5.2.2 Representation Learning Techniques

Autoencoders: These denoising and variational autoencoders extract latent representations that capture shared variance across omics types (Ballard et al., 2024).

Generative adversarial networks: These are used for data augmentation and modality translation across omics layers (Marouf et al., 2020).

GNNs: They model interactions between genes, proteins, and metabolites by incorporating prior biological networks (Zitnik & Leskovec, 2017).

Transformer models: These models capture long-range dependencies across omics features and have shown promise in integrating sequential omics data (Ji et al., 2021).

Multi-view learning: These help in simultaneous modeling of multiple data sources to extract shared and modality-specific features (Serra, 2017).

10.5.3 Case Studies and Applications

10.5.3.1 Cancer Subtyping and Prognosis

DeepMO: It integrates multi-omics data using DL for classification of cancer subtypes and survival prediction (Zhang et al., 2018).

MOGONET: It uses GNNs for multi-omics integration in cancer subtype prediction, showing superior performance compared to conventional models (Wang et al., 2021).

10.5.3.2 Disease Biomarker Discovery

AI-driven integration of genomics, transcriptomics, and proteomics has identified multi-omics biomarkers for complicated diseases like Alzheimer's disease (Choi, 2025) and cardiovascular disorders (Usova et al., 2021).

10.5.3.3 Drug Response Prediction

DeepDR: predicts anticancer drug response by integrating genomic, transcriptomic, and pharmacological data (Jiang & Li, 2024).

10.5.3.4 Personalized Medicine

Multi-omics integration powered by AI helps stratify patients for individualized therapy selection, particularly in oncology and autoimmune diseases (Hasin et al., 2017; Correa-Aguila et al., 2022).

10.5.3.5 Rare Disease Diagnosis

AI models combine clinical phenotypes with multi-omics data to advance rare disease identification where conventional single-omics analysis fails (Smith et al., 2016).

10.6 Ethical, Legal, and Societal Implications

As AI increasingly permeates computational biology and healthcare, Ethical, Legal, and Societal Implications have become serious concerns. It is essential to address these issues to guarantee the responsible and equitable positioning of AI technologies while maintaining public trust.

10.6.1 Data Privacy and Security

Biomedical datasets, including genomic, proteomic, and clinical records, contain highly sensitive personal information. Breaches can lead to privacy violations, discrimination, or misuse of data (Shabani & Marelli, 2019). Challenges and solutions associated with biomedical data privacy and security are discussed in Table 10.2.

Table 10.2: Challenges and solutions associated with biomedical data privacy and security.

Category		Description	References
Challenges	Genomic data identification	Anonymized genomic data can be reidentified.	Gymrek et al. (2013)
	Data sharing and confidentiality	Cross-border data sharing raises privacy concerns.	Bonomi et al. (2020)
	Cybersecurity threats	Biomedical data is vulnerable to cyberattacks.	Christofidou et al. (2021)
Solutions	Federated learning	Trains models without sharing raw data.	Li et al. (2020)
	Differential privacy	Adds noise to protect individual data.	Dwork and Roth (2014)
	Secure computation	Protects data during processing.	Bonawitz et al. (2017)

10.6.2 Bias and Fairness in AI Models

AI models are highly sensitive to partialities in the training data, which may result in inadequate performance across different populations.

10.6.2.1 Sources of Bias

Many datasets are skewed toward individuals of European ancestry, limiting model applicability to other ethnic groups (Popejoy & Fullerton, 2016).

Socioeconomic bias: Clinical datasets may underrepresent marginalized groups, affecting generalizability and fairness (Char et al., 2018).

Technical bias: Variations in data collection methods, platforms, and preprocessing pipelines can introduce systemic biases (Obermeyer et al., 2019).

10.6.2.2 Implications

Biased AI tools may yield inaccurate diagnoses, risk predictions, or treatment recommendations for certain subpopulations (Schönberger, 2019). Perceived unfairness can diminish public confidence in AI-driven healthcare systems (Mittelstadt, 2019).

10.6.2.3 Mitigation Strategies

Proactively include diverse populations in biomedical data collection, regular assessment of model outputs for subgroup disparities, and implement fairness constraints during model training (Mehrabi et al., 2021).

10.6.3 Regulatory Considerations

The regulatory landscape for AI in computational biology is evolving but remains fragmented and inconsistent.

10.6.3.1 Current Regulatory Frameworks

The FDA (USA) has developed specific guidance for AI-based Software as a Medical Device, focusing on transparency, performance monitoring, and patient safety (US FDA, 2021). EMA (Europe) develops AI-specific guidelines as part of broader medical device regulation frameworks. GDPR (Europe) imposes strict requirements on data use, sharing, and consent for personal data, including biomedical information (Voigt & Von dem Bussche, 2017).

10.6.3.2 Emerging Challenges

Continuous-learning AI models pose challenges for regulatory oversight due to their evolving nature (Topol, 2019). Many regulations call for explainable AI to ensure clinical accountability, which remains technically challenging for complex DL models (Gunning & Aha, 2019). Discrepancies between national laws hinder cross-border research and development (Conduah et al., 2025).

10.6.3.3 Future Directions

Several proposals suggest the establishment of specialized agencies to oversee AI applications in healthcare and life sciences. Dynamic monitoring of AI systems post-deployment to ensure ongoing safety and effectiveness. Broad inclusion of patients, clinicians, ethicists, and technologists in the regulatory design process.

10.7 Current Challenges and Limitations

In spite of the transformative prospective of AI in computational biology, several technical and practical challenges remain. These boundaries must be carefully addressed to ensure reliable, ethical, and meaningful applications in life sciences.

10.7.1 Data Quality and Availability

Many biological datasets suffer from missing entries, incomplete metadata, or inconsistent annotations, limiting their utility for robust AI training (Dai et al., 2022). Omics datasets often reflect batch effects, platform differences, and experimental noise that introduce biases into AI models, as discussed in Figure 10.2 (Leek et al., 2010). Collecting large, well-annotated datasets is challenging for rare diseases or expensive assays (Bhardwaj et al., 2022).

10.7.1.1 Heterogeneity of Data Sources

Integrating diverse data types (e.g., RNA-seq, mass spectrometry, imaging) requires harmonization across incompatible formats and scales (Hasin et al., 2017). Legal, ethical, and institutional barriers often prevent open sharing of biomedical datasets, limiting global AI research collaborations (Shabani & Marelli, 2019).

10.7.1.2 Solutions Under Development

Development of large, federated multi-omics consortia, advanced imputation techniques using generative models, and standardized data collection and reporting guidelines (Wilkinson et al., 2016, FAIR principles).

10.7.2 Interpretability of AI Models

DL models, especially neural networks, often operate as "black boxes" where the underlying decision logic is opaque, particularly in complex biomedical applications (Gunning & Aha, 2019). Lack of transparency impedes clinical adoption, regulatory approval, and user trust. AI models may detect statistical associations without revealing mechanistic biological insights, limiting their utility for hypothesis generation and translational research (Doshi-Velez & Kim, 2017). Feature importance analysis: Techniques like SHapley Additive exPlanations and LIME help identify key variables manipulating model predictions (Lundberg & Lee, 2017). In transformer-based models, attention weights can highlight influential data regions (Vaswani et al., 2017). Saliency maps and heatmaps are increasingly used in image-based AI for highlighting critical input regions (Simonyan et al., 2013).

10.7.3 Reproducibility and Standardization

The reproducibility crisis includes variations in data preprocessing, model training pipelines, and hyperparameter tuning result in inconsistent findings across studies (Hutson, 2018). Many AI studies lack comprehensive reporting of methods, hyperparameters, and code, impeding reproducibility (Pineau et al., 2021).

Standardization challenges include the fact that few universally accepted benchmark datasets exist for many computational biology tasks, limiting fair model comparison (Walsh et al., 2021). Cross-validation, external validation, and independent cohort testing are often inconsistently applied (Beam & Kohane, 2018).

Initiatives such as the BioCompute Object, FAIR guidelines, and open-source AI frameworks aim to enhance reproducibility and standardization (Simonyan et al., 2016), which increased emphasis on publishing code, pretrained models, and detailed protocols to enable reproducibility (Hutson, 2018).

10.8 Future Directions and Emerging Trends

The convergence of AI and computational biology is still in its early phases, with tremendous opportunities ahead. As technology, policy, and scientific understanding evolve, several key directions and emerging trends are poised to shape the next decade of this interdisciplinary field. Reduced dependence on labeled data: Self-supervised and few-shot learning approaches intend to train AI models with negligible labeled data, which is highly valuable in biology, where labeled datasets are infrequent (Rives et al., 2021). Inspired by models like AlphaFold and ESM, massive pretrained models can capture general biological principles and be fine-tuned for

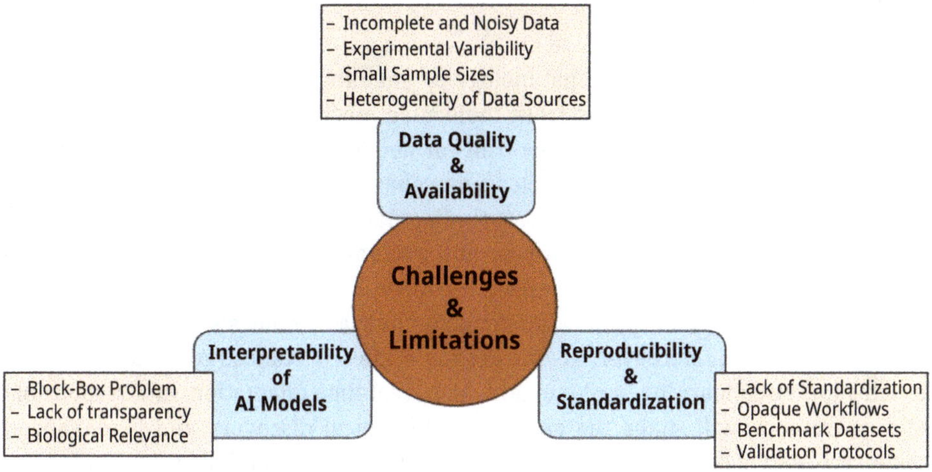

Figure 10.2: Challenges and limitations associated with data quality and availability.

specific tasks (Jumper et al., 2021; Liu et al., 2023). Future AI systems will not only analyze data but also autonomously generate and prioritize hypotheses for experimental validation, accelerating biological discovery (Zitnik et al., 2019). Multimodal fusion: Uniting diverse data types such as genomics, transcriptomics, proteomics, imaging, clinical data, and environmental exposures will enable a truly systems-level understanding of health and disease (Menyhárt & Győrffy, 2021). Comparative AI models leveraging human and model organism data will enhance translational relevance (Bhardwaj et al., 2022). AI systems capable of real-time learning and continuous updates based on newly acquired data can adapt to changing biological knowledge and patient populations (Topol, 2019). Deployment of AI models directly on diagnostic devices and laboratory instruments could enable rapid, on-site decision-making (Li et al., 2019). Cumulative focus will be placed on evolving explainable, equitable, and transparent AI systems that address concerns around bias, fairness, and reproducibility (Mittelstadt, 2019; Mehrabi et al., 2021). Global regulatory harmonization will become critical as AI-based diagnostics and therapeutics are adopted worldwide. Federated learning networks: Collaborative models that preserve privacy while aggregating global data will accelerate biomedical AI research (Li et al., 2020). Growing global efforts toward open-source AI models, reproducible pipelines, and standardized benchmarks will democratize access and foster innovation (Pineau et al., 2021).

10.9 Conclusion

AI is revolutionizing computational biology by enabling unprecedented insights into multifarious biological systems from molecular mechanisms to clinical decision-making. ML and DL algorithms have transformed genomics, proteomics, transcriptomics, drug discovery, structural biology, personalized medicine, and multi-omics integration. However, significant trials, including data quality, model interpretability, reproducibility, ethical considerations, and regulatory oversight, need to be addressed. By addressing these challenges, technical innovation and interdisciplinary collaboration among biologists, data scientists, clinicians, ethicists, and policymakers could be prioritized. Moreover, the integration of these advanced AI paradigms, such as self-supervised learning, multimodal fusion, real-time adaptation, and responsible AI development, will shape the next era of biological discovery and personalized healthcare. As we continue to harness AI's full potential, careful stewardship is critical to ensure that these transformative technologies benefit all individuals and societies, while safeguarding the highest standards of scientific rigor, equity, and transparency.

References

Aibar, S., González-Blas, C. B., Moerman, T., Huynh-Thu, V. A., Imrichova, H., Hulselmans, G., . . . & Aerts, S. (2017). SCENIC: Single-cell regulatory network inference and clustering. *Nature Methods*, *14*(11), 1083–1086.

Andersson, A., Larsson, L., Stenbeck, L., Salmén, F., Ehinger, A., Wu, S., . . . & Lundeberg, J. (2020). Spatial deconvolution of HER2-positive breast tumors reveals novel intercellular relationships. *bioRxiv*, 2020-07.

Ardila, D., Kiraly, A. P., Bharadwaj, S., Choi, B., Reicher, J. J., Peng, L., . . . & Shetty, S. (2019). End-to-end lung cancer screening with three-dimensional deep learning on low-dose chest computed tomography. *Nature Medicine*, *25*(6), 954–961.

Ballard, J. L., Wang, Z., Li, W., Shen, L., & Long, Q. (2024). Deep learning-based approaches for multi-omics data integration and analysis. *BioData Mining*, *17*(1), 38.

Beam, A. L., & Kohane, I. S. (2018). Big data and machine learning in health care. *Jama*, *319*(13), 1317–1318.

Bhardwaj, A., Kishore, S., & Pandey, D. K. (2022). Artificial intelligence in biological sciences. *Life*, *12*(9), 1430.

Bonawitz, K., Ivanov, V., Kreuter, B., Marcedone, A., McMahan, H. B., Patel, S., . . . & Seth, K. (2017 October). Practical secure aggregation for privacy-preserving machine learning. In *Proceedings of the 2017 ACM SIGSAC Conference on Computer and Communications Security* (pp. 1175–1191).

Bonomi, L., Huang, Y., & Ohno-Machado, L. (2020). Privacy challenges and research opportunities for genomic data sharing. *Nature Genetics*, *52*(7), 646–654.

Cable, D. M., Murray, E., Zou, L. S., Goeva, A., Macosko, E. Z., Chen, F., & Irizarry, R. A. (2022). Robust decomposition of cell type mixtures in spatial transcriptomics. *Nature Biotechnology*, *40*(4), 517–526.

Char, D. S., Shah, N. H., & Magnus, D. (2018). Implementing machine learning in health care – Addressing ethical challenges. *New England Journal of Medicine*, *378*(11), 981–983.

Chaudhary, K., Poirion, O. B., Lu, L., & Garmire, L. X. (2018). Deep learning–based multi-omics integration robustly predicts survival in liver cancer. *Clinical Cancer Research, 24*(6), 1248–1259.

Ching, T., Himmelstein, D. S., Beaulieu-Jones, B. K., Kalinin, A. A., Do, B. T., Way, G. P., . . . & Greene, C. S. (2018). Opportunities and obstacles for deep learning in biology and medicine. *Journal of the Royal Society Interface, 15*(141), 20170387.

Choi, J. J. (2025). *Developing and Applying Machine Learning and Artificial Intelligence Approaches to Leverage Multi-Omics to Advance Understanding of Alzheimer's Disease Pathology* (Doctoral dissertation, The University of Wisconsin-Madison).

Christiansen, E. M., Yang, S. J., Ando, D. M., Javaherian, A., Skibinski, G., Lipnick, S., . . . & Finkbeiner, S. (2018). In silico labeling: Predicting fluorescent labels in unlabeled images. *Cell, 173*(3), 792–803.

Christofidou, M., Lea, N., & Coorevits, P. (2021). A literature review on the GDPR, COVID-19 and the ethical considerations of data protection during a time of crisis. *Yearbook of Medical Informatics, 30*(01), 226–232.

Conduah, A. K., Ofoe, S., & Siaw-Marfo, D. (2025). Data privacy in healthcare: Global challenges and solutions. *Digital Health, 11*, 20552076251343959.

Correa-Aguila, R., Alonso-Pupo, N., & Hernández-Rodríguez, E. W. (2022). Multi-omics data integration approaches for precision oncology. *Molecular Omics, 18*(6), 469–479.

Coudray, N., Ocampo, P. S., Sakellaropoulos, T., Narula, N., Snuderl, M., Fenyö, D., . . . & Tsirigos, A. (2018). Classification and mutation prediction from non–small cell lung cancer histopathology images using deep learning. *Nature Medicine, 24*(10), 1559–1567.

Dai, C., Jiang, Y., Yin, C., Su, R., Zeng, X., Zou, Q., . . . & Wei, L. (2022). scIMC: A platform for benchmarking comparison and visualization analysis of scRNA-seq data imputation methods. *Nucleic Acids Research, 50*(9), 4877–4899.

De Fauw, J., Ledsam, J. R., Romera-Paredes, B., Nikolov, S., Tomasev, N., Blackwell, S., . . . & Ronneberger, O. (2018). Clinically applicable deep learning for diagnosis and referral in retinal disease. *Nature Medicine, 24*(9), 1342–1350.

Doshi-Velez, F., & Kim, B. (2017). Towards a rigorous science of interpretable machine learning. *arXiv preprint arXiv:1702.08608*.

Dwork, C., & Roth, A. (2014). The algorithmic foundations of differential privacy. *Foundations and Trends® in Theoretical Computer Science, 9*(3–4), 211–407.

Eduati, F., Utharala, R., Madhavan, D., Neumann, U. P., Longerich, T., Cramer, T., . . . & Merten, C. A. (2018). A microfluidics platform for combinatorial drug screening on cancer biopsies. *Nature Communications, 9*(1), 2434.

Esteva, A., Kuprel, B., Novoa, R. A., Ko, J., Swetter, S. M., Blau, H. M., & Thrun, S. (2017). Dermatologist-level classification of skin cancer with deep neural networks. *nature, 542*(7639), 115–118.

Esteva, A., Robicquet, A., Ramsundar, B., Kuleshov, V., DePristo, M., Chou, K., . . . & Dean, J. (2019). A guide to deep learning in healthcare. *Nature Medicine, 25*(1), 24–29.

Greene, C. S., Krishnan, A., Wong, A. K., Ricciotti, E., Zelaya, R. A., Himmelstein, D. S., . . . & Troyanskaya, O. G. (2015). Understanding multicellular function and disease with human tissue-specific networks. *Nature Genetics, 47*(6), 569–576.

Gulshan, V., Peng, L., Coram, M., Stumpe, M. C., Wu, D., Narayanaswamy, A., . . . & Webster, D. R. (2016). Development and validation of a deep learning algorithm for detection of diabetic retinopathy in retinal fundus photographs. *jama, 316*(22), 2402–2410.

Gunning, D., & Aha, D. (2019). DARPA's explainable artificial intelligence (XAI) program. *AI Magazine, 40*(2), 44–58.

Gymrek, M., McGuire, A. L., Golan, D., Halperin, E., & Erlich, Y. (2013). Identifying personal genomes by surname inference. *Science, 339*(6117), 321–324.

Hannun, A. Y., Rajpurkar, P., Haghpanahi, M., Tison, G. H., Bourn, C., Turakhia, M. P., & Ng, A. Y. (2019). Cardiologist-level arrhythmia detection and classification in ambulatory electrocardiograms using a deep neural network. *Nature Medicine*, *25*(1), 65–69.

Hashemifar, S., Neyshabur, B., Khan, A. A., & Xu, J. (2018). Predicting protein–protein interactions through sequence-based deep learning. *Bioinformatics*, *34*(17), i802–i810.

Hasin, Y., Seldin, M., & Lusis, A. (2017). Multi-omics approaches to disease. *Genome Biology*, *18*, 1–15.

He, K., Gkioxari, G., Dollár, P., & Girshick, R. (2017).Mask r-cnn. In *Proceedings of the IEEE International Conference on Computer Vision* (pp. 2961–2969).

Hu, Y., Zhao, L., Li, Z., Dong, X., Xu, T., & Zhao, Y. (2022). Classifying the multi-omics data of gastric cancer using a deep feature selection method. *Expert Systems with Applications*, *200*, 116813.

Huang, S., Chaudhary, K., & Garmire, L. X. (2017). More is better: Recent progress in multi-omics data integration methods. *Frontiers in Genetics*, *8*, 84.

Hutson, M. (2018). *Artificial Intelligence Faces Reproducibility Crisis*.

Jaganathan, K., Panagiotopoulou, S. K., McRae, J. F., Darbandi, S. F., Knowles, D., Li, Y. I., . . . & Farh, K. K. H. (2019). Predicting splicing from primary sequence with deep learning. *Cell*, *176*(3), 535–548.

Ji, Y., Zhou, Z., Liu, H., & Davuluri, R. V. (2021). DNABERT: Pre-trained bidirectional encoder representations from transformers model for DNA-language in genome. *Bioinformatics*, *37*(15), 2112–2120.

Jiang, Z., & Li, P. (2024). DeepDR: A deep learning library for drug response prediction. *Bioinformatics*, *40*(12), btae688.

Jumper, J., Evans, R., Pritzel, A., Green, T., Figurnov, M., Ronneberger, O., . . . & Hassabis, D. (2021). Highly accurate protein structure prediction with AlphaFold. *nature*, *596*(7873), 583–589.

Kelley, D. R., Snoek, J., & Rinn, J. L. (2016). Basset: Learning the regulatory code of the accessible genome with deep convolutional neural networks. *Genome Research*, *26*(7), 990–999.

King, Z. A., Lu, J., Dräger, A., Miller, P., Federowicz, S., Lerman, J. A., . . . & Lewis, N. E. (2016). BiGG Models: A platform for integrating, standardizing and sharing genome-scale models. *Nucleic Acids Research*, *44*(D1), D515–D522.

Kourou, K., Exarchos, T. P., Exarchos, K. P., Karamouzis, M. V., & Fotiadis, D. I. (2015). Machine learning applications in cancer prognosis and prediction. *Computational and Structural Biotechnology Journal*, *13*, 8–17.

Kraus, O. Z., Grys, B. T., Ba, J., Chong, Y., Frey, B. J., Boone, C., & Andrews, B. J. (2017). Automated analysis of high-content microscopy data with deep learning. *Molecular Systems Biology*, *13*(4), 924.

Lan, K., Wang, D. T., Fong, S., Liu, L. S., Wong, K. K., & Dey, N. (2018). A survey of data mining and deep learning in bioinformatics. *Journal of Medical Systems*, *42*, 1–20.

Lan, L., You, L., Zhang, Z., Fan, Z., Zhao, W., Zeng, N., . . . & Zhou, X. (2020). Generative adversarial networks and its applications in biomedical informatics. *Frontiers in Public Health*, *8*, 164.

Leek, J. T., Scharpf, R. B., Bravo, H. C., Simcha, D., Langmead, B., Johnson, W. E., . . . & Irizarry, R. A. (2010). Tackling the widespread and critical impact of batch effects in high-throughput data. *Nature Reviews Genetics*, *11*(10), 733–739.

Li, T., Sahu, A. K., Talwalkar, A., & Smith, V. (2020). Federated learning: Challenges, methods, and future directions. *IEEE Signal Processing Magazine*, *37*(3), 50–60.

Li, Y., Huang, C., Ding, L., Li, Z., Pan, Y., & Gao, X. (2019). Deep learning in bioinformatics: Introduction, application, and perspective in the big data era. *Methods*, *166*, 4–21.

Libbrecht, M. W., & Noble, W. S. (2015). Machine learning applications in genetics and genomics. *Nature Reviews Genetics*, *16*(6), 321–332.

Liu, M., Li, S., Yuan, H., Ong, M. E. H., Ning, Y., Xie, F., . . . & Liu, N. (2023). Handling missing values in healthcare data: A systematic review of deep learning-based imputation techniques. *Artificial Intelligence in Medicine*, *142*, 102587.

Lundberg, S. M., & Lee, S. I. (2017). A unified approach to interpreting model predictions. *Advances in Neural Information Processing Systems*, 30, 4768–4777.

Marbach, D., Lamparter, D., Quon, G., Kellis, M., Kutalik, Z., & Bergmann, S. (2016). Tissue-specific regulatory circuits reveal variable modular perturbations across complex diseases. *Nature Methods, 13*(4), 366–370.

Marouf, M., Machart, P., Bansal, V., Kilian, C., Magruder, D. S., Krebs, C. F., & Bonn, S. (2020). Realistic in silico generation and augmentation of single-cell RNA-seq data using generative adversarial networks. *Nature Communications, 11*(1), 166.

Mehrabi, N., Morstatter, F., Saxena, N., Lerman, K., & Galstyan, A. (2021). A survey on bias and fairness in machine learning. *ACM Computing Surveys (CSUR), 54*(6), 1–35.

Menyhárt, O., & Győrffy, B. (2021). Multi-omics approaches in cancer research with applications in tumor subtyping, prognosis, and diagnosis. *Computational and Structural Biotechnology Journal, 19*, 949–960.

Mittelstadt, B. (2019). Principles alone cannot guarantee ethical AI. *Nature Machine Intelligence, 1*(11), 501–507.

Moncada, R., Barkley, D., Wagner, F., Chiodin, M., Devlin, J. C., Baron, M., . . . & Yanai, I. (2020). Integrating microarray-based spatial transcriptomics and single-cell RNA-seq reveals tissue architecture in pancreatic ductal adenocarcinomas. *Nature Biotechnology, 38*(3), 333–342.

Obermeyer, Z., Powers, B., Vogeli, C., & Mullainathan, S. (2019). Dissecting racial bias in an algorithm used to manage the health of populations. *Science, 366*(6464), 447–453.

Öztürk, H., Özgür, A., & Ozkirimli, E. (2018). DeepDTA: Deep drug–target binding affinity prediction. *Bioinformatics, 34*(17), i821–i829.

Pal, S., Mondal, S., Das, G., Khatua, S., & Ghosh, Z. (2020). Big data in biology: The hope and present-day challenges in it. *Gene Reports, 21*, 100869.

Pineau, J., Vincent-Lamarre, P., Sinha, K., Larivière, V., Beygelzimer, A., d'Alché-Buc, F., . . . & Larochelle, H. (2021). Improving reproducibility in machine learning research (a report from the neurips 2019 reproducibility program). *Journal of Machine Learning Research, 22*(164), 1–20.

Popejoy, A. B., & Fullerton, S. M. (2016). Genomics is failing on diversity. *Nature, 538*(7624), 161–164.

Poplin, R., Chang, P. C., Alexander, D., Schwartz, S., Colthurst, T., Ku, A., . . . & DePristo, M. A. (2018). A universal SNP and small-indel variant caller using deep neural networks. *Nature Biotechnology, 36*(10), 983–987.

Rives, A., Meier, J., Sercu, T., Goyal, S., Lin, Z., Liu, J., . . . & Fergus, R. (2021). Biological structure and function emerge from scaling unsupervised learning to 250 million protein sequences. *Proceedings of the National Academy of Sciences, 118*(15), e2016239118.

Ronneberger, O., Fischer, P., & Brox, T. (2015). U-net: Convolutional networks for biomedical image segmentation. In *Medical Image Computing and Computer-assisted intervention–MICCAI 2015: 18th International Conference, Munich, Germany, October 5–9, 2015, Proceedings, Part III 18* (pp. 234–241). Springer international publishing.

Schönberger, D. (2019). Artificial intelligence in healthcare: A critical analysis of the legal and ethical implications. *International Journal of Law and Information Technology, 27*(2), 171–203.

Serra, A. (2017). *Multi-view Learning and Data Integration for Omics Data.*

Shabani, M., & Marelli, L. (2019). Re-identifiability of genomic data and the GDPR: Assessing the re-identifiability of genomic data in light of the EU general data protection regulation. *EMBO Reports, 20*(6), e48316.

Simonyan, K., Vedaldi, A., & Zisserman, A. (2013). Deep inside convolutional networks: Visualising image classification models and saliency maps. *arXiv preprint arXiv:1312.6034.*

Simonyan, V., Goecks, J., & Mazumder, R. (2016). Biocompute objects – A step towards evaluation and validation of biomedical scientific computations. *PDA Journal of Pharmaceutical Science and Technology, 71*(2), 136.

Singh, R., Lanchantin, J., Robins, G., & Qi, Y. (2016). DeepChrome: Deep-learning for predicting gene expression from histone modifications. *Bioinformatics*, *32*(17), i639–i648.

Smith, L. D., Willig, L. K., & Kingsmore, S. F. (2016). Whole-exome sequencing and whole-genome sequencing in critically ill neonates suspected to have single-gene disorders. *Cold Spring Harbor Perspectives in Medicine*, *6*(2), a023168.

Tan, J., Ung, M., Cheng, C., & Greene, C. S. (2015). Unsupervised feature construction and knowledge extraction from genome-wide assays of breast cancer with denoising autoencoders. In *Pacific Symposium on Biocomputing. Pacific Symposium on Biocomputing* (Vol. 20, p. 132).

Topol, E. J. (2019). High-performance medicine: The convergence of human and artificial intelligence. *Nature Medicine*, *25*(1), 44–56.

Tran, N. H., Qiao, R., Xin, L., Chen, X., Liu, C., Zhang, X., . . . & Li, M. (2019). Deep learning enables de novo peptide sequencing from data-independent-acquisition mass spectrometry. *Nature Methods*, *16*(1), 63–66.

Usova, E. I., Alieva, A. S., Yakovlev, A. N., Alieva, M. S., Prokhorikhin, A. A., Konradi, A. O., . . . & Baragetti, A. (2021). Integrative analysis of multi-omics and genetic approaches – A new level in atherosclerotic cardiovascular risk prediction. *Biomolecules*, *11*(11), 1597.

Vamathevan, J., Clark, D., Czodrowski, P., Dunham, I., Ferran, E., Lee, G., . . . & Zhao, S. (2019). Applications of machine learning in drug discovery and development. *Nature Reviews Drug Discovery*, *18*(6), 463–477.

Vaswani, A., Shazeer, N., Parmar, N., Uszkoreit, J., Jones, L., Gomez, A. N., . . . & Polosukhin, I. (2017). Attention is all you need. *Advances in Neural Information Processing Systems*, *30*, 6000–6010.

Voigt, P., & von dem Bussche, A. (2017). *The Eu General Data Protection Regulation (Gdpr). A Practical Guide* (Vol. 10(3152676), pp. 10–5555, *1st ed.*). *Cham*: *Springer International Publishing*.

Walsh, I., Fishman, D., Garcia-Gasulla, D., Titma, T., Pollastri, G., Harrow, J., . . . & Tosatto, S. C. (2021). DOME: Recommendations for supervised machine learning validation in biology. *Nature Methods*, *18*(10), 1122–1127.

Wang, T., Shao, W., Huang, Z., Tang, H., Zhang, J., Ding, Z., & Huang, K. (2021). MOGONET integrates multi-omics data using graph convolutional networks allowing patient classification and biomarker identification. *Nature Communications*, *12*(1), 3445.

Weigert, M., Schmidt, U., Boothe, T., Müller, A., Dibrov, A., Jain, A., . . . & Myers, E. W. (2018). Content-aware image restoration: Pushing the limits of fluorescence microscopy. *Nature Methods*, *15*(12), 1090–1097.

Whirl-Carrillo, M., McDonagh, E. M., Hebert, J. M., Gong, L., Sangkuhl, K., Thorn, C. F., . . . & Klein, T. E. (2012). Pharmacogenomics knowledge for personalized medicine. *Clinical Pharmacology & Therapeutics*, *92*(4), 414–417.

Wilkinson, M. D., Dumontier, M., Aalbersberg, I. J., Appleton, G., Axton, M., Baak, A., . . . & Mons, B. (2016). The FAIR Guiding Principles for scientific data management and stewardship. *Scientific Data*, *3*(1), 1–9.

Wong, C. H., Siah, K. W., & Lo, A. W. (2019). Estimation of clinical trial success rates and related parameters. *Biostatistics*, *20*(2), 273–286.

Xu, Y., Dai, Z., Chen, F., Gao, S., Pei, J., & Lai, L. (2015). Deep learning for drug-induced liver injury. *Journal of Chemical Information and Modeling*, *55*(10), 2085–2093.

Zhang, L., Lv, C., Jin, Y., Cheng, G., Fu, Y., Yuan, D., . . . & Shi, T. (2018). Deep learning-based multi-omics data integration reveals two prognostic subtypes in high-risk neuroblastoma. *Frontiers in Genetics*, *9*, 477.

Zhavoronkov, A., Ivanenkov, Y. A., Aliper, A., Veselov, M. S., Aladinskiy, V. A., Aladinskaya, A. V., . . . & Aspuru-Guzik, A. (2019). Deep learning enables rapid identification of potent DDR1 kinase inhibitors. *Nature Biotechnology*, *37*(9), 1038–1040.

Zhou, J., & Troyanskaya, O. G. (2015). Predicting effects of noncoding variants with deep learning–based sequence model. *Nature Methods*, *12*(10), 931–934.

Zitnik, M., & Leskovec, J. (2017). Predicting multicellular function through multi-layer tissue networks. *Bioinformatics*, *33*(14), i190–i198.

Zitnik, M., Nguyen, F., Wang, B., Leskovec, J., Goldenberg, A., & Hoffman, M. M. (2019). Machine learning for integrating data in biology and medicine: Principles, practice, and opportunities. *Information Fusion*, *50*, 71–91.

Vikas Kushwaha, Anuprabha, Varruchi Sharma*, Anupam Sharma, Poonam Bansal, Anil Kumar Sharma*, Nitin Sharma, Ajay Sharma, Imran Sheikh, Punit Puri, Manoj Singh, Milan Bera

Chapter 11
Artificial Intelligence: Pioneering the Future of Synthetic Biology

Abstract: Being an interdisciplinary field, synthetic biology shows synergy between biology, engineering, computer science, and chemistry to design and construct artificial biological systems with enhanced properties and safety features. More recently, synthetic biology has become a powerful tool for manipulating genetic code, cellular processes, and metabolic pathways for creating new biological functionalities, synthetic genomes, valuable compounds, create biofuels, develop novel therapies, or even design organisms with novel applications. The current chapter highlights the recent advancements in the field of synthetic biology with the prominent role of artificial intelligence having many downstream applications particularly in metabolic engineering, in the design and enhancement of microbial strains. The integration of multi-omics data through artificial intelligence encompassing genomics, transcriptomics, proteomics, and metabo-

***Corresponding author: Varruchi Sharma,** Department of Biotechnology and Bioinformatics, Sri Guru Gobind Singh College, Sector 26 Chandigarh 160019, India, e-mail: Varruchi.sharma2013@gmail.com
***Corresponding author: Anil kumar Sharma,** Department of Biotechnology, School of Biological Sciences, Amity University, Mohali 140306, Punjab, India, e-mail: anibiotech18@gmail.com
Vikas Kushwaha, Department of Biotechnology, Goswami Ganesh Dutta Sanatan Dharma College, Sector 32, Chandigarh
Anuprabha, Department of Botany, Goswami Ganesh Dutta Sanatan Dharma College, Sector 32, Chandigarh
Anupam Sharma, Department of Physics, Guru Kashi University, Talwandi Sabo, Bathinda, 151302, Punjab, India
Poonam Bansal, Department of Bio-sciences and Technology, MMEC, Maharishi Markandeshwar (Deemed to be University), Mullana, Ambala, 133207, India.
Nitin Sharma, Department of Biotechnology, Chandigarh College of Technology, CGC, Landran, Mohali 140307, Punjab, India, e-mail: abhinitu30@gmail.com
Ajay Sharma, Department of Chemistry, Career Point University, Tikker-Kharwarian, Hamirpur, 176041, Himachal Pradesh, India [2]Centre for Nano-Science and Technology, Career Point University, Tikker-Kharwarian, Hamirpur 176041, Himachal Pradesh, India
Imran Sheikh, UCRD, Chandigarh University, Mohali
Punit Puri, Department of Zoology, DAV College Jalandhar, Punjab, India
Manoj Singh, Department of Bio-sciences and Technology, MMEC, Maharishi Markandeshwar (Deemed to be University), Mullana, Ambala, 133207, India.
Milan Bera, Department of Physics, MMEC, Maharishi Markandeshwar (Deemed to be University), Mullana, Ambala, 133207, India

 | https://doi.org/10.1515/9783112214275-011

lomics will be extremely helpful in comprehending the functionality of engineered cells across various biological dimensions.

Keywords: Synthetic biology, artificial intelligence, multi-omics, genomics, metabolomics, proteomics, transcriptomics

11.1 Introduction

Synthetic biology is an interdisciplinary field of science which shows synergy between biology, engineering, computer science, and chemistry to design and construct artificial biological systems with enhanced, safe and control functions. Synthetic biology has become a powerful tool for manipulating genetic code, cellular processes, and metabolic pathways for creating new biological functionalities, synthetic genomes, valuable compounds, create biofuels, develop novel therapies, or even design organisms with novel applications (Synthetic Biology, 2025). Systems biology uses computational modeling, high-throughput data analysis (omics technologies), and quantitative approaches to study complex interactions among genes, proteins, metabolites, and environmental factors. By integrating these diverse datasets, it provides a holistic view of how biological systems function and respond to perturbations. The transformative potential of systems biology plays a crucial role in medicine, enabling the exploration of disease mechanisms, the discovery of biomarkers, and the development of personalized therapies specific to an individual's molecular profile. In the realm of biotechnology, the integration of systems biology significantly enhances metabolic engineering and synthetic biology. This approach facilitates the meticulous design of microbial pathways, leading to the optimized production of drugs, enzymes, and biofuels. This innovation is revolutionizing agriculture by utilizing predictive models that enhance crop yield, increase stress tolerance, and optimize nutrient use efficiency. In the field of environmental research, it aids in the design of microbial consortia for bioremediation and ecosystem restoration.

The advancement of artificial intelligence (AI) in biotechnology represents a pivotal breakthrough for understanding, creating, and implementing biological systems in a new era in the life sciences. Initially, AI served as computational models and statistical tools that aided in gene sequence analysis, protein folding predictions, and fundamental bioinformatics applications. With advancements in genomic databases and computing capabilities, a new era of machine learning has arisen, significantly amplifying the impact of AI in biotechnology through the utilization of omics data, including genomics, transcriptomics, proteomics, and metabolomics. This enhancement facilitated the recognition of biomarkers, categorization of diseases, and the development of early predictive diagnostics using algorithms like SVMs and Bayesian networks. Moreover, deep learning is transforming the field of systems biology. Innovative models like GANs, VAEs, and transformers have initiated a groundbreaking era of creativity within the field of biotechnology. These models have the capability to create innovative molecules, synthetic proteins, and even suggest enhanced metabolic pathways, effortlessly merging with automated laboratories and synthetic biology processes.

For instance, AI-powered models now aid in the design of innovative metabolic pathways or in forecasting the efficacy of engineered genes, greatly alleviating the trial-and-error challenges that have historically limited the progress of synthetic biology processes. In the healthcare sector, researchers reconstructed nerve cells and achieved a partial restoration of vision for a blind individual in 2021, utilizing a gene for light-sensitive proteins derived from algae. Altos Labs, a biotechnology firm, is utilizing genetic editing techniques to address hereditary heart disorders and exploring longevity projects by reengineering human cells to enhance their health and durability. Three-dimensional bioprinting serves a crucial role in the field of tissue engineering, particularly in replicating a patient's skin, thereby minimizing the potential risks linked to inflammation and immune-system rejection. Biotech firm Oxitec is addressing malaria through the creation of genetically altered male mosquitoes that do not bite, are nontoxic, and are claimed to be harmless to ecosystems (Fernández, 2024).

The agricultural sector is experiencing a transformative shift as the combined application of advanced technologies and biological sciences is enhancing crop improvement strategies. This innovative approach allows for the optimization of intricate traits by merging diverse datasets, including multi-omics, environmental, and phenotypic information. AI algorithms are now capable of accurately predicting the relation between genes and traits, facilitating the strategic development of crops that are high-yielding, resilient to stress, and rich in nutrients (Najafabadi & Jackson, 2025). Furthermore, biotechnology firms like MNDL Bio are utilizing engineered molecules to improve food flavor profiles and nutritional value, paving the way for healthier and more sustainable food sources. The advancements in this field carry significant consequences for worldwide food stability, offering the potential to alleviate hunger, decrease production expenses, and expedite growing periods, thus transforming the landscape of sustainable farming (Fernández, 2024).

In the pharmaceutical sector, the integration of AI with advanced biological modeling is now aiding in the prediction of molecular interactions, the identification of promising drug candidates, and the optimization of therapeutic pathways. The influence of this collaboration permeates contemporary healthcare, ranging from swift drug lead development and target confirmation to precise gene-editing techniques that facilitate highly tailored and individualized therapies. Collectively, these developments are not just enhancing efficiency they are propelling biomedical innovation forward and broadening the limits of what can be biologically accomplished (da Cunha et al., 2025). This powerful convergence also brings new responsibilities. The fusion of AI and synthetic biology is fundamentally reshaping life-science research shifting the paradigm from "observe and test" to "predict, build, and iterate" and holds immense promise for medicine, agriculture, manufacturing, and beyond.

11.2 Synthetic Biology: Concepts and Applications

Synthetic biology, an interdisciplinary discipline, employs engineering concepts to biological systems, paving the way for the creating, testing, and evaluating new genetic components, various pathways, and even whole organisms (Synthetic Biology, 2025). Synthetic biology uses the concept of Design-Build-Test-Learn (DBTL). Design genetic parts (DNA and proteins), build them in cells or in vitro, test how they function, and learn from the results and iterate. Synthetic biology employs engineering principles such as modularity, which involves the connection of cellular components, standardization for the reuse and sharing of parts, and abstraction. This approach facilitates rapid prototyping and promotes the seamless exchange of designs among practitioners globally (Garner, 2021). Biological components of genetic material like terminators, promoters, and genes are meant to be interchangeable "modules" that may be assembled into devices and then incorporated into larger systems (Chen, Galloway & Smolke, 2012). Incorporating tools such as computer modeling, automated DNA synthesis, and robotic experimentation significantly accelerate and enhance the efficiency of the DBTL cycle. Computer models enable researchers to forecast the behavior of a genetic design prior to its creation. The process of automated DNA synthesis efficiently constructs the necessary DNA sequences with remarkable precision. Automated systems conduct experiments with remarkable consistency and accuracy. Collectively, these instruments enhance the efficiency of designing, evaluating, and refining biological constructs, enabling investigators to innovate new biological systems more rapidly (Garner, 2021).

The field of synthetic biology encompasses numerous significant applications. In the field of medicine, researchers can create biological systems capable of identifying symptoms of disease and autonomously responding to them (Chen, Galloway & Smolke, 2012). For instance, cells can be engineered to identify microRNA signatures associated with cancer and subsequently initiate a therapeutic response. Engineered microorganisms have the potential to synthesize valuable pharmaceuticals. Furthermore, systems that operate without intact living cells composed of cellular components are being advanced for rapid and large-scale disease detection as well as for more efficient protein production (Embra, 2024). In the industrial sector, the application of synthetic biology enables the utilization of microbes to produce beneficial products like biofuels, valuable chemicals, and biodegradable materials. Researchers have the ability to alter the natural metabolic pathways in organisms such as yeast or bacteria, enabling them to produce these compounds with high efficiency. Typically, the production of these items involves intricate and expensive chemical methods; however, specially designed microorganisms can accomplish this task in a more straightforward and environmentally friendly manner (Jiang et al., 2025). This method additionally aids in diminishing reliance on fossil fuels, as numerous industrial products can now be produced through biological processes rather than petroleum-derived techniques.

In regards to the environment, synthetic biology enables to create the engineered organisms that assist in environmental remediation efforts. These engineered organisms are modified to identify and decompose detrimental materials like pollutants, heavy metals, and even plastic debris: a phenomenon referred to as bioremediation. This enables the restoration of contaminated soil and water through methods that are both natural and efficient (Karataş and Ayaz, 2025). The application of synthetic biology is contributing to the advancement of sustainable agricultural practices. For instance, microorganisms can be modified to enhance plant growth by fixing nitrogen, thereby decreasing the reliance on chemical fertilizers. Plants can be engineered to enhance their nutritional value, resist pests, or withstand challenging environments such as drought or suboptimal soil conditions. Collectively, these advancements contribute to enhancing food security and minimizing environmental consequences (Wang, Zang and Zhou, 2022).

In general, synthetic biology occupies a significant intersection of biological sciences and engineering principles, paving the way for innovations that have the potential to revolutionize healthcare, industry, and ecological sustainability. By assembling biological systems from the foundational level utilizing modular, standardized components and cutting-edge automation this discipline transcends the mere observation of life. It enables us to reimagine biological systems in manners that can address some of the most pressing issues of our time, ranging from medical therapies to sustainable energy solutions and the revitalization of ecosystems (Chen, Galloway & Smolke, 2012).

11.3 Artificial Intelligence: Concept and Its Applications in Life Sciences

AI refers to computer-aided designing of tools to perform tasks such as learning, reasoning, problem-solving, and recognizing patterns. AI involves algorithms especially machine learning and deep learning to process large amounts of data and make predictions or decisions (Chatterjee, 2022).

In recent years, the landscape of drug discovery and development has been fundamentally transformed by advancements in AI. In the past, the process of discovering new pharmaceuticals was characterized by significant delays and high expenses, frequently spanning several years. Researchers had to examine thousands of chemical compounds in the laboratory to identify just one that could potentially be effective. Today, AI significantly accelerates and enhances this process. Employing advanced computational techniques, AI can swiftly examine and interpret vast chemical datasets in a short period of time. This approach enables the forecasting of interactions between various molecules and particular biological targets, including proteins that play a role in disease processes. Drawing from these forecasts, advanced algorithms can help in the selection of the most viable drug candidates. This enables researchers to concentrate their efforts and resources on the most promising compounds, minimizing expenses and accelerating the advancement of novel treatments (Seal et al., 2025).

A significant application of AI lies in the advancement of genomics and the development of personalized medicine. Advanced algorithms efficiently process extensive genetic datasets to identify variations associated with diseases. This aids in gaining a deeper understanding of an individual's disease risk and in formulating treatments specifically designed for their distinct genetic profile (Athanasopoulou et al., 2022). AI has the capability to integrate genomic data with additional health information like medical history or lifestyle factors to develop models that forecast disease progression and suggest optimal treatment strategies.

A significant advancement in the field of biology is the prediction of protein structures using AI. AI tools such as DeepMind's AlphaFold, possess the capability to precisely forecast a protein's three-dimensional structure based on its sequence. This advancement aids in comprehending protein functionality and facilitates the design of targeted drugs, ultimately accelerating both research and therapeutic development (National Academies of Sciences, Engineering, and Medicine; Policy and Global Affairs; Committee on International Security and Arms Control; Division on Engineering and Physical Sciences; Computer Science and Telecommunications Board; Division on Earth and Life Studies; Board on Life Sciences; Committee on Assessing and Navigating Biosecurity Concerns and Benefits of Artificial Intelligence Use in the Life Sciences, 2025). Overall, the advancements in AI are revolutionizing the field of life sciences accelerating the process of drug discovery, refining personalized medicine, deepening our understanding of proteins, and optimizing diagnostics and manufacturing. By integrating extensive biological datasets with advanced algorithms, AI is enabling researchers to gain a deeper understanding of life and create quicker, more efficient solutions for human health.

11.4 AI-Driven Innovations in Synthetic Biology

11.4.1 AI in Genome Design and Editing

Gene-editing tools enable precise alterations to DNA within living cells. Various gene editing tools, including the CRISPR-Cas9 system, zinc-finger nucleases (ZFNs), and transcription activator-like effector nucleases (TALENs), are used for genetic modifications. The CRISPR-Cas gene editing tool facilitates accurate modifications in an organism's DNA, employing a system derived from bacteria that has proven to be effective for editing purposes. The mechanism employs a guide RNA to locate a particular DNA sequence, while a Cas9 protein is utilized to cleave the DNA at the designated site. A significant challenge in CRISPR-Cas editing lies in the design of guide RNAs (gRNAs) that achieve both high efficiency (on-target) and high specificity (minimal off-target cuts). AI, particularly in the realms of machine learning and deep learning, contributes by analyzing extensive datasets derived from prior CRISPR experiments (Dixit et al., 2024).

DeepCRISPR represents a significant advancement in AI, employing deep learning techniques to forecast the efficacy of gRNA and assess the probability of unintended cuts in DNA. The system autonomously evaluates sequence and epigenetic characteristics that affect gRNA efficacy (Chuai et al., 2018). Various advanced computational tools assist in the design of gRNA. Models such as CRISTA and DeepHF demonstrate a remarkable ability to predict both on-target and off-target activity with impressive accuracy (Pandey et al., 2025). CRISPRpred(SEQ), a conventional machine learning tool, occasionally outperforms DeepCRISPR in specific cell types (Muhammad Rafid et al., 2020). Hybrid models like CrnnCrispr integrate neural networks with explainable AI techniques to elucidate the sequence features that impact gRNA efficiency (Zhu et al., 2024). These elucidative AI methodologies assist investigators in comprehending the rationale behind the efficacy of certain guides compared to others, thereby enhancing assurance in gRNA selection (Figure 11.1).

11.4.2 Protein Structure Designing

The advancements in AI have significantly revolutionized the forecasting of protein configurations and their three-dimensional representations. An advanced computational tool, AlphaFold (AlphaFold, AlphaFold2, AlphaFold3), created by DeepMind, employs deep neural networks alongside evolutionary data derived from multiple sequence alignments (MSAs) to generate precise atomic-level predictions. RoseTTAFold, developed in David Baker's lab, employs a "three-track" neural network that concurrently analyzes sequence, pairwise distances, and 3D coordinates, resulting in both speed and accuracy (Chen et al., 2024). Furthermore, ESMFold, created by Meta AI, utilizes a substantial protein "language model" that allows it to predict structure from a single amino acid sequence, demonstrating a remarkable speed increase of up to approximately six times compared to methods based on multiple sequence alignments. ESMFold demonstrates the ability to rapidly and precisely predict protein structures. This approach operates at a significantly accelerated rate compared to numerous traditional techniques, thereby facilitating the management of the vast and rapidly expanding array of protein sequences being identified. This facilitates better understanding of protein structure and function, particularly as databases continue to develop (Rives, 2022) (Figure 11.1).

11.4.3 Pathway and Metabolic Network Optimization

AI is significantly enhancing the efficiency of biological pathways and metabolic networks. Researchers forecast the interactions of millions of biochemical reactions, identify crucial enzymes or processes that hinder efficiency through the machine learning (Merzbacher and Oyarzún, 2023). AI systems facilitate the development of novel or enhanced metabolic pathways, enabling cells to produce a larger amount of a targeted product (such as a biofuel or pharmaceutical) while minimizing waste byproducts. Instruments such as active learning facilitate experimentation by recommending specific genetic modifications or environmental adjustments to pursue, thereby accelerating and reducing the costs associated with the design-build-test cycle. AI enhances pathway control by forecasting the dynamic regulation of enzyme expression, utilizing mechanisms such as feedback loops or biosensors, resulting in increased stability and efficiency in production (Jang et al., 2022) (Figure 11.1).

11.4.4 AI in Genetic Circuit Design

The integration of advanced computational techniques in genetic circuit design is becoming more prominent, utilizing machine learning, reinforcement learning, and evolutionary algorithms to enhance the optimization of synthetic gene networks (Müller, Arndt and Hoffmann, 2025). Through the application of reinforcement learning, systems can systematically investigate and enhance genetic circuit parameters; the intelligent entity evaluates various designs, identifies those that yield superior performance, and progressively optimizes the circuit informed by the results obtained. Algorithms that draw inspiration from the principles of natural selection simulate groups of circuit variations, identifying the most successful ones and incorporating mutations to develop designs that are more efficient or resilient. Models utilizing machine learning have the capability to forecast the dynamics of gene expression, providing insights into how circuit behavior evolves over time under varying conditions (Hiscock, 2019). Through the simulation of these dynamics, advanced systems facilitate the creation of circuits that exhibit dependable responses, produce oscillations, or execute functions akin to logical operations. Collectively, these methodologies enhance the investigation of extensive design landscapes and the precise optimization of circuits, thereby rendering synthetic biology more effective and robust (Figure 11.1).

11.5 AI-Enabled Biomanufacturing

The integration of AI-driven automation is transforming the DBTL cycle through the synergy of robotics and machine learning, resulting in a simple, closed-loop workflow.

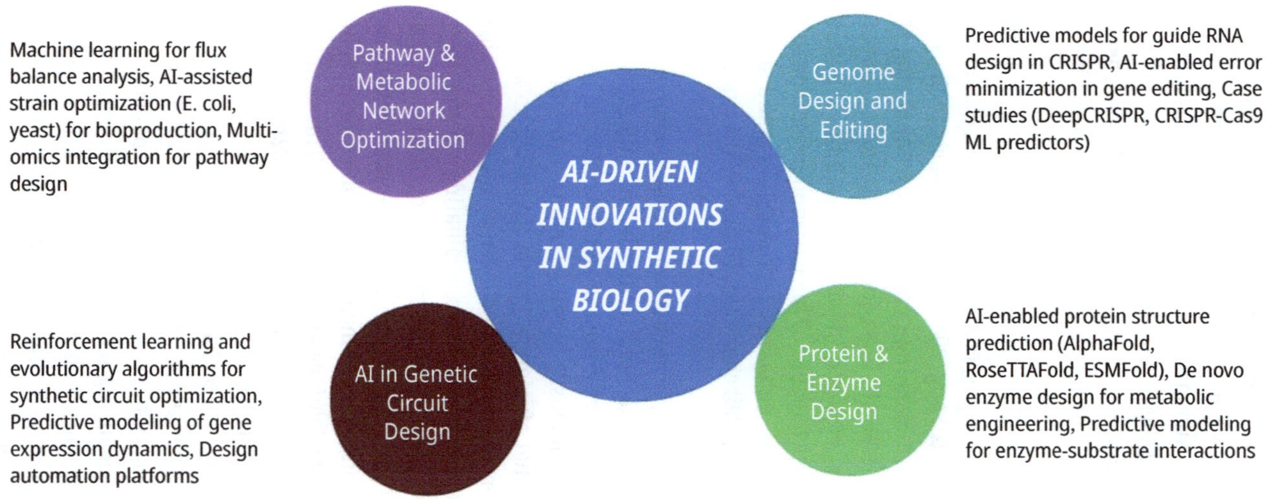

Figure 11.1: AI-driven innovations transforming design, analysis, and optimization in synthetic biology.

This indicates that experimental procedures can be systematically designed, executed, evaluated, and refined autonomously, thereby enhancing the speed, precision, and efficiency of biomanufacturing processes. In automated biofoundries, liquid-handling robots execute a variety of tasks, including the assembly of DNA, the setup of reactions, and the simultaneous execution of hundreds of experiments. All the data generated undergoes immediate analysis through advanced computational models in real time. The AI system is adept at identifying successful strategies, understanding failures, and analyzing the underlying reasons for both outcomes. This analysis facilitates the automatic updating of subsequent designs or experimental conditions, eliminating the need for human intervention (Kim, Bae & Song, 2020). This approach enables the system to autonomously finalize the entire learning cycle, thereby enhancing the speed, precision, and overall efficiency of research. In a robotic configuration employing active-learning algorithms, the system acquired the ability to modify induction conditions in bacteria to enhance protein production. This closed-loop methodology accelerates the DBTL cycle, reduces expenses, enhances reproducibility, and significantly optimizes synthetic biology processes (Mirchandani, Khandhediya and Chauhan, 2025; Spannenkrebs et al., 2025).

The advancements in AI are revolutionizing fermentation within biomanufacturing by enhancing process control, boosting yields, and facilitating real-time monitoring. Advanced models are capable of examining real-time sensor data like pH, temperature, and oxygen levels and can autonomously modify feeding rates or other parameters to maintain optimal operational efficiency. Digital twins, which are virtual replicas of bioreactors, are also generated by machine-learning algorithms. These models are capable of simulating fermentation processes, evaluating various strategies through computational methods, and assisting in determining optimal control decisions for the actual system (Butean et al., 2025). These digital replicas collaborate with AI-driven biosensors that assess challenging-to-measure variables such as biomass or metabolite concentrations using straightforward sensor data. This facilitates immediate oversight of the fermentation process (Zobel-Roos et al., 2021). Together, they ensure that conditions remain ideal, enhance output, minimize waste, and streamline production for greater reliability and scalability. These digital twins function within a closed-loop system utilizing AI-enhanced biosensors, often referred to as "soft sensors." Soft sensors have the capability to estimate critical yet challenging-to-measure variables within the bioreactor, including biomass, nutrient concentrations, or metabolites, by using simple measurements such as temperature, pH, or oxygen levels. This enables the system to comprehend the internal dynamics of the reactor instantaneously and modify conditions without delay. Through collaboration, virtual replicas and flexible measurement devices contribute to maintaining a stable and efficient fermentation process. They facilitate the maintenance of ideal conditions, enhance product yield, minimize waste, and guarantee that the entire production process is more dependable and simpler to expand (Yadav, Marwah & Kumar, 2025) (Figure 11.2).

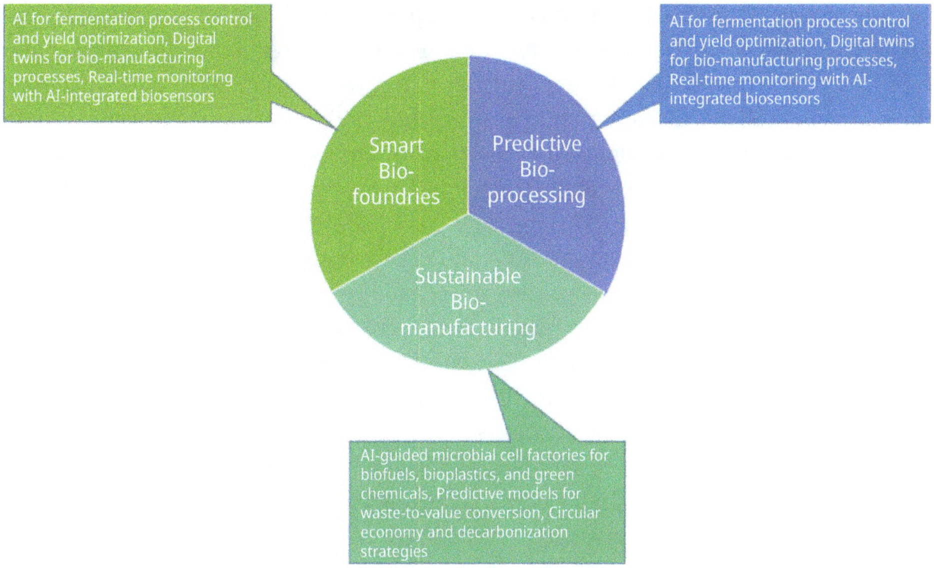

Figure 11.2: AI-enabled bio-manufacturing enhancing efficiency, scalability, and precision in biological production systems.

11.6 AI in Precision and Personalized Medicine Through Synthetic Biology

AI is emerging as an important factor in precision and personalized medicine, facilitating synthetic biology in the development of treatments that are increasingly accurate and customized for each individual. The integration of AI with synthetic biology allows for the simulation and prediction of the behavior of engineered cells, genes, or biological circuits across diverse individuals. This enables the development of treatments customized to an individual's specific genetic composition or molecular characteristics (Nimmagadda, 2022).

For instance, advanced algorithms can analyze a patient's multi-omics data, including their genome, gene expression patterns, and protein levels to identify specific biomarkers or vulnerabilities associated with disease. With this information, researchers can design synthetic gene circuits or engineer therapeutic cells, such as personalized CAR-T cells, that function specifically for that patient's needs. AI can assist in selecting more secure gene targets, forecasting potential side effects, and optimizing the functionality of modified biological systems prior to their application within the organism (Mirchandani, Khandhediya & Chauhan, 2025).

Generative AI models utilizing variational autoencoders (VAEs) or generative adversarial networks (GANs) are currently employed to design novel biological molecules, such as synthetic DNA sequences or proteins, which have the potential to function as personalized therapeutics. These models address data constraints, create innovative biomolecules designed for particular disease scenarios, and predict the potential behavior of these molecules within a patient's physiology (Mishra et al., 2025).

In regenerative medicine, AI significantly contributes by assisting researchers in comprehending and enhancing the functioning of engineered cells or tissues within the biological system. AI systems possess the capability to forecast the growth patterns of these cells, their differentiation into particular cell types, or their responses to stimuli within a patient's internal milieu. This enables experts to develop safer and more efficient therapies based on synthetic biology that can heal or substitute damaged tissues, including in wound healing, organ restoration, or cartilage regeneration (Nosrati et al., 2023). The convergence of synthetic biology and AI accelerates the advancement of therapies capable of simultaneously diagnosing and treating diseases. AI contributes to the development of advanced synthetic circuits capable of identifying disease markers within the body, including atypical proteins or indicators of inflammation. Upon detection of these markers, the system can autonomously initiate a therapeutic response, which may include the release of a drug, activation of immune cells, or deactivation of detrimental pathways (Brosch-Lenz et al., 2021). This results in enhanced therapies that minimize adverse effects, advancing us toward genuinely smart and responsive healthcare solutions (Figure 11.3).

11.7 Computational and Data-Driven Advances

11.7.1 Integration of Multi-omics Data

The integration of multi-omics data genomics, transcriptomics, proteomics, and metabolomics is being revolutionized by advanced computational techniques like AI, enabling the construction of comprehensive models of cellular systems. By integrating these data, AI examines the functionality of genes, their expression, the formation of corresponding proteins, and the interactions and changes of small-molecule metabolites (Zack et al., 2025). Advanced computational tools such as OmiEmbed simplify the intricacies of multi-omics data by mapping it into a unified low-dimensional space, facilitating the effective prediction of phenotypes like disease susceptibility or therapeutic response (Zhang et al., 2021).

The role of AI is becoming more significant in the fields of synthetic biology and metabolic engineering, particularly in the design and enhancement of microbial strains. A significant breakthrough is the integration of multi-omics data through AI encompassing genomics, transcriptomics, proteomics, and metabolomics, which aids in compre-

Figure 11.3: AI enabling precise, data-driven personalization of diagnosis, treatment, and disease prevention in medicine.

hending the functionality of engineered cells across various biological dimensions. In the realm of strain engineering, advanced computational tools can scrutinize proteomic and metabolomic data derived from modified Aspergillus strains to pinpoint limitations within metabolic pathways. Leveraging these insights, researchers can refine or optimize metabolic networks to enhance the synthesis of valuable chemicals. A significant example involves enhancing the production of 3-hydroxypropionic acid, where analytical support from AI revealed critical enzyme constraints and pathway discrepancies, facilitating precise re-engineering of the organism (Kim et al., 2021). Furthermore, the integration of multi-omics through AI enhances precision medicine by pinpointing biomarkers and potential therapeutic targets. In the field of hepatology, advanced computational models integrate data from genomics, transcriptomics, proteomics, and metabolomics to enhance disease classification and facilitate drug development (Renganathan et al., 2020). In the realm of cardiovascular research, the integration of various omics layers through advanced computational techniques enhances the precision of patient outcome predictions, thereby facilitating tailored approaches to diagnosis and treatment (Lin et al., 2025).

11.7.2 Generative AI for Synthetic Biology

The advent of generative AI is catalyzing a significant shift in synthetic biology, enabling the design of entirely new biological components, including DNA sequences, proteins, enzymes, and regulatory elements. These instruments can generate sequences with targeted functions, enhancing the design process to be quicker, more intelligent, and significantly more innovative compared to conventional trial-and-error approaches. GANs serve as a powerful mechanism for analyzing extensive datasets of natural biological sequences, enabling the generation of novel sequences that adhere to established patterns. An innovative model known as FBGAN has been employed to create artificial DNA sequences that encode antimicrobial peptide molecules capable of combating infections (Gupta and Zou, 2018). A significant instrument in this field is the utilization of extensive language models developed from protein or DNA sequences. Instruments like ProGen, Peptide-GPT, and various protein-language models possess the capability to engineer novel proteins characterized by distinct shapes, enhanced stability, and targeted activities (Madani et al., 2020). In a significant advancement, investigators successfully employed a protein-language model to engineer synthetic transposases (modified PiggyBac enzymes) that exhibited superior performance compared to their natural counterparts. The application of generative AI extends to the design of synthetic regulatory DNA, including elements like promoters and enhancers. These components can be customized to activate or deactivate genes in specific tissues or under certain circumstances. For instance, investigators at Yale School of Medicine have employed AI to create regulatory sequences that provide highly accurate control over gene expression in specific cell types.

11.8 Challenges and Ethical Considerations

The application of AI in synthetic biology holds significant potential for advancements, yet it simultaneously raises critical ethical and safety considerations that must be addressed. A significant concern is biosecurity. The acceleration of designing new or modified organisms through AI presents the possibility of inadvertently or deliberately generating harmful pathogens. The integration of automated laboratories with AI technologies raises concerns about the potential for misuse in the efficient design of hazardous biological agents (Hynek, 2025).

Concerns regarding biosafety are significant. Organisms designed through AI could exhibit unforeseen behaviors if they were to break free into the ecosystem. They may disrupt natural ecosystems, propagate in unforeseen manners, or surpass indigenous species in competition. Should the information utilized to develop AI systems be substandard or skewed, the ensuing outcomes could be defective or potentially dangerous (Matthews et al., 2025). There are also significant issues regarding transparency and accountability. An increasingly significant concern is the security of biological data in the digital realm. As synthetic biology becomes more digital and automated, there is a risk that hackers could break into systems, tamper with AI models, or even insert harmful DNA instructions into automated workflows.Ultimately, we encounter deficiencies in regulation. Current biosecurity treaties and AI regulations were not built for this new blend of AI and synthetic biology. Researchers highlight the necessity for revised global protocols, enhanced regulatory supervision, and improved risk-evaluation structures (Farbiash & Puzis, 2020; Hynek, 2025).

In summary, although the integration of advanced technologies in synthetic biology holds the promise to revolutionize healthcare, manufacturing, and ecological systems, it is imperative that this development is approached with rigorous safety protocols, ethical considerations, and international collaboration to mitigate risks and safeguard both society and the environment.

11.9 Global Policy and Governance Landscape

AI-enabled synthetic biology is creating new global governance challenges. Traditional international agreements like the Biological Weapons Convention (BWC) remain important for biosecurity, but experts say they are no longer enough. AI can speed up the design of biological systems, which increases dual-use risks (technologies that can be used for both good and harmful purposes). Because of this, international policymakers are calling for stronger global monitoring systems, better verification tools, improved containment strategies for AI-generated risks, and stricter DNA synthesis screening (Lentzos, 2012; Patrick & Barton, 2024). As AI and synthetic biology rapidly merge, many regulatory gaps are becoming obvious. Existing AI rules, such as the OECD AI principles or various national AI policies, rarely consider the risks of using AI for biological design. That means current governance systems do not adequately address the safety challenges of "SynBioAI" (Hynek, 2025; Synthetic biology, 2025). Global biosecurity experts stress that policies must be flexible and able to evolve quickly. They recommend new types of institutions or "early-warning" systems that can detect emerging risks as technologies advance. The overall aim is to support innovation in synthetic biology while ensuring strong safeguards. This balance is essential to enjoy the benefits of faster bioengineering without increasing the chances of misuse or accidental harm.

11.10 Future Directions

The influence of AI in synthetic biology is anticipated to expand swiftly, propelled by emerging generative frameworks like extensive biological language models (DNA-

LLMs). These advanced models can design completely new DNA sequences and regulatory elements, enabling researchers to construct synthetic organisms with specialized functions for instance, organisms that generate new medicines or remediate environmental pollutants (Fernández, 2024). A notable expanding field is the utilization of AI in cell-free biomanufacturing, where biochemical reactions take place outside of living organisms. Through the application of predictive modeling and reinforcement learning, these systems can be adjusted in real time, facilitating rapid, on-demand synthesis of biomolecules.

A significant trend on the horizon is the advancement of digital replicas for biological systems. These represent digital models of cells, microorganisms, or complete bioprocesses that are grounded in actual experimental data. Digital twins enable professionals to simulate and refine designs digitally prior to conducting experiments, thereby conserving time, expenses, and resources. Simultaneously, the field of geometric deep learning, which comprehends three-dimensional forms and intricate structures, is rising as a crucial instrument for protein engineering, enabling AI to forecast and create proteins with enhanced accuracy (Renganathan et al., 2020).

In the field of agriculture, innovative AI models that merge predictive and generative capabilities are being developed to design crops that can withstand climate challenges, utilizing data from various omics layers (Najafabadi & Jackson, 2025). In summary, the trajectory of AI in the realm of synthetic biology indicates a shift towards increasingly automated, intelligent, and context-sensitive biological design facilitating the development of sustainable, programmable, and scalable living systems.

References

Athanasopoulou, K., Daneva, G. N., Adamopoulos, P. G., Scorilas, A., Athanasopoulou, K., Daneva, G. N., Adamopoulos, P. G., & Scorilas, A. (2022). Artificial intelligence: The milestone in modern biomedical research. *BioMedInformatics*, *2*, 727–744. doi: 10.3390/biomedinformatics2040049.

Brosch-Lenz, J., Yousefirizi, F., Zukotynski, K., Beauregard, J.-M., Gaudet, V., Saboury, B., Rahmim, A., & Uribe, C. (2021). Role of AI in theranostics: Towards routine personalized radiopharmaceutical therapies. *arXiv*, 13913. doi: 10.48550/arXiv.2107.13913.

Butean, A., Cutean, I., Barbero, R., Enriquez, J., Matei, A., Butean, A., Cutean, I., Barbero, R., Enriquez, J., & Matei, A. (2025). A review of artificial intelligence applications for biorefineries and bioprocessing: From data-driven processes to optimization strategies and real-time control. *Processes*, *13*, 2544. doi: 10.3390/pr13082544.

Chatterjee, S. (2022). What is Artificial Intelligence (AI)? Its meaning, applications, and the future. *Emeritus India*, 23 March. Available at: https://emeritus.org/in/learn/what-is-artificial-intelligence/ (Accessed: 17 November 2025).

Chen, L., Li, Q., Nasif, K. F. A., Xie, Y., Deng, B., Niu, S., Pouriyeh, S., Dai, Z., Chen, J., Xie, C. Y., Chen, L., Li, Q., Nasif, K. F. A., *et al.* (2024). AI-driven deep learning techniques in protein structure prediction. *International Journal of Molecular Sciences*, *25*, 8426. doi: 10.3390/ijms25158426.

Chen, Y. Y., Galloway, K. E., & Smolke, C. D. (2012). Synthetic biology: Advancing biological frontiers by building synthetic systems. *Genome Biology*, *13*, 240. doi: 10.1186/gb-2012-13-2-240.

Chuai, G., Ma, H., Yan, J., Chen, M., Hong, N., Xue, D., Zhou, C., Zhu, C., Chen, K., Duan, B., *et al.* (2018). DeepCRISPR: Optimized CRISPR guide RNA design by deep learning. *Genome Biology, 19*, 80. doi: 10.1186/s13059-018-1459-4.

da Cunha, N. B., Fernandes, F. C., Gil-Ley, A., Franco, O. L., Timakondu, N., & Costa, F. F. (2025). Bridging bioSciences and technology: The impact of AI & GenAI in life sciences and agribusiness. *Gene, 964*, 149623. doi: 10.1016/j.gene.2025.149623.

Dixit, S., Kumar, A., Srinivasan, K., Vincent, P. M. D. R., & Ramu Krishnan, N. (2024). Advancing genome editing with artificial intelligence: Opportunities, challenges, and future directions. *Frontiers in Bioengineering and Biotechnology, 11*, 1335901. doi: 10.3389/fbioe.2023.1335901.

Embra, A. (2024). Engineering life: An introduction to synthetic biology and its applications. *12*, 104. doi: 10.35248/2332-0737.24.12.104.

Farbiash, D., & Puzis, R. (2020). Cyberbiosecurity: DNA injection attack in synthetic biology. *arXiv*. doi: 10.48550/arXiv.2011.14224.

Fernández, M. (2024). *Artificial Intelligence Powering Synthetic Biology: The Fundamentals.S&P Global.* Available at: https://www.spglobal.com/en/research-insights/special-reports/artificial-intelligence-powering-synthetic-biology-the-fundamentals(Accessed: 21 November 2025).

Garner, K. L. (2021). Principles of synthetic biology. *Essays in Biochemistry, 65*, 791–811. doi: 10.1042/EBC20200059.

Gupta, A., & Zou, J. (2018). Feedback GAN (FBGAN) for DNA: A novel feedback-loop architecture for optimizing protein functions. *arXiv*. doi: 10.48550/arXiv.1804.01694.

Hiscock, T. W. (2019). Adapting machine-learning algorithms to design gene circuits. *BMC Bioinformatics, 20*, 214. doi: 10.1186/s12859-019-2788-3.

Hynek, N. (2025). Synthetic biology/AI convergence (SynBioAI): Security threats in frontier science and regulatory challenges. *AI & SOCIETY*. doi: 10.1007/s00146-025-02576-4.

Jang, W. D., Kim, G. B., Kim, Y., & Lee, S. Y. (2022). Applications of artificial intelligence to enzyme and pathway design for metabolic engineering. *Current Opinion in Biotechnology, 73*, 101–107. doi: 10.1016/j.copbio.2021.07.024.

Jiang, X., Ji, J., Yang, Q., Dou, Y., Li, Y., Yang, X., Liu, C., Dou, S., & Dong, L. (2025). Key technologies of synthetic biology in industrial microbiology. *Microorganisms, 13*, 2343. doi: 10.3390/microorganisms13102343.

Karataş, P., & Ayaz, F. (2025). Synthetic biology and application areas. *Discover Biotechnology, 2*, 3. doi: 10.1007/s44340-025-00010-5.

Kim, S.-H., Bae, S., & Song, M. (2020). Recent development of aminoacyl-tRNA synthetase inhibitors for human diseases: A Future perspective. *Biomolecules, 10*, E1625. doi: 10.3390/biom10121625.

Kim, Y., Petzold, C. J., Kerkhoven, E. J., & Baker, S. E. (2021). Editorial: Multi-Omics technologies for optimizing synthetic biomanufacturing. *Frontiers in Bioengineering and Biotechnology, 9*. doi: 10.3389/fbioe.2021.818010.

Lentzos, F. (2012). Synthetic biology, security and governance. *BioSocieties, 7*, 339–351. doi: 10.1057/biosoc.2012.28.

Lin, M., Guo, J., Gu, Z., Tang, W., Tao, H., You, S., Jia, D., Sun, Y., & Jia, P. (2025). Machine learning and multi-omics integration: Advancing cardiovascular translational research and clinical practice. *Journal of Translational Medicine, 23*, 388. doi: 10.1186/s12967-025-06425-2.

Madani, A., McCann, B., Naik, N., Keskar, N. S., Anand, N., Eguchi, R. R., Huang, P.-S., & Socher, R. (2020). *ProGen: Language modeling for protein generation, arXiv.org*. Available at: https://arxiv.org/abs/2004.03497v1 (Accessed: 21 November 2025).

Matthews, K. R. W., Johnson, A. L., Iltis, A. S., George, D. R., & Berto, S. (2025). Ethical challenges and concerns in synthetic biology. *Rice University's Baker Institute for Public Policy*. Available at: https://www.bakerinstitute.org/research/ethical-challenges-and-concerns-synthetic-biology (Accessed: 21 November 2025).

Merzbacher, C., & Oyarzún, D. A. (2023). Applications of artificial intelligence and machine learning in dynamic pathway engineering. *Biochemical Society Transactions*, *51*, 1871–1879. doi: 10.1042/BST20221542.

Mirchandani, I., Khandhediya, Y., & Chauhan, K. (2025). Review on advancement of AI in synthetic biology. *Methods in Molecular Biology (Clifton, N.J.)*, *2952*, 483–490. doi: 10.1007/978-1-0716-4690-8_26.

Mishra, A., Majumder, A., Kommineni, D., Anna Joseph, C., Chowdhury, T., & Anumula, S. K. (2025). Role of generative artificial intelligence in personalized medicine: A systematic review. *Cureus*, *17*, e82310. doi: 10.7759/cureus.82310.

Muhammad Rafid, A. H., Toufikuzzaman, M., Rahman, M. S., & Rahman, S. M. (2020). CRISPRpred(SEQ): A sequence-based method for sgRNA on target activity prediction using traditional machine learning. *BMC Bioinformatics*, *21*, 223. doi: 10.1186/s12859-020-3531-9.

Müller, M. M., Arndt, K. M., & Hoffmann, S. A. (2025). Genetic circuits in synthetic biology: Broadening the toolbox of regulatory devices. *Frontiers in Synthetic Biology*, *3*. doi: 10.3389/fsybi.2025.1548572.

Najafabadi, M. Y., & Jackson, S. A. (2025). Hybrid AI in synthetic biology: Next era in agriculture. *Trends in Plant Science*, *S1360-1385*(25), 00237–7. doi: 10.1016/j.tplants.2025.08.011.

National Academies of Sciences, Engineering, and Medicine; Policy and Global Affairs; Committee on International Security and Arms Control. Division on engineering and physical sciences; computer science and telecommunications board; division on earth and life studies; board on life sciences; committee on assessing and navigating biosecurity concerns and benefits of artificial intelligence use in the life sciences (2025). In *The Age of AI in the Life Sciences: Benefits and Biosecurity Considerations*. Washington (DC): National Academies Press (US). Available at: http://www.ncbi.nlm.nih.gov/books/NBK614593/ Accessed: 17 November 2025.

Nimmagadda, V. S. P. (2022). AI in Synthetic biology: Designing and optimizing genetic constructs for bioengineering applications. *Australian Journal of Machine Learning Research & Applications*, *2*, 341–381. Available at: https://www.sydneyacademics.com/index.php/ajmlra/article/view/118 Accessed: 19 November 2025.

Nosrati, H., Nosrati, M., Nosrati, H., & Nosrati, M. (2023). Artificial intelligence in regenerative medicine: Applications and implications. *Biomimetics*, *8*, 442. doi: 10.3390/biomimetics8050442.

Pandey, S., Choudhari, J. K., Tripathi, A., Singh, A., Antony, A., & Chouhan, U. (2025). Artificial intelligence-based genome editing in CRISPR/Cas9. *Methods in Molecular Biology (Clifton, N.J.)*, *2952*, 273–282. doi: 10.1007/978-1-0716-4690-8_16.

Patrick, S., & Barton, J. (2024). Mitigating Risks from Gene Editing and Synthetic Biology: Global Governance Priorities, Carnegie Endowment for International Peace. Available at: https://carnegieendowment.org/research/2024/10/mitigating-risks-from-gene-editing-and-synthetic-biology-global-governance-priorities?lang=en (Accessed: 21 November 2025).

Renganathan, V. G., Vanniarajan, C., Karthikeyan, A., & Ramalingam, J. (2020). Barnyard millet for food and nutritional security: Current status and future research direction. *Frontiers in Genetics*, *11*, 500. doi: 10.3389/fgene.2020.00500.

Rives, A. (2022). Meta AI releases ESMFold, a new protein structure prediction tool – All Health Tech. Available at: https://allhealthtech.com/esmfold/ (Accessed: 18 November 2025).

Seal, S., Uynh, D. L., Chelbi, M., Khosravi, S., Kumar, A., Thieme, M., Wilks, I., Davies, M., Mustali, J., Sun, Y., & Edwards, N. (2025). AI agents in drug discovery. arXiv. doi: 10.48550/arXiv.2510.27130.

Spannenkrebs, J. B., Eiermann, A., Zoll, T., Hackenschmidt, S., & Kabisch, J. (2025). Closing the loop: Establishing an autonomous test-learn cycle to optimize induction of bacterial systems using a robotic platform. *Frontiers in Bioengineering and Biotechnology*, *12*, 1528224. doi: 10.3389/fbioe.2024.1528224.

Synthetic Biology. (2025). National Institute of Biomedical Imaging and Bioengineering. Available at: https://www.nibib.nih.gov/science-education/science-topics/synthetic-biology (Accessed: 22 November 2025).

Wang, L., Zang, X., & Zhou, J. (2022). Synthetic biology: A powerful booster for future agriculture. *Advanced Agrochem*, *1*. doi: 10.1016/j.aac.2022.08.005.

Yadav, J., Marwah, H., & Kumar, C. (2025). Synthetic biology and metabolic engineering paving the way for sustainable next-gen biofuels: A comprehensive review. *Energy Advances*, *4*, 1209–1228. doi: 10.1039/D5YA00118H.

Zack, M., Stupichev, D. N., Moore, A. J., Slobodchikov, I. D., Sokolov, D. G., Trifonov, I. F., & Gobbs, A. (2025). Artificial intelligence and multi-omics in pharmacogenomics: A new era of precision medicine. *Mayo Clinic Proceedings. Digital Health*, *3*, 100246. doi: 10.1016/j.mcpdig.2025.100246.

Zhang, X., Xing, Y., Sun, K., & Guo, Y. (2021). OmiEmbed: A unified multi-task deep learning framework for multi-omics data. *Cancers*, *13*, 3047. doi: 10.3390/cancers13123047.

Zhu, W., Xie, H., Chen, Y., Zhang, G., Zhu, W., Xie, H., Chen, Y., & Zhang, G. (2024). CrnnCrispr: An interpretable deep learning method for CRISPR/Cas9 sgRNA on-target activity prediction. *International Journal of Molecular Sciences*, *25*(8). doi: 10.3390/ijms25084429.

Zobel-Roos, S., Schmidt, A., Uhlenbrock, L., Ditz, R., Köster, D., & Strube, J. (2021). Digital twins in biomanufacturing. *Advances in Biochemical Engineering/Biotechnology*, *176*, 181–262. doi: 10.1007/10_2020_146.

Diksha Pathania*, Shivani Saklani, Sonu Sonu, Priti Duhan

Chapter 12
AI-Driven Environmental Biotechnology for a Sustainable Future

Abstract: Our planet is grappling with a multitude of sustainability challenges in the modern era that pose a threat to the fragile equilibrium of ecosystems around the globe. Recent developments in artificial intelligence (AI) hold revolutionary promise for improving current conservation initiatives. AI technologies, such as the use of machine learning and statistical analysis, are increasingly being used to enhance the recognition of species, habitat surveillance, and threat evaluation with remarkable efficacy and precision. AI has the potential to improve conservation initiatives and aid in ensuring the survival of biodiversity worldwide.

Keywords: Bioremediation, artificial intelligence, machine learning, sustainability, environment

12.1 Introduction

A recent World Bank analysis states that cities produce more than 2 billion tons of municipal solid waste yearly, with that amount expected to rise to 3.4 billion tons by 2050 (Lakhouit, 2025). Attempts for active conservation, like restoration projects, are frequently made on smaller budgets than passive restoration; nevertheless, this is where present management initiatives are frequently less successful and cost-effective. At the moment, active conservation initiatives come at a hefty expense (Díaz-García et al., 2020). For instance, restoring one hectare of marine habitat costs 80,000 USD on average worldwide, but the true costs are more likely to be two to four times more because of the unknowns in the restoration process (Bayraktarov et al., 2016).

Urban garbage treatment plant design and operation can be completely transformed by integrating robotics and artificial intelligence (AI), which will result in

*Corresponding author: Diksha Pathania, Dairy Technology Division, National Dairy Research Institute (ICAR), Karnal, Haryana, India,
e-mail: dikshapathania198@gmail.com
Shivani Saklani, School of Biological and Environmental Sciences, Shoolini University of Biotechnology and Management Sciences, Solan 173229, India
Sonu Sonu, School of Advanced Chemical Sciences, Shoolini University of Biotechnology and Management Sciences, Solan 173229, India
Priti Duhan, Department of Biotechnology, K. L. Mehta Dayanand College for Women, Faridabad 121001, Haryana, India

 | https://doi.org/10.1515/9783112214275-012

cleaner waste management techniques and higher operational efficiency (Mukherjee et al., 2021; Yigitcanlar & Mehmood, 2021). To optimize resource use, efficiency, and recycling alternatives throughout the solid waste management cycle, a number of developed nations, including Austria, Germany, New Zealand, the USA, the UK, Japan, Singapore, Switzerland, South Korea, and Canada, have already started implementing AI innovations (Soni et al., 2019).

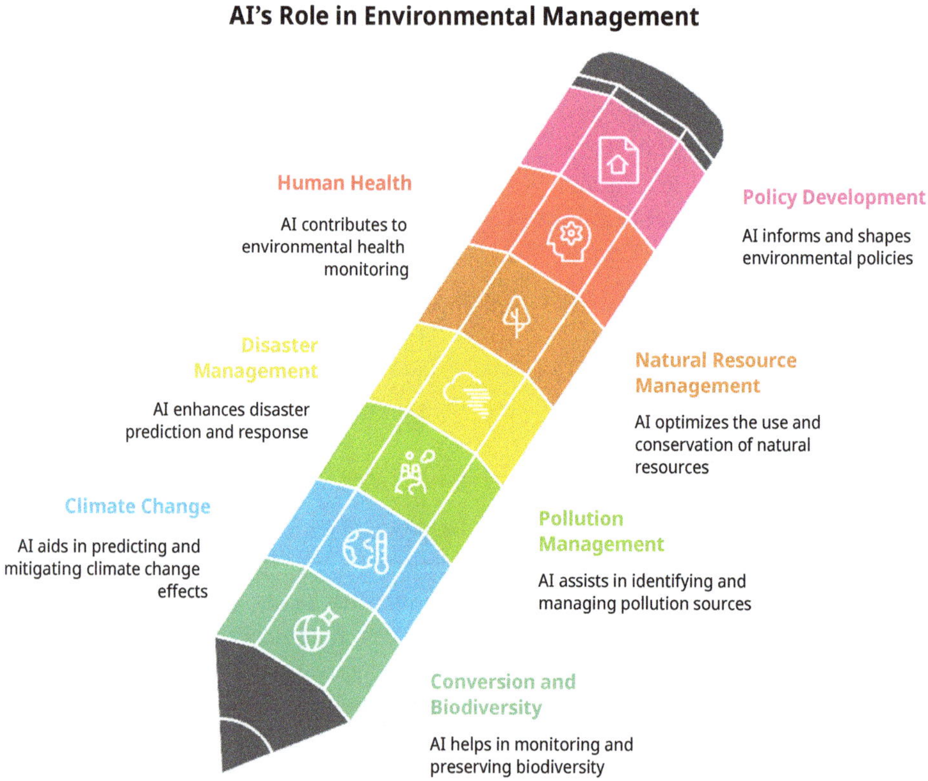

Figure 12.1: Artificial intelligence enhancing monitoring, prediction, and sustainable decision-making in environmental management.

AI-driven flood forecasting systems offer warnings, enabling communities to better prepare and evacuate. Challenges, including data integrity and accessibility, algorithmic bias, and infrastructural restrictions, have been seen as chances for progress. The ethical application and accountable development of AI are highlighted as fundamental tenets, promising that the incorporation of AI into environmental preservation is consistent with the values of equity, transparency, and social advantage (Chisom et al., 2024). AI provides a toolset to conservation managers through automated method number of problems can be mitigated, and AI lowers the long-term costs of monitoring and monitoring bottlenecks. Users can get access to more information through automating data collection, transport, and processing, which helps with time management (Ditria et al., 2022).

The data collection and evaluation of this information for several environmental factors, such as soil, water, and air quality, are part of environmental monitoring. The main goals of environmental monitoring are identifying and checking environmental changes, evaluating the effects of human activity, and assessing the efficiency of environmental regulations (Divine et al., 2024) (Figure 12.1). A lot of techniques were used for environmental monitoring, such as satellite images, remote sensing, and physical sampling (Mao et al., 2019).

12.2 Bioremediation

Environmental pollution has become a major worldwide concern due to the increase in industrialization, urbanization, and agricultural activities. Pollutants like heavy metals, hydrocarbons, and synthetic compounds affect the soil, water, and air. These pollutants have a harmful impact on human health and the ecosystem. For example, heavy metal contamination can disrupt the metabolic processes in living organisms, which lowers biodiversity and impairs the performance of ecosystems (Akintola, 2024). These types of pollutants can be detected and evaluated by AI-modified software. AI is the term used for the technology, especially computer systems that can simulate human intelligence processes. AI has a lot of processes like learning, reasoning, problem-solving, observation, and language comprehension. The main objective of AI is to build machines that are capable of monitoring activities like speech recognition, linguistic comprehension, and decision-making that normally need human intelligence. AI can be divided into two main categories: general AI to mimic human cognitive abilities across a variety of tasks, and narrow AI, which is made to carry out certain tasks, e.g., virtual assistants like Siri and Alexa (Russell & Norvig, 2021). Machine learning (ML) is a branch of AI that focuses on generating algorithms for computers to analyze data, learn from it, and make judgments or predictions. ML algorithms are utilized for statistical techniques to find patterns in data rather than being obviously coded to carry out a task, and AI will be better as exposed to more data (Jordan & Mitchell, 2015).

The search for innovative pollution-reduction techniques has been spurred by increasing concerns about the state of the planet and its effects on human health. AI technology includes the decisions or predictions of data generation and learning of computational techniques. AI is gaining popularity in environmental research because of its capacity to handle enormous quantities of complex data, identify their trends, and produce vital insights for addressing environmental concerns. AI provides possibilities to increase the efficiency of managing environmental activities through monitoring pollution and its remediation (Wani et al., 2024). This application of sewage treatment and bioremediation by AI technologies has given it a fresh perspective on life in the modern era (Bansal et al., 2023).

AI can mimic the behavior of nanomaterials using massive datasets and sophisticated algorithms by predictive models to identify the best features for the removal of pollutants. Researchers can quickly assess and optimize nanomaterials using AI with automated screening technologies for efficient bioremediation and speeding up the creation of environmentally friendly solutions. AI can provide important insights into the connections between the characteristics of nanomaterials and pollutant removal from the environment by utilizing ML and multi-scale modeling techniques (Kumar, 2025).

Future models will be created using AI and statistical techniques based on input factors. Critical analysis is done through the AI-driven bioremediation, such as the necessity for accountable deployment, ethical issues, and technological limitations. The focus of AI is on encouraging interdisciplinary cooperation to get beyond these obstacles. Future directions and practical suggestions will be completed through combining AI with conventional methods by filling in technological and legislative gaps, and making sure ecosystems are managed sustainably. According to recent studies, AI-driven optimization can enhance novel bioremediation systems by enabling adaptive control, real-time monitoring, and futureproofing of remediation procedures. In a new time of environmental management, experts can enhance the efficacy of pollution remediation, lessen environmental consequences, and maximize resource utilization by blending AI algorithms with biological remediation methods (Blessing et al., 2025).

AI can improve the efficiency of microbial processes in bioremediation through automated modeling and microbial activity optimization. AI can help in identifying the ideal conditions for the microbial breakdown of contaminants by assessing environmental factors like pH, temperature, and supply of nutrients (Akintola, 2024). AI can help create novel bioremediation techniques, such as creating genetically modified microorganisms that are suited to particular pollutants. The ability of these genetically modified microorganisms can be assessed in real-time, enabling prompt modifications to remediation strategies (Blessing & Olateru, 2025).

12.3 Ecosystem Monitoring and Management

AI has extended Ecological Services evaluations by integrating the huge data with things on the internet. This has enabled the successful integration of information from various data sources to ease resource management and ecological protection. The advantage of AI in ecological monitoring is its capability to apply its sophisticated algorithms to record the real-time environmental changes, ensuring the stability of

Ecological Services and presenting the scientific underpinnings for sustainable management and ecological renewal (Han et al., 2025).

Till now, a number of conventional methods have been employed to preserve biodiversity, but the time and lack of availability of data necessitate the use of new technology (Ullah et al., 2025). It would be valuable to gain extensive knowledge regarding the distribution, behavior, and quantity of animals in their natural habitat to better manage them and comprehend their intricacies (Harris et al., 2010; Norouzzadeh et al., 2018).

These aircraft platforms provide high-resolution photographs, data, and live data while providing researchers, environmentalists, and policymakers access to remote regions. Drones can smoothly analyze huge amounts of data when combined with AI technology, assisting in thorough habitat mapping, species monitoring, and ecosystem evaluations (Bekmurzaeva et al., 2024). The restoration of an ecosystem with accessible natural resources would be fruitful for maintaining biodiversity. In this regard, AI-based structures ecosystem planning would establish zones rich in biodiversity, which considers soil type, climate, and temperature range for that zone (Jayanthi & Kumar, 2024).

12.4 Waste Management and Pollution Control

Trash cans, anticipate waste collection, and efficiency of waste processing facilities can be monitored by AI-based technologies such as intelligent trash cans, categorization machines, forecasting algorithms, and wireless detection (Fang et al., 2023) (Figure 12.2). By utilizing AI technologies, municipalities can lower their expenses, increase safety, and lessen the environmental effects of trash management.

Traditional trash cans only hold rubbish, and sanitation personnel must manually check the bins to determine how much trash is inside. For routine trash disposal inspections, this method is ineffective. Additionally, because the containers are frequently filled, insects and infectious diseases have a tendency to breed on them (Noiki et al., 2021). By using a system on a chip made by Espressif Systems (ESP 8266) module, an intelligent trash can can be made that automatically detects things and establishes thresholds inside the bin. After being collected, the data can be sent to another node for additional processing and analysis (Praveen et al., 2020). The primary goal of smart garbage system research is to check the level of waste filling and alert users. Data is collected through their main source sensors, and then, the data will be uploaded to the system. AI bin systems have the capacity to improve the atmosphere of the city, lower the rate of spread of disease, and boost the efficiency of garbage collection (Fang et al., 2023).

For municipal solid waste management, garbage arrangement and employing robots can significantly increase garbage categorization efficiency. However, to classify

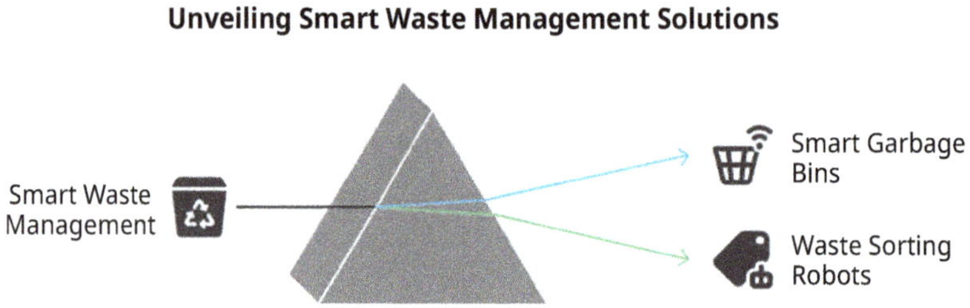

Figure 12.2: Leveraging artificial intelligence to optimize waste collection, improve recycling efficiency, and reduce environmental impact through smart data-driven decisions.

waste in extremely complex, diverse, and unpredictable industrial contexts, robots need highly developed visual and operational skills (Koskinopoulou et al., 2021). Scientists are now exploring ways to incorporate waste-sorting robots into current waste management systems, like using them to separate waste before it is dumped in a landfill. Studies have proposed a parallel robot model in this context, with the main idea centered on a grasping device that is completely incorporated into a comparable robot structure with four degrees of freedom (Leveziel et al., 2022). A method known as sensor-based trash monitoring makes use of sensors to quantify the quantity of garbage produced, pinpoint its sources, and assess how well waste management plans are working in a particular location Gurram et al., 2022). There are different types of sensors that detect different parameters such as humidity, odor, sound, infrared, and gas. Researchers suggested monitoring nuclear waste glassware melts in cold crucible induction melting furnaces in situ using a noncontact microwave sensor. They also created moisture, temperature, and sound sensors to keep an eye on noise pollution, infrared sensors to determine the carriages' filling level, and gas sensors to identify potentially dangerous pollutants (Sivaprakasam et al., 2020). Proper disposal of waste has many benefits, as summarized in Figure 12.3.

12.5 Biodiversity Preservation

Human activities have brought substantial damage to our planet's ecosystems, prompting creative approaches to conserve biodiversity and address ecological challenges. AI has become a revolutionary tool, delivering unprecedented capacity for environmental preservation and monitoring (Sharma & Jamwal, 2024). AI can assist in ecological surveillance and conservation through tracking wildlife, assessing habitats, conserving resources, analyzing biodiversity, and identifying species (Ayoola et al., 2024). AI-powered algorithms utilize GPS information, drone imagery, and camera trap recordings to identify and estimate the size of the population, which will enhance the anti-poaching efforts, protecting a diverse group of species. AI-driven picture anal-

Figure 12.3: Effective waste management supports environmental protection, public health, resource conservation, and sustainable development across urban and industrial systems.

ysis has been employed in habitat monitoring and resource preservation to assist with identifying regions that require restoration, deforestation detection, and evaluating the wellness of forests (Chisom et al., 2024).

Possible applications of technology integrated with AI are highlighted in Figure 12.4.

12.6 Challenges

The spread of automated surveillance may also be hampered by social issues. Groups' unwillingness or incapacity to exchange data may hinder rapid and efficient conservation initiatives and slow joint research on a local or even global scale. For this reason, improving the application of AI in environmental research requires making relevant data easily accessible (Ditria et al., 2022). Knowledge sharing among researchers may be facilitated by collaboration and the adoption of a systematic, standardized reporting framework for marine monitoring. By combining AI technology with uniform reporting, managers may be able to exchange helpful information and make well-informed decisions that will promote and enhance management performance (Eger et al., 2022). Democratizing data necessitates working to make relevant data available, including managing and storing it appropriately (Pimm et al., 2015). Regarding AI to detect unlawful dumping, it was suggested that the convolutional neu-

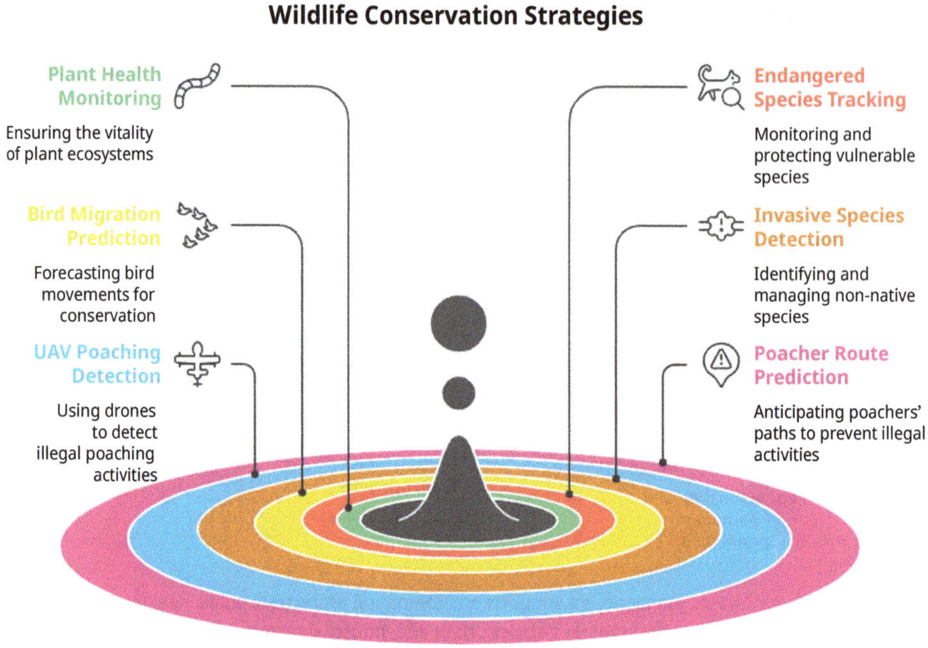

Figure 12.4: AI helps monitor wildlife, analyze ecosystems, predict threats, and support data-driven strategies to protect and restore biodiversity.

ral network features model's faster regions could only detect garbage close to rivers; they could not detect waste in other intricate locations, such as cities (Takahashi et al., 2022). Furthermore, there are some restrictions on the single-shot detection algorithm's ability to detect waste in drone photos. Waste in covered places, such as forests, cannot be detected by the algorithm (Youme et al., 2021). Adherence to strict environmental standards is frequently necessary for the application of AI-driven bioremediation techniques. It can be difficult to follow these rules, particularly when adopting new technologies. To evaluate the security and effectiveness of AI applications in bioremediation, regulatory bodies could demand a great deal of testing and documentation (Akintola, 2024).

Some of the major challenges of using AI for a sustainable future are highlighted in Figure 12.5.

12.7 Future Recommendations

For accurate environmental monitoring and predictive ecosystem monitoring, it is essential to integrate AI-enhanced surveillance technologies with IoT networks. IoT sensor networks and satellite and aerial remote sensing should be combined in a hybrid

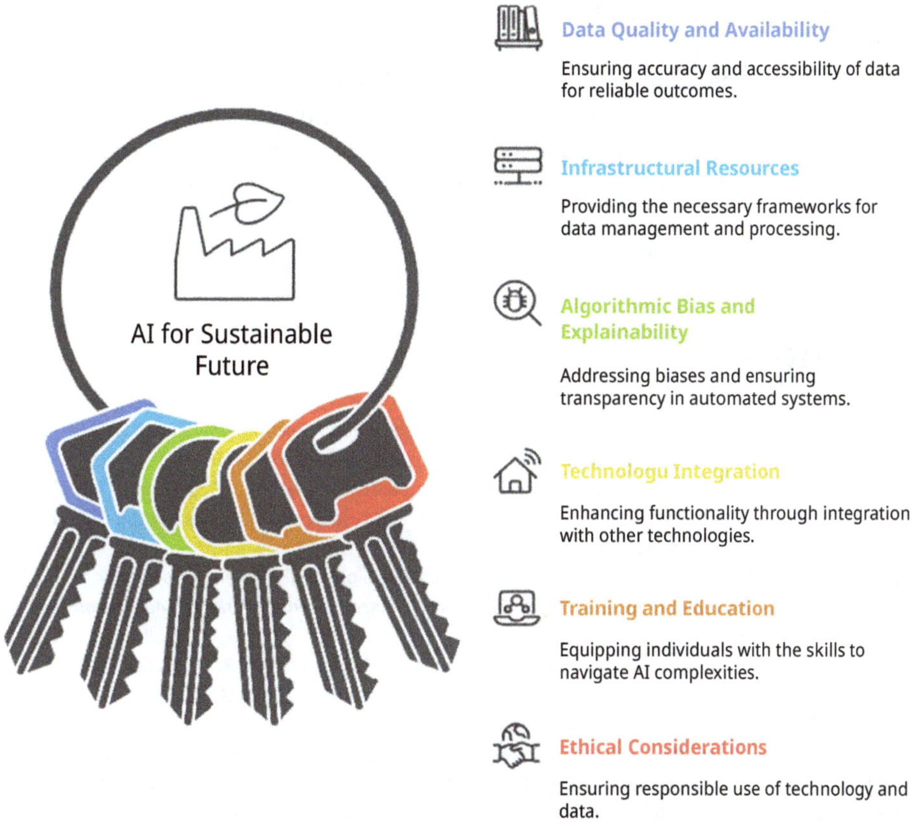

Figure 12.5: High costs, data limitations, ethical concerns, energy consumption, and lack of technical infrastructure hinder the effective and equitable adoption of AI for sustainability.

architecture that uses AI algorithms to combine and analyze data from many sources. IoT data should be processed in real-time using edge computing to improve responsiveness and lower latency. To guarantee smooth data transfer between IoT devices and remote sensing systems, standardized communication protocols (such as MQTT and CoAP) ought to be implemented. Proactive decision-making is made possible by the integration of ML algorithms, which can detect patterns and forecast environmental trends.

12.8 Conclusion

AI has the potential to significantly alter pollution management efforts in river ecosystems, according to an analysis of current bioremediation techniques, case studies, challenges, and opportunities. Using AI algorithms offers chances to increase the effectiveness of pollutant removal and lessen environmental effects, from data-driven pollution tracking to AI-driven restoration process optimization.

Automation is becoming a more appealing option for protection and management in marine environments as finance for long-term monitoring programs becomes more limited, and the cost of acquiring, deploying, and maintaining AI technology continues to drop. Before these technologies are integrated into completely end-to-end automated monitoring systems, there are still a number of study topics that require more examination on their viability and scalability.

References

Akintola, A. A. (2024). AI-driven monitoring systems for bioremediation: Real-time data analysis and predictive modelling. *World Journal of Advanced Research and Reviews, 24*(1), 788–803.

Ayoola, V. B., Idoko, I. P., Eromonsei, S. O., Afolabi, O., Apampa, A. R., & Oyebanji, O. S. (2024). The role of big data and AI in enhancing biodiversity conservation and resource management in the USA. *World Journal of Advanced Research and Reviews, 23*(2), 1851–1873.

Bansal, H., & Jha, S. (2023). Implementation of AI techniques for bioremediation and wastewater treatment. In *The Internet of Medical Things (IoMT) and Telemedicine Frameworks and Applications* (pp. 124–134). IGI Global Scientific Publishing.

Bayraktarov, E., Saunders, M. I., Abdullah, S., Mills, M., Beher, J., Possingham, H. P., . . . & Lovelock, C. E. (2016). The cost and feasibility of marine coastal restoration. *Ecological Applications, 26*(4), 1055–1074.

Bekmurzaeva, R., Kalimullin, R., & Aguzarova, F. (2024). Application of drones and artificial intelligence to monitor and protect natural ecosystems. In *BIO Web of Conferences* (Vol. 140, p. 01008). EDP Sciences.

Blessing, A. A., & Olateru, K. (2025). AI-driven optimization of bioremediation strategies for river pollution: A comprehensive review and future directions. *Frontiers in Microbiology, 16*, 1504254.

Chisom, O. N., Biu, P. W., Umoh, A. A., Obaedo, B. O., Adegbite, A. O., & Abatan, A. (2024). Reviewing the role of AI in environmental monitoring and conservation: A data-driven revolution for our planet. *World Journal of Advanced Research and Reviews, 21*(1), 161–171.

Díaz-García, J. M., López-Barrera, F., Pineda, E., Toledo-Aceves, T., & Andresen, E. (2020). Comparing the success of active and passive restoration in a tropical cloud forest landscape: A multi-taxa fauna approach. *PloS One, 15*(11), e0242020.

Ditria, E. M., Buelow, C. A., Gonzalez-Rivero, M., & Connolly, R. M. (2022). Artificial intelligence and automated monitoring for assisting conservation of marine ecosystems: A perspective. *Frontiers in Marine Science, 9*, 918104.

Divine, I. L. O. H., Aguma, C. P., & Olagunju, A. O. (2024). Integrating AI enhanced remote sensing technologies with IOT networks for precision environmental monitoring and predicative ecosystem management. *World Journal of Advanced Research and Reviews, 23*(02), 2156–2166.

Eger, A. M., Earp, H. S., Friedman, K., Gatt, Y., Hagger, V., Hancock, B., . . . & Reeves, S. (2022). The need, opportunities, and challenges for creating a standardized framework for marine restoration monitoring and reporting. *Biological Conservation, 266*, 109429.

Fang, B., Yu, J., Chen, Z., Osman, A. I., Farghali, M., Ihara, I., . . . & Yap, P. S. (2023). Artificial intelligence for waste management in smart cities: A review. *Environmental Chemistry Letters, 21*(4), 1959–1989.

Gurram, G. V., Shariff, N. C., & Biradar, R. L. (2022). A secure energy aware meta-heuristic routing protocol (SEAMHR) for sustainable IoT-wireless sensor network (WSN). *Theoretical Computer Science, 930*, 63–76.

Han, H., Yuan, X., Zhang, Y., & Jian, Y. (2025). Application of artificial intelligence in ecosystem services research: Current status, challenges, and prospects. *Contemporary Problems of Ecology, 18*(4), 589–602.

Harris, G., Thompson, R., Childs, J. L., & Sanderson, J. G. (2010). Automatic storage and analysis of camera trap data. *Bulletin of the Ecological Society of America, 91*(3), 352–360.

Jayanthi, J., & Kumar, K. A. (2024). AI-Driven restoration: Enhancing biodiversity conservation and ecosystem resilience. In *Explainable AI (XAI) for Sustainable Development* (pp. 180–193). Chapman and Hall/CRC.

Jordan, M. I., & Mitchell, T. M. (2015). Machine learning: Trends, perspectives, and prospects. *Science, 349*(6245), 255–260.

Koskinopoulou, M., Raptopoulos, F., Papadopoulos, G., Mavrakis, N., & Maniadakis, M. (2021). Robotic waste sorting technology: Toward a vision-based categorization system for the industrial robotic separation of recyclable waste. *IEEE Robotics and Automation Magazine, 28*(2), 50–60.

Kumar, B. (2025). AI-based predictive models for engineering nanostructures for bioremediation. *Journal Publication of International Research for Engineering and Management (JOIREM), 5*(05), 1–4.

Lakhouit, A. (2025). Revolutionizing urban solid waste management with AI and IoT: a review of smart solutions for waste collection, sorting, and recycling. *Results in Engineering,* 25, 104018.

Leveziel, M., Laurent, G. J., Haouas, W., Gauthier, M., & Dahmouche, R. (2022). A 4-DoF parallel robot with a built-in gripper for waste sorting. *IEEE Robotics and Automation Letters, 7*(4), 9834–9841.

Mao, S., Wang, B., Tang, Y., & Qian, F. (2019). Opportunities and challenges of artificial intelligence for green manufacturing in the process industry. *Engineering, 5*(6)), 995–1002.

Mukherjee, A. G., Wanjari, U. R., Chakraborty, R., Renu, K., Vellingiri, B., George, A., . . . & Gopalakrishnan, A. V. (2021). A review on modern and smart technologies for efficient waste disposal and management. *Journal of Environmental Management, 297,* 113347.

Noiki, A., Afolalu, S. A., Abioye, A. A., Bolu, C. A., & Emetere, M. E. (2021, February). Smart waste bin system: A review. In *IOP Conference Series: Earth and Environmental Science* (Vol. 655, 1, p. 012036). IOP Publishing.

Norouzzadeh, M. S., Nguyen, A., Kosmala, M., Swanson, A., Palmer, M. S., Packer, C., & Clune, J. (2018). Automatically identifying, counting, and describing wild animals in camera-trap images with deep learning. *Proceedings of the National Academy of Sciences, 115*(25), E5716–E5725.

Pimm, S. L., Alibhai, S., Bergl, R., Dehgan, A., Giri, C., Jewell, Z., . . . & Loarie, S. (2015). Emerging technologies to conserve biodiversity. *Trends in Ecology and Evolution, 30*(11), 685–696.

Praveen, A., Radhika, R., Rammohan, M. U., Sidharth, D., Ambat, S., & Anjali, T. (2020, July). IoT based smart bin: A swachh-bharat initiative. In *2020 International Conference on Electronics and Sustainable Communication Systems (ICESC)* (pp. 783–786). IEEE.

Russell, S. J., & Norvig, P. (2021). *Artificial Intelligence: A Modern Approach*. Global Edition 4e.

Sharma, V., & Jamwal, A. (2024). Artificial intelligence's role in environmental conservation: A study on harnessing artificial intelligence for planetary preservation. *Issue 5 International Journal of Management and Human., 7,* 297.

Sivaprakasam, B. T., Sugilal, G., Shah, J. G., Kaushik, C. P., Krishnamurthy, C. V., & Arunachalam, K. (2020). In situ process monitoring of nuclear waste glass melts using non-contact microwave sensor. *Measurement, 160,* 107793.

Soni, U., Roy, A., Verma, A., & Jain, V. (2019). Forecasting municipal solid waste generation using artificial intelligence models – A case study in India. *SN Applied Sciences, 1*(2), 162.

Takahashi, Y., Fujii, J., & Amakata, M. (2022). Training a robust uav river patrol AI for different river and an analysis of the training data set. In *Proceedings of the National Congress of the Society for Artificial Knowledge* (Vol. JSAI2022, pp. 3S3IS2e01–3S3IS2e01).

Ullah, F., Saqib, S., & Xiong, Y. C. (2025). Integrating artificial intelligence in biodiversity conservation: Bridging classical and modern approaches. *Biodiversity and Conservation, 34*(1), 45–65.

Wani, A. K., Rahayu, F., Ben Amor, I., Quadir, M., Murianingrum, M., Parnidi, P., . . . & Latifah, E. (2024). Environmental resilience through artificial intelligence: Innovations in monitoring and management. *Environmental Science and Pollution Research*, *31*(12), 18379–18395.

Yigitcanlar, T., Mehmood, R., & Corchado, J. M. (2021). Green artificial intelligence: Towards an efficient, sustainable and equitable technology for smart cities and futures. *Sustainability*, *13*(16), 8952.

Youme, O., Bayet, T., Dembele, J. M., & Cambier, C. (2021). Deep learning and remote sensing: Detection of dumping waste using UAV. *Procedia Computer Science*, *185*, 361–369.

Shallu Saini*, Manuj Saini*, Hardeep Singh Tuli, Priti Duhan

Chapter 13
Emerging Technologies at the Intersection of AI and Biotechnology

Abstract: Artificial intelligence (AI) is rapidly transforming biotechnology, enabling predictive, autonomous high-resolution analyses across biological systems. The chapter highlights that deep learning, neural architectures, and biological foundation models underpin tasks such as protein structure prediction, multi-omics integration, drug discovery, and clinical diagnostics. AI accelerates target identification, de novo molecule generation, and biologics engineering while improving preclinical safety assessment and adaptive clinical trials. Laboratory automation has advanced to self-driving labs where robotic execution and AI-driven decision-making operate in closed feedback loops to optimize experiments. Multimodal models integrate genomic, imaging, and clinical data to enhance biomarker discovery and personalized medicine. Digital twins further support real-time bioprocess control and precision manufacturing. However, these advances raise concerns over data quality, algorithmic bias, reproducibility, dual-use risks, and insufficient regulatory frameworks. The chapter emphasizes responsible governance, transparency, and ethical oversight as indispensable for ensuring safe and beneficial bio-AI innovation.

Keywords: Bioprocess, precision, multi-omics, biomarker, robotic

13.1 Introduction

The rapid evolution of computational technologies has brought artificial intelligence (AI) to the forefront of biotechnology, fundamentally transforming biological research, biomedical discovery, and healthcare innovation. The integration of data-driven algorithms with biological sciences is enabling capabilities that were previ-

***Corresponding author: Manuj Saini,** Department of Genetics and Plant Breeding, Chaudhary Charan Singh Haryana Agricultural University, Hisar 125004, Haryana, India, e-mail: doc.manujsaini@gmail.com
***Corresponding author: Shallu Saini,** Department of Bio-Sciences and Technology, Maharishi Markandeshwar Engineering College, Maharishi Markandeshwar (Deemed to be University), Mullana, Ambala 133207, India
Hardeep Singh Tuli, Department of Bio-Sciences and Technology, Maharishi Markandeshwar Engineering College, Maharishi Markandeshwar (Deemed to be University), Mullana, Ambala 133207, India
Priti Duhan, Department of Biotechnology, K. L. Mehta Dayanand College for Women, Faridabad 121001, Haryana, India

 | https://doi.org/10.1515/9783112214275-013

ously considered unattainable, including automated molecular analysis, large-scale biological pattern recognition, and intelligent experimental design. This paradigm shift is driven not only by advances in machine learning and computational power but also by the unprecedented growth of biological data generated through emerging high-throughput technologies. As biotechnology continues to embrace AI, the research ecosystem is transitioning from conventional analytical approaches toward autonomous, predictive, and self-optimizing scientific systems.

13.1.1 The Convergence of AI and Biotechnology: From Rule-Based Bioinformatics to Deep Learning and Foundation Models

Traditionally, computational biology centered on bioinformatics systems based on rules that were derived from prior knowledge and biological rules that were manually maintained. Although these systems provided a way to do early genomic and proteomic studies, they did not provide much flexibility or generalization, limiting scientific advancement. With the development of machine learning and later the deep learning algorithm, there was a significant shift because large-scale biological data could be used to extract patterns without necessarily relying on hard-coded assumptions. This has enabled investigators to discover intricate biological connections, forecast the structural-function interactions, and better represent nonlinear cell processes.

Recent advances in science show that deep learning has become integrated into all layers of biological research probes: gene regulatory networks prediction to protein structure prediction, and phenotype prediction. Hein et al. (2025) demonstrate that the current AI designs are speeding up the process of discovering biology at the continuum of genes to proteins and providing high-resolution analysis where previous models cannot. In addition to deep learning, there is the emergence of foundation models in the biotechnology landscape, which are large pretrained models that can transfer knowledge to many tasks. Guo et al. (2025) emphasize that the biological foundation models, which are trained on the sequences of multi-omic, biomolecular structures, and scientific corpora, are a significant step toward general-purpose and adaptive AI systems in bioinformatics. The shift to these scalable models demonstrates the coming of age of AI-based biology, in which algorithmic thought supplements scientific investigation as opposed to being directly dependent on it. On a larger scale, this convergence becomes a structural shift in the ecosystem of life sciences research, which is prioritized by Hassoun et al. (2021), who assert AI to be a central discipline that can redefine experimental processes, biological explanations, and the development of biological theories. In addition to this, Abu-El-Ruz et al. (2025) show that AI has already entered virtually all fields of biomedical inquiry, which reinforces the idea that the intersection of AI and biotechnology is no longer a fantasy but a necessity in research in the field of biology.

13.1.2 Drivers of Convergence: Data Explosion, Multi-omics Growth, Cheaper Sequencing, and High-Performance Computing

There is a series of forces within the system that have hastened the incorporation of AI in the field of biotechnology. The first and most impactful is the surge of biological data, which has taken place as a result of the developments in multi-omics systems, high-resolution images, wearable biosensors, electronic health records, and real-time biomedical monitoring systems. According to da Silva (2024), the explosive growth of biomedical data, in particular the data in the emerging economies, has made AI an analytical engine that is indispensable in deriving meaningful patterns in both population-level and molecular-level biological information.

Simultaneous improvements in technological infrastructure have supported this tendency. The availability of reduced costs of genome sequencing, as well as the enhancement of mass spectrometry, cryo-EM, and spatial transcriptomics, has made large-scale mapping in biology tremendously more feasible. Moreover, via cloud computing and distributed computing architecture, real-time collaborative use of high-performance resources is achievable without the requirement of in-house supercomputing resources. According to Koppad et al. (2021), the current capability of cloud-enabled platforms to realize large-scale multi-omics analyses renders them both computationally accessible and economically viable to a wide range of research institutions. The increasing reliance of the research community on accelerated computing, including GPUs and TPUs, has also maximized the process of training and deploying deep learning models in the biology domain.

These drivers are supported by the educational trends. As Pickard et al. (2025) emphasize, the current state of biomedical researchers is demanding computational competency as one of the fundamental necessities of their work, with their performance being measured by data analytics, high-performance computing, and AI-enabled research processes, thereby making computational competency an inevitable expectation, and not a highly specialized skill anymore. Together, the spread of biological big data, the low cost of multi-omic technologies, and the accessibility of cloud-based high-performance computing are not simply redefining the processes of biotechnology but are also making sure that the integration of AI and life sciences will not only keep gaining momentum but that the pace of its growth will keep accelerating.

13.1.3 Core AI Technologies Underpinning Modern Biotechnology

The rapid progress of AI has reshaped the methodological foundations of biotechnology by transforming how biological data are acquired, analyzed, and interpreted. Over the last decade, the field has shifted from traditional statistical modeling to adaptive computational systems that learn directly from high-dimensional and heterogeneous biological information. Breakthroughs in deep learning frameworks, specialized neural architectures, and powerful biological foundation models have enabled the life sciences to operate at scales and speeds once considered unachievable. These technologies now underpin a new experimental paradigm in which biological discovery relies on computational intelligence as much as laboratory observation. The core AI technologies driving the acceleration of biological discovery depicted in Figure 13.1.

13.2 Deep Learning Frameworks and Software Ecosystems

The framework of contemporary biotechnology, upheld by AI, is deep learning platforms. Out of these, TensorFlow and PyTorch are the most popularly followed infrastructures with regard to their scalability, modularity, and compatibility with distributed computing. TensorFlow provided a platform where complicated biological mechanisms can be taught on heterogeneous computing outfits, empowering high-performance education of tasks including protein structure apprehension, cellular picture assessment, and multi-omics forecasting (Abadi et al., 2016). Later, PyTorch extended them with an increasingly flexible, imperative programming model with a better understanding of the neural operations and dynamic computation graphs, enabling quick experimentation and optimization of biological models (Paszke et al., 2019). Several libraries, including Keras, JAX, and Hugging Face, have developed around these two structures and pushed the barrier to AI tool development even lower, allowing biologists to develop their own tailor-made AI tools.

Different toolkits tailored to particular domains were created as the breadth and depth of biological queries have grown, and general-purpose deep learning systems are being adapted to particular bioengineering processes. Such toolkits facilitate very technical processes like biomolecular simulation, large-scale sequence representation, and molecular-property prediction. The more recent history of AI-native laboratory systems demonstrates that computer platforms are able to organize the entire process of biological experimentation, both in silico hypothesis generation and in vitro experimentation. An example of a demonstration of an autonomous biomolecular engineering laboratory that combines deep learning models with robotic experimentation shows how computational infrastructures are transforming into tools of analytical functions toward active scientific discovery (Wu et al., 2025).

There is an equal pace in the development of hardware in line with biotechnology, accelerated with the help of software ecosystems. Current biological deep learning models are built on the concept of parallelism using GPUs and are becoming progressively dependent on accelerators of TPU scale in order to process multiscale biological data. Pharmaceutical partnerships related to industrial cooperation show how AI supercomputing is now strategically vital to competitive drug discovery, and massive companies invest heavily in AI-powered platforms to reduce molecular screening timelines and enhance therapeutic hit rates (Philippidis, 2021). Together, this digital infrastructure – structures, toolkit, domain-specific, and special hardware – constitutes the technological basis on which modern AI-based biotechnology functions.

13.2.1 Neural Network Architectures in Biological Data Analysis

The triumph of AI in biotechnology cannot be attributed to the existence of the software and hardware only, but rather to the development of the neural design, which can learn the natural structure of biological events. Convolutional neural networks (CNNs) are one of the first structures to redefine biological studies. They have made digital microscopy, automated pathology, and radiomics high-precision diagnostic instruments due to their capability to identify hierarchical spatial patterns. CNN-based systems are currently regularly used to analyze histopathology slides, medical imaging, and subcellular fluorescence patterns with degrees of consistency and sensitivity that make them useful in clinical decision-making (Esteva et al., 2021).

Biological sequence analysis has been facilitated by neural networks modeled sequentially. There are numerous recurrent and hybrid sequence models, which add attention mechanisms to learn the long-range dependencies across genomic, transcriptomic, and proteomic data-point dependencies, which traditional sequence alignment methods are unable to predict. Recent architectures that operate on hybrid deep sequential networks have shown high predictive performance in multi-omics tasks, which involve survival and disease profiling and need integration of regulatory and clinical features across time (Hazarika et al., 2026).

This evolution is continued with the introduction of the transformer-based genomic language models. Such models consider the sequence of DNA, RNA, and proteins as biological languages and are pretrained on large quantities of unlabeled sequences, and then transfer knowledge to a wide variety of downstream bioinformatics tasks. The models based on transformer have demonstrated the state of the art in regulatory genomics, predicting variant impacts and molecular phenotyping tasks (Leema et al., 2025). Notably, besides making progress in sequence learning, transformers are also closing the gap between biological modalities (text, structures, gene networks, and molecular images) and single predictive models.

Graph neural networks (GNNs) are a second architecture of biological data analysis due to their capacity to capture the non-Euclidean structure of molecules. GNN methods consider atoms, proteins, or biological entities as nodes and learn the interactions spread across the molecular graphs. It has been shown that this structural representation is quite successful in predicting protein-protein interactions, molecular binding affinities, and stability of biological complexes (Jha et al., 2022). With biological problems often being associated with relational complexity, GNNs are actively combined with transformers and sequential models to be able to represent biochemical hierarchy at all spatial, structural, and temporal scales.

13.2.2 Foundation Models and "Scientific Superintelligence."

A new stage of biotechnology is continuing to emerge whereby foundation models, large, self-supervised AI systems that are trained on large and heterogeneous biological data, are now driving discovery. The difference between these models and the traditional AI methods is inherent, as these models are not tailored to a specific task; they are constructed to generalize knowledge on wide biological and biochemical issues. The foundation models have the ability to absorb genetic sequences, protein structure, scientific literature, and phenotypic measures, which allows it to be used as a single engine to provide biological inferences. Their theoretical and practical value to computational biology is fully developed, which is a shift to more generalizable and reusable computational knowledge systems (Guo et al., 2025).

Various types of biological foundation models have been proposed in proteins, DNA, RNA, single-cell datasets, and multimodal omics. Models that are protein-based focus on conformation and stability prediction of a molecule, whereas genomic foundation models have the capacity to predict regulatory grammar and genetic variation. Unimodal models have developed into multimodal architectures based on the possibility of integrating structural, textual, and experimental data and developing a holistic perspective of biological systems. Their growing ability has spurred the creation of dedicated big language models for biological and chemical reasoning, and more are being utilized in protein engineering, ligand design, and automated biomolecular annotation (Leema et al., 2025).

It is the direction that foundation models are taking that biotechnology is headed to a future sometimes called scientific superintelligence. Within this paradigm, systems of AI not only analyze biological data but also generate hypotheses autonomously, coordinate experiments, and self-optimize their knowledge in closed cycles of scientific discoveries. Recent polls on autonomous scientific discovery detail the transformation of AI-based research pipelines in accordance with computational assistants to agentic collaborators of laboratory equipment and knowledge systems (Wei et al.,

2025). Gabriel, with the emergence of AI-native experimental labs (Wu et al., 2025), this trend implies that lifetime discovery processes and their sequence, starting with molecular ideation and ending with experimental validation, can soon be performed by the computational agent with little human oversight.

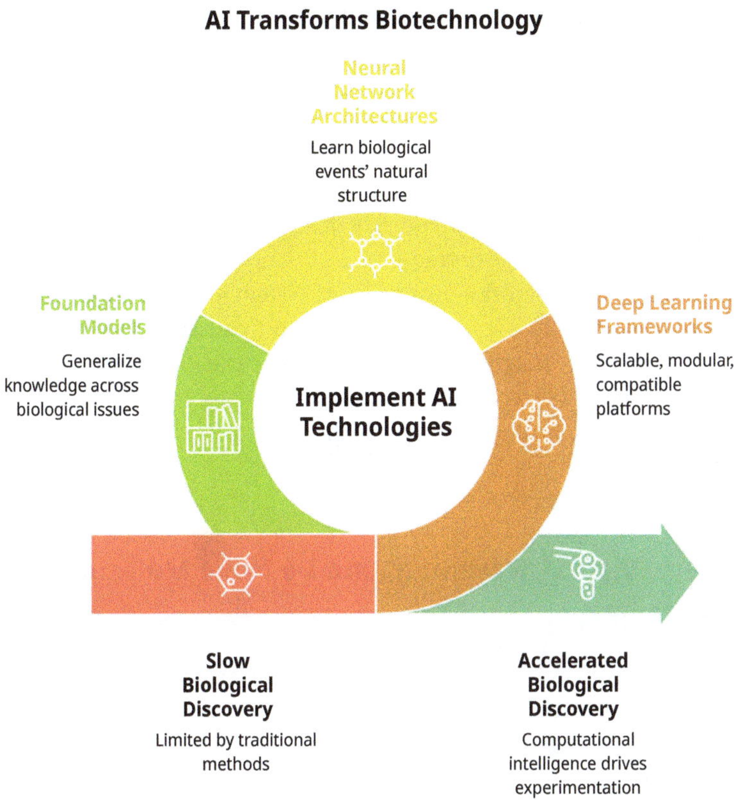

Figure 13.1: Advanced AI methods are accelerating biological research by predicting molecular structures, modeling complex systems and uncovering new insights in genomics and drug discovery.

13.3 AI-Driven Drug Discovery and Design

Pharmaceutical research has accepted AI to become an aspect of the research through the transformation of how drugs are discovered and designed, through the development of biologics, and the screening of drug candidates before clinical trials. In contrast to the previous computational drug-discovery techniques, where the computational techniques were highly dependent on preestablished molecular assumptions, the current AI methods are now trained on biological and chemical data. This enables predictive models to extrapolate the disease biology, molecular libraries, and variability. Consequently, AI is becoming viewed as a strategic pipeline accelerator of drugs and not an additional research resource.

13.3.1 Target Identification and Validation

Effective drug discovery starts with the identification of biological targets that are pharmacologically accessible and that are mechanistically significant. Contemporary AI has been able to assist with this step by combining multi-omics data and literature via large-scale deep learning and language-model reasoning. They present useful models in complex therapeutic fields that can identify hidden gene-disease or protein-disease interactions not presented in more traditional statistical processes. Network-based analysis is also starting to be implemented as an AI system for target prediction, and allows researchers to study biological pathways, not individual molecules. This tendency is supported by recently designed generative frameworks that combine molecular interaction prediction with drug-target scoring to reduce the search space during the initial discovery phase and decrease the probability of failure at the later stages of development (Kotkondawar et al., 2025).

13.3.2 Hit Discovery, Virtual Screening, and De Novo Molecule Generation

AI can be used to hasten the search for chemical matter once promising targets are known. Advances in de novo molecule design are due to the emergence of generative models, e.g., variational autoencoders, generative adversarial networks, reinforcement learners, and diffusion frameworks, where new molecule structures are generated instead of sampling one of the existing libraries. Such models are not only maximizing chemical novelty, but also trying to match molecular generation and therapeutic purpose, and physicochemical characteristics. The recent research reviews point to the fact that there is a significant shift in the paradigm of the traditional computational screening to the generative learning approaches that can suggest optimized lead candidates (Martinelli, 2022; Mouchlis et al., 2021).

The industrial drug-design strategies are also being affected by generative AI. Gangwal and others also noted in 2024 that generative methods are leaving the experimental phases and entering real-life platforms in pharmaceutical companies, since they are able to combine biological interest with chemical producibility (Gangwal et al., 2024). One of these directions is the creation of policy-driven generative systems that include feasibility rules and synthetic constraints. Such systems are aimed at generating molecules that are not only pharmacologically significant but also realistic to generate and scale, and aid in shortening the gap between the generation of leads in computational models and experimental validation.

The other aspect of hit discovery is the AI-based structure design. After three-dimensional protein-structure prediction has become a part of drug screening, virtual screening can be used to assess binding affinity and interaction geometry based on the computational models, not on pure experimental work. The research of AI-assisted molecular design demonstrates that the synergistic use of expected structural data with generative pipelines has allowed prioritizing more compounds to be experimentally tested, more rapidly (Mak et al., 2024).

13.3.3 Biologics, Antibodies, and Protein Engineering

AI has also been of equal significance in the discovery of biologics. Whereas the classical methods of engineering antibodies relied on lengthy iterative cycles of mutation and testing, deep learning is currently offered as a rational method to search the protein sequence structure function space. Recent studies demonstrate that the protein-design models are capable of suggesting sequence alterations that would result in an increase in affinity, a decrease in immunogenicity, or enhanced thermostability (Chungyoun & Gray, 2023). The design of antibody libraries is already being performed using machine-learning techniques, binding hot spots are being identified, and mutations that enhance antigen recognition are proposed. Surveys in this area suggest that AI-assisted workflows are deployed in both the discovery and optimization and developability evaluation of biologics pipelines (Zheng et al., 2024; Kim et al., 2023).

13.3.4 AI in Preclinical Development and Clinical Trials

Once the choice of leads or biologics has been made, AI models keep reinventing preclinical and clinical testing. One of the biggest problems with drug development is the failure of the candidate because of inadequate pharmacokinetic or toxicological characteristics. Machine-learning systems like ADMET-oriented predictors confirmed that simulations with high throughput can identify toxicity and off-target effects before animal testing, which saves financial and ethical expense (Swanson et al., 2024). Simultaneously, AI will be a major facilitator of more adaptive clinical trials. Instead of having homogenous patient enrolment, AI-aided designs encourage trial personalization by assessing disease subtypes, biomarker signatures, and anticipated treatment response. It is an adaptive method, which is supposed to reduce timescales, enhance safety, and achieve higher statistical power when conducting therapeutic programs by patient stratification (Ogidi et al., 2025).

13.4 AI-Integrated Laboratory Automation and Self-Driving Labs

The automation of laboratories has been taken to such an extent that AI does not just help in experiments, but it manages them. This automation, robotics, and data-driven reasoning convergence is changing the traditional laboratory into a computationally sensitive ecosystem, whereby experiments can now be designed, run, assessed, and improved through further refinement with slight human intervention. Three interacting forces are helpful in this transition, including robotics and liquid-handling hardware, workflow coordination, and data management software systems, as well as intelligent models that can make decisions during experimentation. Collectively, these developments are indicative of a novel research paradigm, namely the self-driving laboratories capable of achieving self-optimization of scientific workflows.

13.4.1 Evolution of Laboratory Automation

The first robots in automation were liquid-handling robots and laboratory information management systems (LIMS) to enhance efficiency and minimize human error. These systems, however, involved manual interpretation of data by researchers as well as planning of further experiments. This method has been changed with the integration of AI into the laboratory setting, with robots providing the precision of a system and algorithmic thought. This transition is recorded in clinical and biomedical laboratory studies, where the recent evolution of AI-driven automation has become the decisive factor in the workflow instead of merely assisting in its implementation (Naugler & Church, 2019). Intelligent instrumentation is being quickly adopted, which indicates changes at the market level, where laboratories are becoming more and more invested in automation not only to achieve throughput, but also to acquire features that previously were the prerogative of scientists: interpretation, planning, and optimization. Further expansion of AI-powered lab infrastructure is an indication that the research community now regards automation as a strategic driver in speeding up discovery and not just an upgrade in operations. The evolution of this shift is the initiation of a cultural shift in industrial and academic science as a result of computational thinking.

13.4.2 Architecture of AI-Integrated Laboratory Automation Systems

The automation system based on AI in laboratories is developed on three layers that are mutually dependent, including hardware, software, and data.

13.4.2.1 Hardware Layer

Modular robots have become increasingly popular in laboratories that can communicate with each other in a singular workflow. The modern lab now consists of a physical execution network that is made up of robotic arms, liquid handlers, mobile carriers, and automated analyzers. Redesigning automation platforms of high-throughput mammalian and bacterial cell-culture processes demonstrate how modular hardware is becoming flexible instead of being fixed (Hartman, 2024).

13.4.2.2 Software Layer

The role of automation software becomes one of the central ones. As opposed to basic scheduling solutions, advanced software combines LIMS, workflow planners, and digital-twin simulations that can simulate the working process of an experiment before its performance. The software-based orchestration is considered by professional analyses to be among the key factors that define reliability in automated laboratory settings (Naugler & Church, 2019). The digital twin also allows minimizing the waste of resources and greatly decreasing optimization cycles.

13.4.2.3 Data Layer

The third automation pillar is a structured and cohesive data infrastructure. One records temperature, humidity, process parameters, and reaction outputs in sensors and Internet-of-Things devices. Metadata standards are incorporated into these streams, enabling to comparison the instruments and tracking the experiments in a longitudinal way. Probably more important than the level of sophistication of the individual hardware is the integrity and accessibility of data, which determines the value of an automated laboratory. Clear data models are also useful for traceability in order to support historians, as well as training machine-learning algorithms in order to support pipelines.

13.4.3 Self-Driving Laboratories (SDLs)

The next phase of automated science is self-driving laboratories, which operate based on a closed feedback loop, where AI chooses experiments, robotics performs them, and the results are automatically analyzed to make the next decision. SDLs replace experimentation as a sequential, human-directed experimentation process with an adaptive algorithmic cycle that is much more efficient at exploring scientific search spaces compared to traditional workflows (Tobias & Wahab, 2025). Synthetic biology evidence demonstrates that SDLs enhance genetic construct and cellular responses optimization to a series of design-build-test-learn cycles (Martin et al., 2023). SDLs have demonstrated complete autonomous discovery and optimization of functional materials in materials science without being screened by humans exhaustively (Soldatov et al., 2021). Performance metrics of SDL, such as discovery rate, convergence rate, and resource efficiency, are currently being formalized, which is an indicator of growing maturity in the field (Volk & Abolhasani, 2024). An important aspect that facilitates SDL efficiency is the application of adaptive algorithms. Instead of exploring all combinations of variables, active-learning and Bayesian optimization methods can be used to steer the system to the most informative regions of the experiment and can be very fast in converging (Minocha & Pandey, 2025). This makes SDLs not only fully automated laboratories, but smart discovery engines.

13.4.4 Cloud Labs and the Rise of "AI Science Factories"

Cloud laboratories are used to provide robotic automation, coupled with remote accessibility, such that researchers can design and submit experiments electronically, and the robotic infrastructure is executed. This model gets rid of the geographic limitations, and it democratizes high-end laboratory resources. The national implementation of cloud laboratories within Japan is an example of remote automation that allows for the performance of entire scientific processes without physical instrumentation (Yoshikawa et al., 2025). The conceptual phase that follows the cloud labs is the AI science factory. Under this paradigm, autonomous laboratories are not only paired with robotic capability but also with AI-powered systems that can create hypothesis, process and reason over data, as well as model discovery priorities. Technological futurists refer to this as a trend in the direction of scientific intelligence with the ability to speed up the production of knowledge more rapidly than what experiments constrained by human limits can provide (Kurzweil, 2022). The model of a science-factory symbolizes the strategy of making scientific discovery continuous – it is working like an independent engine instead of a series of projects.

An even more ambitious direction is the development of "AI science factories." In this model, high-throughput automated labs are paired with generalist scientific AI systems capable of generating hypotheses, designing experiments, and analyzing outcomes autonomously. Commercial technology developers are now advancing this paradigm, aiming to build laboratories that operate as discovery engines rather than as service facilities. While cloud labs focus primarily on accessibility and throughput, science-factory models prioritize autonomous knowledge generation and continuous discovery.

13.5 Multimodal and Systems-Level AI in Biotechnology

AI has progressed from analyzing isolated biological datasets to interpreting the biological system as an integrated whole. Modern biotechnology increasingly depends on AI models that merge multimodal information genomic sequences, transcriptomic signatures, spatial imaging, clinical records, behavioral markers, and environmental influences – to generate a unified understanding of biological complexity. This transition is not merely computational; it represents a conceptual shift in how disease mechanisms, therapeutic targets, and biological behaviors are interpreted at scale.

13.5.1 Multimodal AI for Integrated Biological Understanding

Multimodal AI computational models learn through several different types of biological data, as opposed to one. This is interpreted as the ability to combine genomics, transcriptomics, proteomics, imaging, metabolomics, and clinical measurements and enables interpretation of health and disease at a resolution never before experienced. As one example, molecular features do not necessarily reflect the entire phenotype of a disease; however, with spatial imaging and indicators of a patient, models can be used to identify subtle disease progressivity signatures. The latest studies focus on the idea that combining multimodal signals will lead to stronger diagnostic biomarkers and increased biological interpretability (Parvin et al., 2025).

Multimodal AIs are becoming a necessity in personalized and precision medicine. The manifestation of the disease in different individuals is different since it depends on genetic background, lifestyle, environment, and comorbidities. Risk stratification, prediction of treatment response, and personalized/individual therapeutic choice are being revolutionized by AI systems that synthesize patient-specific molecular measurements using continuous clinical records. A significant one is the introduction of multimodal pipelines into the pharmaceutical decision process itself, especially in chronic, metabolic, and neurodegenerative diseases, in which population-scale biomarkers are lacking (Somara et al., 2025). Such advancements mark the beginning of an era where the treatment approach is based on the population to patient-specific therapeutic models that are facilitated by ongoing and data-driven learning.

Scientific generalization is also enhanced with multimodality. The single-omic models are not effective when used with new datasets since biological systems are frequently dependent on coordinated signaling as opposed to the isolated change. This problem can be addressed through multimodal AI, which models co-regulation across biological layers; thus, increasing reproducibility in a wide range of populations and in different centers (Parvin et al., 2025).

13.5.2 AI in Synthetic Biology and Genetic Circuit Design

Synthetic biology aims at building biological worlds with predictable and programmable behaviors. AI has now become a crucial accelerator of this process since the performance of genetic circuits cannot be entirely deduced using linear intuition. Models based on deep learning are now able to predict the impact of promoter strength, transcriptional regulator, ribosome binding location, and circuit architecture before being experimentally demonstrated. Research proves that automated design pipelines significantly make fewer trips to the laboratory before working constructs are found, and multiparametric design of circuits, once thought too complex circuits is now possible (Mirchandani et al., 2025).

The role of AI in the safety and precision of genome editing is also high. CRISPR-based therapeutics are associated with the difficulties of off-target mutagenesis and different cell types, with unpredictable efficiency of editing. Learning-based frameworks do not require wet-lab editing or exposure to the risk of such unintended interactions, and thus can be used to predict unintended interactions earlier, making therapeutic translation in the wet lab quicker. The literature reviews have indicated that AI-based CRISPR predictive models are currently being implemented at the initial phases of pipeline design, and not solely at later stages as either validation tools (Mirchandani et al., 2025). With the increasing scope of genetic engineering, AI is being placed at the center of the toolbox, appearing as a replacement for a tool rather than a supplement to biological design.

13.5.3 Digital Twins in Bioprocessing and Biomanufacturing

Digital twins are used to model and optimize bioprocessing and biomanufacturing processes. One of the most disruptive points of AI and biotechnology is digital-twin systems. A digital twin is a constantly evolving virtual simulator of a biological or industrial process that obtains real-time measurements of sensors and returns computational projections to physical activities. Digital twins can be used to simulate metabolic flux, biomass growth, product yield, risk of contamination, and bioreactor dynamics during the manufacture of biopharmaceuticals. It results in control-feedback environment that can regulate the use of nutrients, oxygenation, temperature, feed rates, or agitation to achieve the best conditions (Panjwani et al., 2025).

The digital twins based on AI are currently used in more extensive bioprocessing domains like microbial fermentation, algae-based production, and enzyme production. A recent survey shows that optimization with digital twins can reduce scale-up times by a significant margin and resource usage by effectively aiding in the refinement of manual processes (Butean et al., 2025). The discipline is shifting quickly toward human-machine collaborative intelligence, with engineers controlling the results and AI taking care of the repetitive control loops, sensitivity testing, and anomaly detecting (Shahab et al., 2025).

Industrial research also indicates that digital twins have a significant role to play when it comes to regulatory compliance and quality by design, since they ensure that variability is predicted ahead of its impact on the quality of products. Their use is not restricted to factories: digital-twin models have been suggested as the models of computational health twins in personalized medicine, as the connection between omics, clinical history, and behavioral tracking to predict disease development and progression (Li et al., 2025). This patient-manufacturing twin connection is an extension of an increasing trend in biotechnology, which is systems-level modeling of biological function at the laboratory, the factory, and the clinic level.

13.6 AI in Diagnostics, Biomarker Discovery, and Personalized Medicine

The use of AI has played a major role in the decision-making process in the contemporary clinical environment, allowing for faster disease recognition, identification of biomarkers that humans cannot, and the provision of customized treatment approaches. Compared to conventional diagnostic pipelines, which use a single type of data or a set of clinical rules, AI enables the analysis of complex patterns in imaging, molecular measurements, and longitudinal patient profiles. These computing functions are putting a new spin on the course of healthcare toward reactive treatment to a preventive and precision-oriented medicine.

13.6.1 AI-Powered Diagnostics

The first and most prominent application of AI has been in the field of diagnostics, specifically medical imaging computer vision systems. The models of deep learning can now match or surpass the performance of an expert in radiology, pathology, and point-of-care imaging. CNNs and multimodal image fusion in radiology resulted in the automated detection of lesions, fracture classification, the possibility of anatomical mapping of organs, and tumor segmentation with high reliability (Katal et al., 2024). Current practice reviews underline that AI is not going to oust radiologists, but increase their diagnostic visibility by highlighting a subtle abnormality and making critical cases visible (Obuchowicz et al., 2025).

Mergers between pathology and radiology in the form of single pipelines of analysis are getting faster. Classification systems based on deep learning are now capable of analyzing both cross-sectional scans and histopathology images to create more accurate disease signatures than either of the two alone (He et al., 2025). There is also the improvement in diagnostics in low-resource, point-of-care environments through AI by providing automated image interpretation assistance in cases where expert skills are limited (Al-Waeli & bin Abdul Hameed, 2025). In addition to imaging, molecular diagnostics, especially cancer profiling and infectious disease surveillance, are enjoying the advantage of AI-assisted interpretation of complex assay results, in which molecular signatures can be very heterogeneous.

13.6.2 Biomarker Discovery from High-Dimensional Data

The analysis of high-dimensional datasets with the capacity to capture cellular, structural, and physiological complexity is becoming more and more important in the discovery of clinically meaningful biomarkers. Single-cell omics and spatial transcriptomics can be used to map the heterogeneity of cells in a comprehensive manner, but the size and resolution of such data exceed human analytic ability. AI-based systems have become available to uncover trends connecting molecular phenotypes and disease conditions. The research on AI-enabled biomarker pipelines states that algorithmic feature learning eliminates the fear of false discoveries and increases sensitivity to infrequent biological signatures (Javaid et al., 2025). Proteomic and metabolomic biomarker discovery, too, has focused attention on AI as the dimensionality and stochastic variability of data make the traditional methods of analysis problematic. Beyond the association level prediction, deep learning makes it possible to predict the risk of disease dynamically and time-varyingly. A recent example of these is a model predicting the pancreatic cancer development based on longitudinal health trajectories through AI that shows how AI can identify preclinical risk patterns within a long time before radiologic manifestation (Placido et al., 2023). The idea of imaging biomarkers to predict disease risk is also becoming increasingly popular, and AI capacity to rank people by their risk of future disease validates itself with massive population-scale imaging data (Seletkov et al., 2025).

13.6.3 Personalized and Stratified Therapies

One of the most notable contributions of AI to healthcare is the transition of population-level medicine to the selection of patient-specific therapy. Multimodal models are based on the concept that precision medicine platforms classify patients into biologically relevant subsets instead of culminating in broad diagnostic categories. Such AI-based stratification models enable clinicians to align therapies to genetic characteristics, disease progression, comorbidity, and anticipated responsiveness by calculative forecast (Khalifa & Albadawy, 2024). This is contrary to traditional treatment guidelines that believe that there is a consistent response in large groups of people.

There is also the development of companion diagnostics or algorithms that detect the groups of patients who may respond to a particular therapy and the people who might be susceptible to adverse events, an AI trend that continues to grow. Digital phenotyping combined with imaging biomarkers and longitudinal data can be used to provide adaptive treatment guidance where therapeutic decisions are updated as the patient changes in physiology. This dynamically controlled adjustment of treatment algorithmically is particularly promising in chronic diseases, oncology and metabolic, where the dynamically advancing biological progress is demanding personalized assessment of response and not fixed treatment choices.

As multimodal biomarker discovery, diagnostic AI, and adaptive therapeutic systems meet, healthcare is transforming into continuously learning clinical spaces that combine patient data in real time and optimize treatment plans through repeated cycles throughout the life cycle of the patient.

13.7 Trust, Ethics, Governance, and Biosafety

The integration of AI into biotechnology has brought profound scientific benefits, yet it has also introduced new challenges relating to trust, safety, and global responsibility. AI systems that analyze biological data, design genetic constructs and operate automated laboratories raise questions about data integrity, algorithmic fairness, dual-use risk, and the adequacy of regulatory frameworks. Ensuring that advances in bio-AI remain safe, ethical, and socially beneficial requires deliberate policies and technical safeguards that evolve alongside innovation.

13.7.1 Data Quality, Bias, and Reproducibility in Bio-AI Systems

Biotechnological data is extremely heterogeneous, and includes genomics, imaging, clinical measures, biobanking data, and experimental outcomes. In case data sources are missing, unbalanced or inadequately standardized, AI models might learn spurious correlations and produce biased results. Ethical analyses of AI in biomedical science emphasize the problem of biases being transferred to the clinical decision tool because of the presence of biases in biological data, which ultimately leads to the deterioration of diagnostic reliability and patient care (Kargl et al., 2022). Such a threat is further increased when multi-omic data and global biobanks are integrated with machine learning pipelines.

The other essential part of credible bio-AI is model interpretability. When algorithms are opaque black boxes, it is hard to understand whether clinicians and regulators are able to know whether predictions have a biological basis or are based on artifacts in the training data. Ethical considerations stress that both validation and benchmarking are processes that need to be performed continuously, as opposed to demonstrating the accuracy once (Dara & Azarpira, 2025). This type of caution should especially be observed when dealing with AI systems, which are supposed to provide information to make critical decisions during diagnostics and therapeutics.

The transparency of training data, metadata, hyperparameters, and evaluation datasets is also important in terms of reproducibility. In the absence of documentation and disclosure, when experimental results are observed as potentially clinically promising, they cannot be generalized in practice when applied in a real-world environment (Lorenc-Kukuła, 2025). Consequently, to achieve trustworthy performance, accurate algorithmic performance is not sufficient, and stringent laboratory, clinical, and computational reproducibility practices are also necessary.

13.7.2 Safety and Dual-Use Concerns

The dual-use of AI in biotechnology has turned out to be of primary concern. The safe and positive discovery can be expedited by the use of AI, and the same level of computer capabilities can be used to create harmful biological entities or to optimize the breeding environment of dangerous organisms. Research on AI-powered bioengineering warns that, unless controlled, the availability of biological design tools might give malicious users the ability to do what used to take a large amount of lab work to perform (Sandbrink, 2023).

Investigations on risk analysis suggest that AI-assisted laboratory automation increases this dual-use dilemma. Autonomous planning and execution of biology-related experiments could be abused in the case of a lack of appropriate safeguards (De Haro, 2024). Simultaneously, AI-based innovation can be enormous in relation to the creation of antimicrobial designs, protein engineering, and vaccine development, creating a fine line between fostering discovery and encouraging abuse (Wheeler, 2025).

The international bodies have started studying these risks at hand. The National Academies held a massive workshop that cautioned the increased availability of dangerous functionalities of AI-powered automated laboratories and design technologies without effective oversight mechanisms (National Academies of Sciences, Engineering, and Medicine, 2024). Control issues spread to the new fields of AI-assisted organism design, gene drives, and synthetic genomes, and even hypothetical computational knowledge may drive a shift in the biological security environments (Zhang et al., 2025).

13.7.3 Regulatory and Policy Landscape

The bio-AI regulatory framework is yet to be developed, and its development pace remains slower than the development of technology. Existing policies are centered on conventional biotechnology, and there is an absence of clarity regarding AI-based processes in drug discovery, diagnostics, and laboratory automation. The policy environment reviews indicate that the governance should be dynamic but not fixed since the rules can become obsolete with the development of AI methods (Undheim, 2024). The issue is to facilitate innovation, without losing its focus on safety and social responsibility.

Responsible innovation relies on algorithm auditability standards, data integrity, design decisions, and training source traceability standards. Such protection measures play a vital role in detecting all the unwanted model behavior and in ensuring that biological candidates made by AI satisfy biosafety regulations (Bloomfield et al., 2024). The moral commentaries argue the necessity of control measures that integrate technical control and professional codes of conduct to make sure that market motives do not take precedence over safety concerns (Dara & Azarpira, 2025).

Policy analysts claim that an internationalized model of governance is needed because biotechnology has a transnational nature. In the absence of international coordination, high-risk experimentation would migrate to other areas where it might not be controlled, thus sabotaging safety in other areas (Wheeler, 2025). The future of safe and useful bio-AI must be in the form of regulatory ecosystems that promote discovery but impose transparency, accountability, and risk reduction.

13.8 Conclusion

The accelerated introduction of AI into the biotechnology field has brought about a transformation in the whole gamut of biological science, from one end of initiation of molecular discovery to clinical decision-making and biomanufacturing. AI is now not an assistant computing tool but a backbone that facilitates scientific research, translation, and innovation, as discussed in this chapter. Deep learning models, neural networks, dedicated bio-AI software, and biological basis models have all taken biological knowledge generation and use to new levels of capability and predictability in pattern recognition, structural inference, and predictive modeling to transform the mode of biological knowledge production and use.

The effects of AI are the most notable in areas where biological complexity had created barriers that were impossible to break. Generative learning models, multimodal sequence-structure predictors, and AI-guided protein engineering have played a major role in drug discovery and molecular design to shorten the time that is spent between hypothesis generation and actionable therapeutic candidates. Equally, highly sensitive methods of AI-assisted diagnostics and biomarker discovery have guided the medical field toward personalized and proactive therapy by using multimodal clinical, molecular, and imaging data to identify disease at earlier stages and map individual responses to therapy. Further evidence of the merging of automation, robotics, and computational intelligence is found in the emergence of self-driving laboratories and digital-twin bioprocessing, which allows a closed-loop, autonomous experimentation that can more and more operate above human throughput limits.

With these achievements, this chapter highlights that advancements should be accompanied by accountability. The strength of AI in biological design, laboratory automation, and synthetic engineering is associated with new threats to the integrity of data, algorithmic fairness, reproducibility, and dual-use risks. There must be governance, transparency, ethical control, and international coordination that ensures that innovation is safe and fair. The responsible implementation of bio-AI must be validated continuously, the computational choices and structures must be traced, and they must serve both the interests of society and communal development.

Together, AI and biotechnology have reached the point of convergence that signifies the onset of a new scientific era known as predictive, autonomous, and computationally guided biological knowledge. When directed with care and commitment, the instruments and models introduced in this chapter can hasten the discovery process, democratize scientific innovation, and help to create healthier and more robust world communities. What lies ahead will be influenced not only by technical advancements but also by the ability of all parties to bring AI-enabled biotechnology in line with ethics, safety, and human good.

References

Abadi, M., Agarwal, A., Barham, P., Brevdo, E., Chen, Z., Citro, C., . . . & Zheng, X. (2016). Tensorflow: Large-scale machine learning on heterogeneous distributed systems. *arXiv preprint arXiv:1603.04467.*

Abu-El-Ruz, R., Hasan, A., Hijazi, D., Masoud, O., Abdallah, A. M., Zughaier, S. M., & Al-Asmakh, M. (2025). Artificial intelligence in biomedical sciences: A scoping review. *British Journal of Biomedical Science, 82*, 14362.

Al-Waeli, S. H. A., & bin Abdul Hameed, H. N. (2025). AI in medical imaging & diagnostics. *Wasit. Journal for Pure Sciences, 4*(3), 35–45.

Bloomfield, D., Pannu, J., Zhu, A. W., Ng, M. Y., Lewis, A., Bendavid, E., . . . & Inglesby, T. (2024). AI and biosecurity: The need for governance. *Science, 385*(6711), 831–833.

Butean, A., Cutean, I., Barbero, R., Enriquez, J., & Matei, A. (2025). A review of artificial intelligence applications for biorefineries and bioprocessing: From data-driven processes to optimization strategies and real-time control. *Processes, 13*(8), 2544.

Chungyoun, M. F., & Gray, J. J. (2023). AI models for protein design are driving antibody engineering. *Current Opinion in Biomedical Engineering, 28*, 100473.

da Silva, R. G. L. (2024). The advancement of artificial intelligence in biomedical research and health innovation: Challenges and opportunities in emerging economies. *Globalization and Health, 20*(1), 44.

Dara, M., & Azarpira, N. (2025). Ethical considerations emerge from Artificial Intelligence (AI) in biotechnology. *Avicenna Journal of Medical Biotechnology, 17*(1), 80.

De Haro, L. P. (2024). Biosecurity risk assessment for the use of artificial intelligence in synthetic biology. *Applied Biosafety, 29*(2), 96–107.

Esteva, A., Chou, K., Yeung, S., Naik, N., Madani, A., Mottaghi, A., . . . & Socher, R. (2021). Deep learning-enabled medical computer vision. *NPJ Digital Medicine, 4*(1), 5.

Gangwal, A., Ansari, A., Ahmad, I., Azad, A. K., Kumarasamy, V., Subramaniyan, V., & Wong, L. S. (2024). Generative artificial intelligence in drug discovery: Basic framework, recent advances, challenges, and opportunities. *Frontiers in Pharmacology*, *15*, 1331062.

Guo, F., Guan, R., Li, Y., Liu, Q., Wang, X., Yang, C., & Wang, J. (2025). Foundation models in bioinformatics. *National Science Review*, *12*(4), nwaf028.

Hartman, T. W. (2024). *Development of a Modular Lab Automation System with Applications to Animal and Bacteria Cell Culture* (Master's thesis, University of South Dakota).

Hassoun, S., Jefferson, F., Shi, X., Stucky, B., Wang, J., & Rosa, E., Jr (2021). Artificial intelligence for biology. *Integrative and Comparative Biology*, *61*(6), 2267–2275.

Hazarika, A., Kumar, A., & Halder, A. (2026). AHDSN: An attention-enabled hybrid deep sequential network for cancer survivability prediction from multi-omics data. *Mammalian Genome*, *37*(1), 2.

He, L., Luan, L., & Hu, D. (2025). Deep learning-based image classification for AI-assisted integration of pathology and radiology in medical imaging. *Frontiers in Medicine*, *12*, 1574514.

Hein, Z. M., Guruparan, D., Okunsai, B., Che Mohd Nassir, C. M. N., Ramli, M. D. C., & Kumar, S. (2025). AI and machine learning in biology: From genes to proteins. *Biology*, *14*(10), 1453.

Javaid, H., Petrescu, C. C., Schmunk, L. J., Monahan, J. M., O'Reilly, P., Garg, M., . . . & Martin-Herranz, D. E. (2025). The impact of artificial intelligence on biomarker discovery. *Emerging Topics in Life Sciences*, *8*(2), 89–105.

Jha, K., Saha, S., & Singh, H. (2022). Prediction of protein–protein interaction using graph neural networks. *Scientific Reports*, *12*(1), 8360.

Kargl, M., Plass, M., & Müller, H. (2022). A literature review on ethics for AI in biomedical research and biobanking. *Yearbook of Medical Informatics*, *31*(01), 152–160.

Katal, S., York, B., & Gholamrezanezhad, A. (2024). AI in radiology: From promise to practice– A guide to effective integration. *European Journal of Radiology*, *181*, 111798.

Khalifa, M., & Albadawy, M. (2024). Artificial intelligence for clinical prediction: Exploring key domains and essential functions. *Computer Methods and Programs in Biomedicine Update*, *5*, 100148.

Kim, J., McFee, M., Fang, Q., Abdin, O., & Kim, P. M. (2023). Computational and artificial intelligence-based methods for antibody development. *Trends in Pharmacological Sciences*, *44*(3), 175–189.

Koppad, S., . B., . A., Gkoutos, G. V., & Acharjee, A. (2021). Cloud computing enabled big multi-omics data analytics. *Bioinformatics and Biology Insights*, *15*, 11779322211035921.

Kotkondawar, R. R., Sutar, S. R., Kiwelekar, A. W., Kadam, V. J., & Jadhav, S. M. (2025). A generative framework for enhancing drug target interaction prediction in drug discovery. *Scientific Reports*, *15*(1), 35588.

Kurzweil, R. (2022). Superintelligence and singularity. *Machine Learning and the City: Applications in Architecture and Urban Design*, 579–601. https://doi.org/10.1002/9781119815075.ch48

Leema, A., Leema, D. A. A., Saad, M., & Babu, M. (2025). Gene-LLMs: A comprehensive survey of transformer-based genomic language models for regulatory and clinical genomics. *Frontiers in Genetics*, *16*, 1634882.

Li, X., Loscalzo, J., Mahmud, A. F., Aly, D. M., Rzhetsky, A., Zitnik, M., & Benson, M. (2025). Digital twins as global learning health and disease models for preventive and personalized medicine. *Genome Medicine*, *17*(1), 11.

Lorenc-Kukuła, K. (2025). Cutting-edge AI tools revolutionizing scientific research in life sciences. *BioTechnologia*, *106*(1), 77.

Mak, K. K., Wong, Y. H., & Pichika, M. R. (2024). Artificial intelligence in drug discovery and development. *Drug Discovery and Evaluation: Safety and Pharmacokinetic Assays*,1461–1498. https://doi.org/10.1007/978-3-031-35529-5_92

Martin, H. G., Radivojevic, T., Zucker, J., Bouchard, K., Sustarich, J., Peisert, S., . . . & Singh, A. (2023). Perspectives for self-driving labs in synthetic biology. *Current Opinion in Biotechnology*, *79*, 102881.

Martinelli, D. D. (2022). Generative machine learning for de novo drug discovery: A systematic review. *Computers in Biology and Medicine*, *145*, 105403.
Minocha, N., & Pandey, P. (2025). self-driving labs for materials sciences. *Artificial Intelligence in Polymer Science and Nanotechnology: Autonomous Labs, Polymer Design, Materials Simulation*, *69*, 69–96.
Mirchandani, I., Khandhediya, Y., & Chauhan, K. (2025). Review on advancement of AI in synthetic biology. *Artificial Intelligence (AI) in Cell and Genetic Engineering*, *2952*, 483–490.
Mouchlis, V. D., Afantitis, A., Serra, A., Fratello, M., Papadiamantis, A. G., Aidinis, V., . . . & Melagraki, G. (2021). Advances in de novo drug design: From conventional to machine learning methods. *International Journal of Molecular Sciences*, *22*(4), 1676.
National Academies of Sciences, Engineering, and Medicine. (2024). Artificial Intelligence and Automated Laboratories for Biotechnology: Leveraging Opportunities and Mitigating Risks: Proceedings of a Workshop – in Brief.
Naugler, C., & Church, D. L. (2019). Automation and artificial intelligence in the clinical laboratory. *Critical Reviews in Clinical Laboratory Sciences*, *56*(2), 98–110.
Obuchowicz, R., Lasek, J., Wodziński, M., Piórkowski, A., Strzelecki, M., & Nurzynska, K. (2025). Artificial intelligence-empowered radiology – Current status and critical review. *Diagnostics*, *15*(3), 282.
Ogidi, O. I., Alfred-Ugbenbo, D., & Poripo, B. E. (2025). AI-Driven drug discovery, design, and development in immunological disorders. In *AI-Assisted Computational Approaches for Immunological Disorders* (pp. 143–180). IGI Global Scientific Publishing.
Panjwani, S., Wei, H., & Mason, J. (2025). Artificial intelligence and machine learning-assisted digital applications for biopharmaceutical manufacturing. *Biotechnology Progress*, *42*, e70089.
Parvin, N., Joo, S. W., Jung, J. H., & Mandal, T. K. (2025). Multimodal AI in biomedicine: Pioneering the future of biomaterials, diagnostics, and personalized healthcare. *Nanomaterials*, *15*(12), 895.
Paszke, A., Gross, S., Massa, F., Lerer, A., Bradbury, J., Chanan, G., . . . & Chintala, S. (2019). Pytorch: An imperative style, high-performance deep learning library. *Advances in Neural Information Processing Systems*, *32*, 8026–8037.
Philippidis, A. (2021). From model to molecule: NVIDIA doubles down on ai drug discovery: GPU inventor partners with Schrödinger, AstraZeneca, UF Health on computational approaches to therapy development. *GEN Edge*, *3*(1), 238–246.
Pickard, J., Sturgess, V. E., McDonald, K. O., Rossiter, N., Arnold, K. B., Shah, Y. M., . . . & Beard, D. A. (2025). A hands-on introduction to data analytics for biomedical research. *Function*, *6*(2), zqaf015.
Placido, D., Yuan, B., Hjaltelin, J. X., Zheng, C., Haue, A. D., Chmura, P. J., . . . & Sander, C. (2023). A deep learning algorithm to predict risk of pancreatic cancer from disease trajectories. *Nature Medicine*, *29*(5), 1113–1122.
Sandbrink, J. B. (2023). Artificial intelligence and biological misuse: Differentiating risks of language models and biological design tools. *arXiv preprint arXiv:2306.13952*.
Seletkov, D., Starck, S., Mueller, T. T., Zhang, Y., Steinhelfer, L., Rueckert, D., & Braren, R. (2025). AI-driven preclinical disease risk assessment using imaging in UK biobank. *NPJ Digital Medicine*, *8*(1), 480.
Shahab, M. A., Destro, F., & Braatz, R. D. (2025). Digital twins in biopharmaceutical manufacturing: Review and perspective on human-machine collaborative intelligence. *arXiv preprint arXiv:2504.00286*.
Soldatov, M. A., Butova, V. V., Pashkov, D., Butakova, M. A., Medvedev, P. V., Chernov, A. V., & Soldatov, A. V. (2021). Self-driving laboratories for development of new functional materials and optimizing known reactions. *Nanomaterials*, *11*(3), 619.
Somara, S., Joshi, A. M., Mitra, K., Desai, S., Lundberg, M. S., Bhasin, S., & Hunsberger, J. (2025). Artificial intelligence in biotechnology and pharmaceuticals: Evolution, applications, and regulatory frontiers. *Current Stem Cell Reports*, *11*(1), 1–10.

Swanson, K., Walther, P., Leitz, J., Mukherjee, S., Wu, J. C., Shivnaraine, R. V., & Zou, J. (2024). ADMET-AI: A machine learning ADMET platform for evaluation of large-scale chemical libraries. *Bioinformatics, 40*(7), btae416.

Tobias, A. V., & Wahab, A. (2025). Autonomous 'self-driving'laboratories: A review of technology and policy implications. *Royal Society Open Science, 12*(7), 250646.

Undheim, T. A. (2024). The whack-a-mole governance challenge for AI-enabled synthetic biology: Literature review and emerging frameworks. *Frontiers in Bioengineering and Biotechnology, 12*, 1359768.

Volk, A. A., & Abolhasani, M. (2024). Performance metrics to unleash the power of self-driving labs in chemistry and materials science. *Nature Communications, 15*(1), 1378.

Wei, J., Yang, Y., Zhang, X., Chen, Y., Zhuang, X., Gao, Z., . . . & Zhou, B. (2025). From ai for science to agentic science: A survey on autonomous scientific discovery. *arXiv preprint arXiv:2508.14111.*

Wheeler, N. E. (2025). Responsible AI in biotechnology: Balancing discovery, innovation and biosecurity risks. *Frontiers in Bioengineering and Biotechnology, 13*, 1537471.

Wu, M., Wang, Z., Wang, J., Dong, Z., Yang, J., Li, Q., . . . & Chen, H. (2025). An AI-native experimental laboratory for autonomous biomolecular engineering. *arXiv preprint arXiv:2507.02379.*

Yoshikawa, N., Asano, Y., Futaba, D. N., Harada, K., Hitosugi, T., Kanda, G. N., . . . & Ushiku, Y. (2025). Self-driving laboratories in Japan. *Digital Discovery, 4*(6), 1384–1403.

Zhang, Z., Chakraborty, S., Bedi, A. S., Mathew, E., Saravanan, V., Cong, L., . . . & Wang, M. (2025). Generative AI for biosciences: Emerging threats and roadmap to biosecurity. *arXiv preprint arXiv:2510.15975.*

Zheng, J., Wang, Y., Liang, Q., Cui, L., & Wang, L. (2024). The application of machine learning on antibody discovery and optimization. *Molecules, 29*(24), 5923.

Tanya, Pooja Saini, Avinash Godara, Vivek Kumar, Ajeev Kumar*,
Sonali Sangwan, Priti Duhan

Chapter 14
Revolutionizing Biotechnology: The Real-World Impact of AI

Abstract: Artificial intelligence (AI) has significantly evolved with advancements in digital technologies, emerging as a transformational force across biotechnology. Biotechnology focuses on molecular processes, genetic engineering, and biological systems. The integration of AI enables rapid analysis of huge, complex biological datasets, including genomics, proteomics, metabolomics, and biomedical imaging. By shifting biotechnology from hypothesis-driven to data-driven discovery, this digital transformation improves research speed, accuracy, and efficiency while reducing cost and uncertainty. Innovation within modern biotechnology is significantly supported by AI, reshaping how biological data is analyzed, interpreted and applied. AI-driven methods offer unprecedented capabilities for pattern identification and predictive modeling with the rapid growth of high-throughput technologies producing vast datasets in genomes, transcriptomics, proteomics, metabolomics, and biomedical imaging. Machine learning and deep learning algorithms support complex biological tasks such as gene expression analysis, protein structure prediction, biomarker identification, and drug-target interaction modeling. In addition to improving research accuracy and reproducibility, these computational methods significantly reduce experimental time and cost. In drug discovery, AI facilitates target identification, virtual screening, de novo molecule design, toxicity prediction and ADMET profiling are all accelerating the transition from early discovery to clinical validation. Rational drug design and protein structure prediction have been transformed by deep learning frameworks like AlphaFold and transformer-based models like convolutional neural networks. In healthcare, AI improves diagnostic precision, disease classification, and

***Corresponding author: Ajeev Kumar,** Department of Agricultural Biotechnology, College of Biotechnology, Chaudhary Charan Singh Haryana Agricultural University, Hisar, Haryana, India, e-mail: ajeevsangwan@hau.ac.in

Tanya, Department of Bioinformatics and Computational Biology, College of Biotechnology, Chaudhary Charan Singh Haryana Agricultural University, Hisar, 125001, Haryana, India

Pooja Saini, Department of Bio-Sciences and Technology, MMEC, Maharishi Markandeshwar (Deemed to be University), Mullana, Ambala, 133207, Haryana, India

Avinash Godara, Vivek Kumar, Department of Bioinformatics and Computational Biology, College of Biotechnology, Chaudhary Charan Singh Haryana Agricultural University, Hisar, 125001, Haryana, India

Sonali Sangwan, Department of Biotechnology, Guru Nanak Khalsa College, Yamuna Nagar, 135001, Haryana, India

Priti Duhan, Department of Biotechnology, K.L. Mehta Dayanand College for Women, Faridabad

 | https://doi.org/10.1515/9783112214275-014

personalized treatment strategies through advanced image analysis and clinical data modeling. AI-enabled genomic selection in agriculture supports sustainable crop improvement by improving crop yield, drought tolerance, and disease resistance prediction. Overall, the integration of AI across biotechnology has transformed traditional hypothesis-driven approaches into powerful data-driven systems capable of solving complex biological issues more quickly, extensively, and intelligently. As AI continues to evolve, its synergy with biotechnology develops significant advancements in healthcare, agriculture, environmental management, and industrial bioprocessing.

Keywords: Machine learning, bioprocessing, protein structure prediction, intelligently

14.1 Introduction: AI and Biotechnology

Artificial intelligence (AI) was introduced by a group of computer scientists in 1956 during a workshop at Dartmouth College. The expansion of AI relies on digital technology. AI has become a disruptive force across multiple scientific domains, including biotechnology. Though biotechnology depends on molecular processes, genetic engineering, and biological systems for improved enterprises, farming, and human health, the incorporation of AI has greatly widened its potential. Rapid analysis of intricate, high-dimensional biological information, including imaging, proteomics, metabolomics, and genomes, is made possible by AI. In various research institutes, universities, organizations, and companies, the introduction of digital technologies and processes to enhance accuracy, efficiency, pace of research, and development is digital transformation (Holzinger et al., 2023). This has reduced time, expense, and uncertainty in research results by moving biotechnology away from hypothesis-driven experimentation and toward data-driven discovery (Jordan & Mitchell, 2015). AI, machine learning, and deep learning are interconnected but not identical. AI is a broad term, refers to designing intelligent systems to perform tasks usually requiring human intelligence, i.e., problem solving, decision making, learning, and reasoning. Machine learning (ML) is a branch of AI in which computers are trained to perform tasks by identifying patterns and observations from data. Deep learning is a subfield of ML, utilizes artificial neural networks to perform tasks, gather information and make decisions. It handles large and complex types of data sets such as images (DALL-E2), text (ChatGPT), or audio-speech (VALL-E) (Holzinger et al., 2019).

Biotechnology and AI share a symbiotic relationship; biotechnological advancements create enormous datasets, and AI offers computational power to extract valuable insights from them. For instance, AI techniques can mine the terabytes of data generated by next-generation sequencing (NGS) technology to identify innovative medicinal targets, regulatory mechanisms, or crop improvement approaches (Ching et al., 2018). Recently, the use of AI technologies in bioprocess optimization and control has increased, upgrading the performance of the bioprocess. A hybrid multi-objective approach has been utilized to enhance biomass and microorganisms' production yield for industrial applications (Saini et al., 2021). In this chapter, we have discussed key areas of biotechnology revolutionized by AI, comprising agricultural biotechnology, drug discovery and development, biomolecular engineering, bioprocess optimization, and genome editing techniques such as CRISPR (Table 14.1). Hence, it can drive innovation in the research and industrial domains. The powerful interaction between AI and biotechnology is accelerating various discoveries and addressing global challenges such as environmental protection, waste management, healthcare, and sustainable food production.

14.2 AI in Healthcare and Drug Discovery

Conventionally, drug discovery has provided various treatments yet faces challenges that delay in the development and delivery of new therapies for patients. Drug research and development would take 10–15 years and cost around $2.6 billion to bring a new drug with <10% success (Ocana et al., 2015). Hence, there was a need to find ways to reduce and optimize the drug discovery and development process. AI has revolutionized the pharmaceutical industry in various ways; it allows the analysis and interpretation of large amounts of data simultaneously, which was difficult to handle manually with statistical methods. Furthermore, AI has the potential to present innovative approaches for effective drug development and finding novel agents.

Structure-based drug design (SBDD) relies crucially on the 3D structure of the protein to develop molecules that can bind efficiently to the target. Previously, these protein and drug binding prediction were done using X-ray crystallography or cryogenic electron microscopy (Cheng, 2015).

AI tools can be used to explore different stages of drug discovery and development. For instance, deep learning can examine multi-omics and network data to discover novel targets, predict protein structures (e.g., AlphaFold). Machine learning algorithms can learn from existing data, identify patterns, and make accurate predictions. For instance, machine learning approaches, such as Kronecker regularized least squares (KronRLS), study the similarities between protein molecules and drugs to calculate drug-target binding affinity (DTBA). Congruently, SimBoost predict DTBA using regression trees while considering both feature-based and similarity-based interaction (Paul et al., 2021). Hence, AI accelerates each discovery stage (from target ID to molecule design) and also improves clinical trial planning (Kumar et al., 2018; Romeo-Guitart et al., 2018; Yaseen et al., 2023).

Table 14.1: Application of AI in different fields of biotechnology and bioinformatics.

Fields	Sub-field	Applications	Methods/tools	Challenges	Examples	References
AI in agriculture	Crop breeding	Genomic selection, trait prediction, and drought/salt tolerance prediction	ML regression, random forests, and deep learning	Scarcity of genomic-phenotypic datasets and environmental variability	AI-driven genomic section in rice and pearl millet	Liakos et al. (2019)
	Pest detection	CNNs, YOLO, and transfer learning	CNNs and transfer learning	Dataset diversity and lighting variability	Mobile apps for maize/rice leaf disease detection	Kamilaris and Prenafeta-Boldu (2018) and Pantazi et al. (2019)
	Soil, water, and climate management	Regression models	Regression models	Small farm data scarcity and sensor costs	AI-based irrigation decision support	Liakos et al. (2018)
AI in animal biotechnology	Breeding and genetics	Genomic prediction, disease resistance and fertility traits	ML regression, Bayesian ML, and DNNs	Breed bias and multi-trait prediction limits	AI genomic selection in cattle/swine	Harlow et al. (2018) and Neethirajan (2020)
	Disease detection	Early detection of zoonoses and health screening	CNNs, anomaly detection, and NLP	Sparse animal health datasets	AI for mastitis in dairy cattle	Wathes et al. (2008) and Neethirajan (2020)
AI in forest biotechnology	Tree health monitoring	Canopy stress and pest/disease detection	CNNs and remote sensing ML	Sparse labeled datasets	AI-based oak decline detection	Fassnacht et al. (2016)
	Species identification and mapping	Tree species classification and biodiversity	Object detection and ecological ML	AI tree census in urban forests	Seasonal variability	Fassnacht et al. (2016)
	Carbon and ecosystem modeling	Biomass prediction and carbon stock estimation	Random forests and geospatial DL	Model uncertainty	AI-based Amazon carbon monitoring	Fassnacht et al. (2016)

AI in bioinformatics	Genomics	Variant calling and genome annotation	CNNs, GNNs, and ensemble ML	Class imbalance and interpretability	DeepVariant and AI gene annotation tools	Zou et al. (2019) and Eraslan et al. (2019)
	Proteomics	Protein folding and PTM predictions	Transformers and GNNs	Sparse validation data	AlphaFold2 for protein structures	Min et al. (2017) and Eraslan et al. (2019)
	Transcriptomics	Expression clustering and trajectory inference	Autoencoders and DL clustering	Batch effects and reproducibility	scRNA-seq AI clustering tools	Zou et al. (2019)

14.2.1 Target Identification, Validation, and Molecular Design

AI allows high throughput in silico screening to prioritize the most promising targets and compounds. Machine learning applied to genomics, proteomics, and network data enables the prediction of critical genes or proteins involved in disease and their potential as drug targets. After that, interpreting the roles of these targets in diseases and validating these targets to confirm their direct involvement in a disease mechanism and therapeutic effect. Convolutional and graph neural networks assess 3D structures and molecular graphs to precisely estimate binding affinity and ADMET properties as compared to traditional methods. Deep learning enables de novo drug design through generative models, such as generative adversarial networks (GANs) and reinforcement learns, that suggest novel chemical structures optimized for desired properties, quickening hit-to-lead optimization.

AlphaFold is a tool developed by Google DeepMind that predicts protein 3D structures with near-experimental accuracy by employing a deep neural network. It facilitates the structure-based design of inhibitors that would be quite challenging to conceive manually. GANs technique can be utilized to create novel biomolecules with desired characteristics. This innovative technique has revolutionized biomolecular engineering and design (Sousa et al., 2021). Docking simulation is a crucial technique for computer-aided drug development (Bohr, 2020). This works by anticipating binding affinity and guiding drug design by simulating the interaction between small molecules (drug) and target macromolecule (protein) (Gentile et al., 2022). This implies that AI models can virtually scan chemical libraries, predict molecular interactions with targets and generate new candidate compounds.

Biomolecular engineering is a technique that develops and modifies biomolecules such as proteins, nucleic acids, and lipids to perform specific functions. Later, these biomolecules were used in biotechnology in various applications, including the production of biomaterials, gene therapy, drug delivery, and biocatalysts (Nagamune et al., 2017).

Target validation is considered an important step in drug discovery. It helps in ensuring that a selected molecular target is directly linked with the disease mechanism, and the target's modulation will give therapeutic effect (Mathai et al., 2020). Target validation involved determining the structure-activity relationship

14.2.2 Clinical Trials and Personalized Medicines

AI has been used to analyze electronic health records and genetic profiles to correlate patients with relevant studies and endpoints. The predictive algorithms can pretend "synthetic control arms" or virtual patient cohorts to examine trial outcomes. Natural language processing (NLP) is a technique that can be used to scan medical literature to get information regarding current treatment possibilities. NLP-powered AI models can study patient data and help to tailor treatments by reducing risk and cost.

AI applications are divided into two types: physical approach and virtual approach (Figure 14.1). The physical approach of AI includes real-time patient monitoring, robot-assisted surgery, AI-enhanced prosthetics, and automated laboratory processes. AI virtual approach assists in drug discovery and development, and more accurate disease diagnosis. ML algorithms can detect abnormalities and help in early disease detection by analyzing medical imaging data from X-rays, CT scans, and MRI images (Yoon et al., 2019).

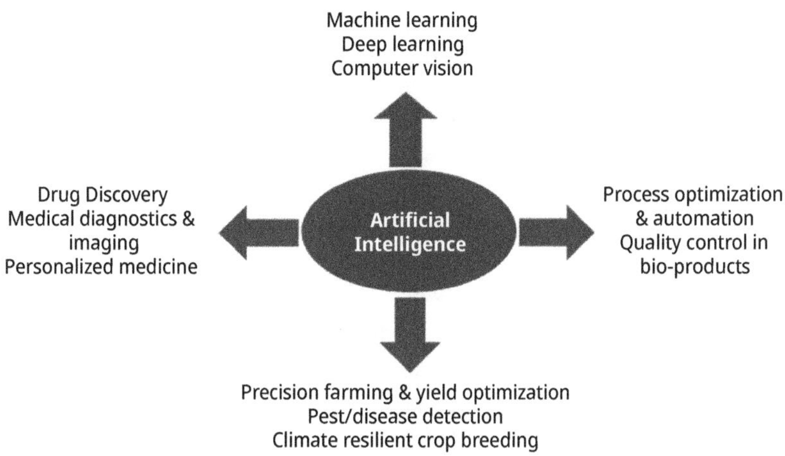

Figure 14.1: Artificial intelligence drives innovation across healthcare, agriculture, finance, transportation, education, and environmental management.

14.2.3 AI-Based Programs and Platforms

Various AI tools/platforms are available that are useful in drug discovery (Table 14.2). DeepChem is a Python library for machine learning and deep learning. It is a useful tool for drug discovery and quantum chemistry. Another Python-based computational platform is DeepNeuralNetQSAR, which aids in the detection of the molecular activity of compounds. The Chemputer program was used to report the standardized format procedure for chemical synthesis (Keenan, 2018). DeepTox is a DL-based software for predicting toxicity (Mayr, 2016). ORGANIC is a molecular generation tool based on generative ML approaches and deep neural networks. It is used to create molecules with desired properties. Another useful ML-based platform is Hit Dexter for hit discovery, such as identifying early molecules that are likely to bind and modulate biological targets (Dara et al., 2022). It utilizes ADME prediction models, deep learning models, boosting, and QSAR approaches to analyze the chemical structures. It also helps in bioactivity prediction, protein-protein interaction analysis, digital pathology, and protein folding prediction. DeltaVina is an ML scoring function using XGBoost and random forest to score protein-ligand binding affinities. It incorporates diverse molecular features to improve binding affinity predictions compared to classical scoring functions. The performance of DeltaVina has been validated on it benchmark datasets and is further available for research.

Table 14.2: List of AI tools/platforms used in drug discovery.

Program/tools	Function	AI technique	Accession
DeepChem	Drug discovery and quantum chemistry	Python library for machine learning and deep learning	https://deepchem.io/
AlphaFold	Predicts 3D protein structures	AI system	https://alphafold.ebi.ac.uk/
DeepTox	Predicts the toxicity of chemical compounds	Deep learning	www.bioinf.jku.at/research/DeepTox
DeepNeuralNetQSAR	Predict compound molecular structure and their biological activity	Deep neural network	https://github.com/Merck/DeepNeuralNet-QSAR (accessed on 1 February 2024
Neural graph fingerprint	Predicts properties of novel molecules	Graph neural networks (GNNs)	https://github.com/HIPS/neural-fingerprint
Chemputer	Synthesize organic molecules	An automated robotic system with ML algorithms	https://zenodo.org/record/1481731
DeltaVina	ML algorithms include XGBoost and random forest	Score protein-ligand binding affinity	https://github.com/chengwang88/deltavina

The neural graph fingerprint technique can predict the properties of novel molecules, which involves generating molecular sentences. It is encoded using a language model, BERT, to obtain high-dimensional embeddings that represent molecules in vector form. It can be used for downstream tasks, such as molecular property prediction, utilizing ROC-AUC (classification), RMSE, and R^2 (regression) metrics. GastroPlus is a predictive modeling platform that uses AI to simulate absorption, distribution, metabolism, and excretion (ADME) in animals and humans, enabling researchers to optimize drug formulations more efficiently.

14.2.4 Case Studies of AI's Successful Application in Healthcare

14.2.4.1 AI in Diabetic Retinopathy Diagnoses

Retinal microangiopathy/diabetic retinopathy is the most common ill effect faced by diabetic people that results in a gradual decrease in vision and, in the worst situation, blindness. According to the World Health Organization, one-third of diabetic patients will be affected by diabetic retinopathy (Kropp et al., 2023). Therefore, early diagnosis and treatment are essential for preventing blindness. DeepMind, a Google subsidiary, is developing an AI-powered system that can analyze retinal images to deliver diagnoses comparable to specialists, showing how AI can aid in the early diagnosis of problems. DeepMind's AI system for diabetic retinopathy diagnosis is built on deep learning techniques, which involve training an artificial neural network to identify patterns in data (Raman et al., 2019). The model is trained on thousands of retinal images annotated by a human expert to "learn" the traits linked with diabetic retinopathy. There are numerous benefits to using AI in diagnosis compared to conventional methods. AI-based systems can quickly and accurately scan a large number of retinal images, saving time and money. AI-powered technologies could be utilized to expand diabetic retinopathy screening and diagnosis to rural and underserved areas. The system can diagnose new retinal images with speed and accuracy once it has been trained. In a groundbreaking study published in JAMA, researchers from DeepMind and Moorfields Eye Hospital demonstrated the effectiveness of the AI system in identifying diabetic retinopathy. The algorithm achieved an accuracy of 94.5% in identifying referable diabetic retinopathy from a dataset of 128,175 retinal images, which matched the diagnosis of human specialists.

14.3 AI in Agricultural Biotechnology

14.3.1 Plant Breeding (Yield Prediction and Trait Selection)

AI has revolutionized crop breeding by enabling faster identification of cultivars with high-yielding or stress-tolerant characteristics. ML/DL is frequently used in genomic selection, which predicts plant characteristics (height, grain quality, disease resistance) from DNA markers. Montesinos-López et al., (2025) compared deep learning vs. traditional GBLUP models on diverse breeding datasets. Deep nets can effectively capture the complex, nonlinear genetic patterns and give higher accuracy for complex traits (disease resistance, drought tolerance, and yield) as compared to linear methods (Montesinos-López et al., 2025). AI additionally enhances simulation and phenotyping. AI provides data from high-throughput phenotyping systems using cameras, drones, and sensors to build models using ML. These models can predict traits like plant height, leaf color, or canopy cover to yield or stress tolerance. ML agricultural models can combine genetic data with weather and soil to forecast performance in future climates. A study conducted in 2024 used explainable AI (XAI) with tree-based learning algorithms (RF and LGBM) to merge climate and soil data. The model obtained an R^2 of 0.92 for predicting crop yield, with XAI tools (SHAP/LIME) identifying rainfall and nutrient levels as key factors (Mohan et al., 2025). These models can assist plant breeders by providing deeper insights into genetic traits that contribute to yield under varying environmental conditions.

14.3.2 AI in Soil Health Assessment and Management

AI-based tools can be used to analyze soil sensor data and geospatial (spatial) information to assess and improve the quality of soils. Soil sensors measure various factors: moisture, pH level, electrical conductivity, organic carbon levels, and nutrient concentrations and are often combined with handheld or Internet of Things (IoT) devices. IoT sensors: soil-moisture probes, weather stations, nutrient probes, and camera traps, etc., continuously stream data for analysis and decision making. The resultant data can be used to train ML models to assess fertility and recommend action for increasing fertility (Gunasekaran et al., 2025). AI can also aid soil mapping and monitoring. Machine learning algorithms are applied to satellite data or proximal sensor data to estimate soil parameters such as organic carbon, moisture, or degraded soil in fields. Algorithms like Random Forest and Gradient Boosting were used to assess the levels of soil nutrient components, such as nitrogen, phosphorus, and potassium (NPK), through sensor data or spectral data (Gunasekaran et al., 2025).

14.3.3 AI in Crop Disease Detection

AI uses computer vision technology to identify crop diseases from leaf images. Various deep learning models have been developed to identify diseases based on images as input data. A convolutional neural network (CNN) was trained on 54,306 images of crop leaves with both disease and nondisease conditions of 14 different types of crops and 26 disease types; the model achieved 99.35% accuracy.

14.3.4 AI in CRISPR

CRISPR, which stands for "Clustered Regularly Interspaced Short Palindromic Repeats," is a strong gene-editing tool that has revolutionized the biotechnology sector. AI tools can improve the accuracy and effectiveness of genome editing by CRISPR (Nidhi et al., 2021). Machine learning algorithms optimize the results of genome editing studies, forecast off-target effects, and direct RNA design. Earlier, creating guide RNAs, the molecules that target specific DNA regions for editing, used to be a tedious and time-consuming operation. By analyzing large genetic datasets, AI can accurately predict and select highly efficient guide RNA sequences for genome editing (Lin et al., 2023). This improves the success rate of targeted gene editing while saving researchers time and effort. CRISPR has one concern: accidental modification at sites (off-target effects) other than the targeted DNA sequence. AI technologies can facilitate the assessment of proposed guide RNA by identifying predicted off-target sites. Hence, it enhances the accuracy and safety of CRISPR editing. CRISPR-P is an online AI-based platform for precise gRNA design, and DeepCpf1 is a CNN-based AI tool used to predict on-target activity and off-target effects of CRISPR-Cpf1 gRNA (Liu et al., 2023, Listgarden et al., 2018).

14.4 AI in Environmental Biotechnology

The integration of AI and biotechnological techniques and methods increases the efficiency of solving environmental problems and global challenges. Biofuels development from algae or other biomass has contributed to the development of sustainable and renewable energy sources, thereby reducing the use of fossil fuels and the associated carbon footprint (Demirbas, 2011). The growth conditions of biofuel-producing organisms can be optimized by ML algorithms to boost yield and decrease cost. Certain bacteria and plants have the capability of accumulating and detoxifying the heavy metals from contaminated water and soil bodies, thereby reducing industrial waste and pollution. AI and deep learning models help in predicting the efficiency of absorbing specific heavy metals by certain plants or microorganisms, thus improving bioremediation strategies for contaminated sites. The integration of AI and geographic information systems (GIS) can identify and monitor contaminated sites (Fischer et al., 2021).

Biotechnology helps in minimizing the need for landfills and reduces greenhouse gas by converting organic waste into useful resources like compost or biofuels emissions. AI can monitor and control the environment constantly during anaerobic digestion or composting processes, guaranteeing the optimal performance of the microorganisms for faster waste breakdown.

In bioreactor systems, sensors and machine learning algorithms are used to real-time adjust ambient conditions to improve microbial activity, hence accelerate waste decomposition. The effectiveness of biodegradation procedures is increased by using AI driven robots (Waste Sorting Robots) with sophisticated visual capabilities to separate organic waste from other types. Carbon dioxide capturing and storing from the atmosphere, helps combat climate change by reducing greenhouse gases emission (Ejimofor et al., 2022).

AI increases enhance the work of biotechnology in climate change mitigation by identifying the best biological carbon capture systems through the analysis of biological and environmental data. AI facilitates the genetic modifications to enhance photosynthetic efficiency to increase carbon capture by plants. The integration of AI and biotechnology is used in development of eco-friendly agricultural practices and lowering chemical runoff and soil degradation (Wilberforce et al., 2021).

AI models can design bio-based solutions that lessen the nutrient leaching by forecasting the runoff hazards associated with various agricultural methods. To reduce runoff, AI-driven systems can recommend precise application of biopesticides and biofertilizers by evaluating the crop health and soil conditions. Using machine learning algorithms and historical data, crop rotation AI models can forecast the optimal rotation approaches to maintain soil health and minimize chemical inputs (Ahirwar et al., 2020).

14.5 Challenges

For medication development, pharmaceutical organizations may have access to enormous datasets that include millions of molecules, which could present difficulties for conventional machine learning systems. The use of ANNs in personalized medicine is not without its difficulties, despite their potential. The quantity and quality of the data used to train ANNs have a significant impact on their accuracy. Biomolecule engineering and designing present moral dilemmas. To ensure that these technologies are developed and used responsibly, much thought is required. Frameworks for AI-driven bio-manufacturing processes are still being developed by regulatory bodies. Validating AI models and providing accurate documentation are necessary to comply with these rules. Implementing and maintaining AI systems requires a skilled workforce with expertise in both bioprocessing and data science. CRISPR technology raises ethical debates due to its ability to alter human genes and the potential for misuse. Not all research facilities may have easy access to advanced AI tools due to their high cost or the required technical expertise.

14.6 Conclusion

At the forefront of innovation, AI is revolutionizing scientific research and transforming various industries. Drug development, customized treatment, and genome editing techniques like CRISPR technology have been fundamentally transformed by AI in the biotechnology industry. Furthermore, by encouraging the production of renewable energy, bioremediation, and strengthening plant resilience, AI-powdered solutions are actively addressing environmental problems. The continuing evolution of AI and biotechnology promises revolutionary applications that could lead to improvements in health and well-being, medication discovery, and environmental management, despite ethical and accessibility concerns arising from the integration of AI in biotech. AI and biotechnology working together have enormous potential to address global issues and promote sustainable innovation.

14.7 Future Directions

The advancements of AI and robotics technologies, further ground-breaking applications in biotechnology can be developed. This integration has the capability to revolutionize the sector and lead to advancements in healthcare, biotechnology, and drug creation. One emerging trend in personalized medicine is preventive medicine. The development of CRIPSR technique for complex gene editing tasks, such as insertion or removal of lengthy DNA fragments, may be aided by AI. AI can assist genetic sequence analysis; data interpretation and experiment design with AI are just a few of the ways the CRISPR procedure can be automated. This increases the rate of discovery and decreases research time for difficult tasks. However, it's important to recognize the current constraints and possible drawbacks of AI technologies in biotechnology. These techniques frequently show a bias towards previous discoveries, which may hinder "out of the box" thinking and creative forecasts, despite their predictive potential through regression analysis. Furthermore, if AI algorithms are given false or misleading information, they may draw incorrect conclusions since they do not automatically evaluate the reliability of the data they analyze. For AI in biotechnology to advance in the future, several potential problems must be solved. By enhancing data validation procedures and creating more complex algorithms that can take data quality into account, we can reduce these risks and fully utilize AI to drive innovation in the field.

References

Ahirwar, N. K., Singh, R., Chaurasia, S., Chandra, R., & Ramana, S. (2020). Effective role of beneficial microbes in achieving the sustainable agriculture and eco-friendly environment development goals: A review. *Frontiers in Microbiology, 5*, 111–123.

Bohr, H. (2020). Drug discovery and molecular modeling using artificial intelligence. In *Artificial Intelligence in Healthcare* (pp. 61–83). Elsevier.

Cheng, Y. (2015). Single-particle cryo-EM at crystallographic resolution. *Cell, 161*(3), 450–457. https://doi.org/10.1016/j.cell.2015.03.049.

Ching, T., Himmelstein, D. S., Beaulieu-Jones, B. K., Kalinin, A. A., Do, B. T., Way, G. P., & Greene, C. S. (2018). Opportunities and obstacles for deep learning in biology and medicine. *Journal of the Royal Society Interface, 15*(141), 20170387.

Dara, S., Dhamercherla, S., Jadav, S. S., Babu, C. M., & Ahsan, M. J. (2022). Machine learning in drug discovery: A review. *Artificial Intelligence Review, 55*, 1947–1999. https://doi.org/10.1007/s10462-021-10039-z.

Demirbas, M. F. (2011). Biofuels from algae for sustainable development. *Applied Energy, 88*(10), 3473–3480.

Ejimofor, M. I., Aniagor, C. O., Oba, S. N., Menkiti, M. C., & Ugonabo, V. I. (2022). Artificial intelligence in the reduction and management of land pollution. In *Current Trends and Advances in Computer-Aided Intelligent Environmental Data Engineering* (pp. 319–333). Elsevier.

Eraslan, G., Avsec, Ž., Gagneur, J., & Theis, F. J. (2019). Deep learning: New computational modelling techniques for genomics. *Nature Reviews Genetics, 20*(7), 389–403. https://doi.org/10.1038/s41576-019-0122-6.

Fassnacht, F. E., Latifi, H., Stereńczak, K., Modzelewska, A., Lefsky, M., Waser, L. T., . . . Koch, B. (2016). Review of studies on tree species classification from remotely sensed data. *Remote Sensing of Environment, 186*, 64–87. https://doi.org/10.1016/j.rse.2016.08.013.

Fischer, A., Lee, M. K., Ojeda, A. S., & Rogers, S. R. (2021). GIS interpolation is key in assessing spatial and temporal bioremediation of groundwater arsenic contamination. *Journal of Environmental Management, 280*, 111683.

GastroPlus® PBPK and PBBM Modeling and Simulation. (2023). Simulations Plus. https://www.simulations-plus.com/software/gastroplus/

Gentile, F., Yaacoub, J. C., Gleave, J., Fernandez, M., Ton, A.-T., Ban, F., Stern, A., & Cherkasov, A. (2022). Artificial intelligence–enabled virtual screening of ultra-large chemical libraries with deep docking. *Nature Protocols, 17*, 672–697. https://doi.org/10.1038/s41596-021-00643-5.

Gunasekaran, K., Sreevardhan, P., & SamiKannu, R. (2025). Real-time soil fertility analysis, crop prediction, and insights using machine learning and deep learning algorithms. *Frontiers in Soil Science, 5*, 1652058.

Harlow, C. R., *et al.*. (2018). Genomic selection and machine learning for livestock breeding. *Animal Genetics, 49*(6), 569–583. https://doi.org/10.1111/age.12760.

Holzinger, A. (2019). Introduction to machine learning and knowledge extraction (MAKE). *Machine Learning and Knowledge Extraction, 1*(1), 1–20.

Holzinger, A., Keiblinger, K., Holub, P., Zatloukal, K., & Müller, H. (2023). AI for life: Trends in artificial intelligence for biotechnology. *New Biotechnology, 74*, 16–24.

Jordan, M. I., & Mitchell, T. M. (2015). Machine learning: Trends, perspectives, and prospects. *Science, 349*(6245), 255–260. https://doi.org/10.1126/science.aaa8415.

Kamilaris, A., & Prenafeta-Boldú, F. X. (2018). Deep learning in agriculture: A survey. *Computers and Electronics in Agriculture, 147*, 70–90. https://doi.org/10.1016/j.compag.2018.02.016.

Keenan, G. (2018). *ChemputerSoftware: Chemputer First Release (Version 0.1.1) [Software]*. Zenodo. https://doi.org/10.5281/zenodo.1472194.

Kropp, M., Golubnitschaja, O., Mazurakova, A., Koklesova, L., Sargheini, N., Vo, T. T. K. S., & Thumann, G. (2023). Diabetic retinopathy as the leading cause of blindness and early predictor of cascading complications – Risks and mitigation. *Epma Journal, 14*(1), 21–42.

Kumar, R., Sharma, A., Siddiqui, M. H., & Tiwari, R. K. (2018). Prediction of drug–plasma protein binding using artificial intelligence-based algorithms. *Combinatorial Chemistry and High Throughput Screening, 21*(1), 57–64.

Liakos, K. G., Busato, P., Moshou, D., Pearson, S., & Bochtis, D. (2018). Machine learning in agriculture: A review. *Sensors, 18*(8), 2674. https://doi.org/10.3390/s18082674.

Lin, J., & Ngiam, K. Y. (2023). How data science and AI-based technologies impact genomics. *Singapore Medical Journal, 64*(1), 59–66.

Listgarten, J., Weinstein, M., Kleinstiver, B. P., Sousa, A. A., Joung, J. K., Crawford, J., & Fusi, N. (2018). Prediction of off-target activities for the end-to-end design of CRISPR guide RNAs. *Nature Biomedical Engineering*, *2*(1), 38–47.

Liu, T., Zhang, X., Li, K., Yao, Q., Zhong, D., Deng, Q., & Lu, Y. (2023). Large-scale genome editing in plants: Approaches, applications, and future. *Perspectives. Current Opinion in Biotechnology*, *79*, 102875.

Mathai, N., Chen, Y., & Kirchmair, J. (2020). Validation strategies for target prediction methods. *Briefings in Bioinformatics*, *21*(3), 791–802.

Mayr, A., Klambauer, G., Unterthiner, T., & Hochreiter, S. (2016). DeepTox: Toxicity prediction using deep learning. *Frontiers in Environmental Science*, *3*, 80.

Merck/DeepNeuralNet-QSAR. Merck Sharp & Dohme Corp. a Subsidiary of Merck & Co., Inc. Available online: https://github.com/Merck/DeepNeuralNet-QSAR (accessed on 20 October 2023).

Min, S., Lee, B., & Yoon, S. (2017). Deep learning in bioinformatics. *Briefings in Bioinformatics*, *18*(5), 851–869.

Mohan, R. J., Rayanoothala, P. S., & Sree, R. P. (2025). Next-gen agriculture: Integrating AI and XAI for precision crop yield predictions. *Frontiers in Plant Science*, *15*, 1451607.

Montesinos-López, A., Montesinos-López, O. A., Ramos-Pulido, S., Mosqueda-González, B. A., Guerrero-Arroyo, E. A., Crossa, J., & Ortiz, R. (2025). Artificial intelligence meets genomic selection: Comparing deep learning and GBLUP across diverse plant datasets. *Frontiers in Genetics*, *16*, 1568705.

Nagamune, T. (2017). Biomolecular engineering for nanobio and bionanotechnology. *Nano Convergence*, *4*(1), 9. https://doi.org/10.1186/s40580-017-0108-y.

Neethirajan, S. (2020). The role of artificial intelligence in precision livestock farming ecosystems. *International Journal of Dairy Technology*, *73*(2), 362–367.

Nidhi, S., Anand, U., Oleksak, P., Tripathi, P., Lal, J. A., Thomas, G., & Tripathi, V. (2021). Novel CRISPR–Cas systems: An updated review of the current achievements, applications, and future research perspectives. *International Journal of Molecular Sciences*, *22*(7), 3327.

Ocana, A., Pandiella, A., Privat, C., Bravo, I., Luengo-Oroz, M., Amir, E., & Gyorffy, B. (2025). Integrating artificial intelligence in drug discovery and early drug development: A transformative approach. *Biomarker Research*, *13*(1), 45.

Pantazi, X. E., Moshou, D., & Bravo, C. (2019). Automated leaf disease detection in different crop species through image features analysis and machine learning. *Computers and Electronics in Agriculture*, *156*, 96–104.

Paul, D., Sanap, G., Shenoy, S., Kalyane, D., Kalia, K., & Tekade, R. K. (2021). Artificial intelligence in drug discovery and development. *Drug Discovery Today*, *26*, 80–93.

Raman, R., Srinivasan, S., Virmani, S., Sivaprasad, S., Rao, C., & Rajalakshmi, R. (2019). Fundus photograph-based deep learning algorithms in detecting diabetic retinopathy. *Eye*, *33*(1), 97–109.

Romeo-Guitart, D., Forés, J., Herrando-Grabulosa, M., Valls, R., Leiva-Rodríguez, T., Galea, E., . . . Casas, C. (2018). Neuroprotective drug for nerve trauma revealed using artificial intelligence. *Scientific Reports*, *8*, 1879. https://doi.org/10.1038/s41598-018-20361-1.

Saini, D. K., Rai, A., Devi, A., Pabbi, S., Chhabra, D., Chang, J. S., & Shukla, P. (2021). A multi-objective hybrid machine learning–based optimization for enhanced biomass and bioactive phycobiliproteins production in *Nostoc* sp. CCC-403. *Bioresource Technology*, *329*, 124908.

Sousa, T., Correia, J., Pereira, V., & Rocha, M. (2021). Generative deep learning for targeted compound design. *Journal of Chemical Information and Modeling*, *61*(11), 5343–5361.

Wathes, C. M., Kristensen, H. H., Aerts, J. M., & Berckmans, D. (2008). Is precision livestock farming an engineer's daydream or nightmare? *Computers and Electronics in Agriculture*, *64*(1), 2–10.

Wilberforce, T., Olabi, A., Sayed, E. T., Elsaid, K., & Abdelkareem, M. A. (2021). Progress in carbon capture technologies. *Science of the Total Environment*, *761*, 143203. https://doi.org/10.1016/j.scitotenv.2020.143203.

Yaseen, B. T., & Kurnaz, S. (2023). RETRACTED ARTICLE: Drug–target interaction prediction using artificial intelligence. *Applied Nanoscience*, *13*(5), 3335–3345.

Yoon, H. J., Jeong, Y. J., Kang, H., Jeong, J. E., & Kang, D.-Y. (2019). Medical image analysis using artificial intelligence. *Progress in Medical Physics*, *30*, 49–58.

Zou, J., Huss, M., Abid, A., Mohammadi, P., Torkamani, A., & Telenti, A. (2019). A primer on deep learning in genomics. *Nature Genetics*, *51*(1), 12–18.

Shallu Saini, Manuj Saini*, Hardeep Singh Tuli, Jatin Saini, Shreya Nagpal, Priti Duhan, Poonam Bansal

Chapter 15 Deep Learning and Artificial Intelligence in Biotechnology: A Comprehensive Analysis of Current Applications and Future Perspectives

Abstract: The convergence of artificial intelligence (AI) and biotechnology represents one of the most transformative developments in modern science, fundamentally reshaping how we approach biological research, drug discovery, and healthcare innovation. This comprehensive analysis explores the multifaceted applications of deep learning frameworks, neural network architectures, and emerging AI technologies across the biotechnology landscape. From the revolutionary impact of protein structure prediction models like AlphaFold to the integration of quantum computing and federated learning systems, AI is enabling unprecedented capabilities in understanding and manipulating biological systems. The field has evolved from basic pattern recognition tasks to sophisticated autonomous laboratory systems capable of hypothesis generation, experimental design, and iterative optimization. As we advance toward 2025 and beyond, the integration of generative models, transformer architectures, and graph neural networks continues to accelerate discovery timelines while enhancing the precision and scope of biotechnological applications.

Keywords: Artificial intelligence, biotechnology, quantum, machine learning, deep learning

***Correspondence author: Manuj Saini**, Department of Genetics and Plant Breeding, Chaudhary Charan Singh Haryana Agricultural University, Hisar 125004, Haryana, India, e-mail: doc.manujsaini@gmail.com
Shallu Saini, Department of Bio-Sciences and Technology, Maharishi Markandeshwar Engineering College, Maharishi Markandeshwar (Deemed to be University), Mullana, Ambala 133207, India
Hardeep Singh Tuli, Department of Bio-Sciences and Technology, Maharishi Markandeshwar Engineering College, Maharishi Markandeshwar (Deemed to be University), Mullana, Ambala 133207, India
Jatin Saini, Department of Agriculture, Maharishi Markandeshwar (Deemed to be University), Mullana, Ambala 133207, India
Shreya Nagpal, Department of Genetics and Plant Breeding, Chaudhary Charan Singh Haryana Agricultural University, Hisar 125004, India
Priti Duhan, Department of Biotechnology, K. L. Mehta Dayanand College for Women, Faridabad 121001, Haryana, India
Poonam Bansal, Department of Bio-Sciences and Technology, Maharishi Markandeshwar Engineering College, Maharishi Markandeshwar (Deemed to be University), Mullana, Ambala 133207, India

 | https://doi.org/10.1515/9783112214275-015

15.1 Introduction

The intersection of artificial intelligence (AI) and biotechnology has emerged as a critical frontier in scientific innovation, fundamentally transforming approaches to biological research and therapeutic development (Goshisht, 2024; Zhang et al., 2023). Machine learning and deep learning methodologies have revolutionized traditional biotechnology workflows by providing sophisticated tools for pattern recognition, predictive modeling, and automated analysis of complex biological datasets (Kulikova & Adcock, 2021; Srinivasu et al., 2022). The remarkable success of AI systems such as AlphaFold in protein structure prediction has demonstrated the transformative potential of these technologies, achieving near-experimental accuracy in structural predictions that previously required years of laboratory effort (Gligorijević et al., 2021; Hu et al., 2022).

Contemporary biotechnology generates vast quantities of multidimensional data from genomic sequencing, proteomic analyses, metabolomic profiling, and high-throughput screening experiments (Fushimi et al., 2025; Chakravarthi et al., 2024). Traditional analytical approaches often struggle with the scale, complexity, and heterogeneity of these datasets, creating significant bottlenecks in biological discovery processes (Alipanahi et al., 2015; Min et al., 2017). AI addresses these challenges by providing scalable, automated solutions for data integration, pattern discovery, and predictive modeling across diverse biological domains (Shen et al., 2017).

The application of AI in biotechnology spans multiple interconnected areas, including drug discovery and development, protein engineering, synthetic biology, genomics, and personalized medicine (Goshisht, 2024; Chakravarthi et al., 2024). Deep learning architectures such as convolutional neural networks excel at processing biological images and structural data, while recurrent neural networks and transformer models effectively analyze sequential biological information including DNA, RNA, and protein sequences (Kulikova & Adcock, 2021; Zhang et al., 2023). Graph neural networks have proven particularly valuable for molecular modeling and drug design applications, enabling direct representation of chemical structures and their properties (Gligorijević et al., 2021).

Recent advances in transformer-based language models have opened new possibilities for biological sequence analysis and prediction (Zhang et al., 2023). These models, adapted from natural language processing, treat biological sequences as languages and learn complex patterns through self-supervised training on large-scale biological databases (Zhang et al., 2023; Srinivasu et al., 2022). The success of models such as ProtBERT, ESM, and DNABERT demonstrates the effectiveness of this approach across diverse biological prediction tasks (Zhang et al., 2023).

15.2 Comparison of Deep Learning Frameworks in Biotechnology Applications

The autonomous laboratory represents another frontier where AI is making significant contributions to biotechnology (Fushimi et al., 2025). Robotic systems integrated with AI algorithms can perform complex experimental workflows, from sample preparation and analysis to hypothesis generation and iterative optimization (Fushimi et al., 2025; Chakravarthi et al., 2024). These systems promise to accelerate the pace of biological discovery while reducing human error and experimental bias (Fushimi et al., 2025).

However, the integration of AI in biotechnology also presents significant challenges including data quality issues, model interpretability concerns, regulatory compliance requirements, and ethical considerations surrounding privacy and bias (Alipanahi et al., 2015; Min et al., 2017; Shen et al., 2017). Addressing these challenges requires interdisciplinary collaboration between computer scientists, biologists, clinicians, and policymakers to develop robust, trustworthy, and ethically sound AI systems for biotechnology applications (Shen et al., 2017).

The economic impact of AI in biotechnology continues to expand rapidly, with market projections indicating substantial growth across multiple application domains (Goshisht, 2024). Investment in AI-driven biotechnology companies has surged, reflecting growing confidence in the commercial viability of these technologies (Goshisht, 2024). As AI capabilities continue to advance and integration barriers diminish, the biotechnology industry is positioned for unprecedented transformation in research methodologies, product development timelines, and therapeutic outcomes (Goshisht, 2024; Kulikova & Adcock, 2021).

15.3 Deep Learning Frameworks in Biotechnology

The selection and implementation of appropriate deep learning frameworks represent a critical decision point for biotechnology applications, significantly influencing model performance, scalability, and integration capabilities (Meng et al., 2025; Wheeler et al., 2025). Modern biotechnology research requires frameworks that can handle diverse data types, from molecular structures and genomic sequences to medical images and time-series biological measurements (Mehedi et al., 2025; Chakraborty et al., 2024). The evolution of specialized frameworks and the adaptation of general-purpose platforms for biological applications have created a rich ecosystem of computational tools supporting biotechnology innovation (Table 15.1) (Gangwal et al., 2024).

TensorFlow has emerged as a dominant framework in biotechnology applications, particularly for large-scale production systems and research projects requiring extensive scalability (Taylor et al., 2025; Montague & Morris, 1994). Its comprehensive ecosystem includes specialized libraries such as TensorFlow Probability for uncertainty quantification in biological predictions and TensorFlow Extended for production ML pipelines in pharmaceutical companies (Taylor et al., 2025). Notable biotechnology applications include Google DeepMind's AlphaFold protein structure prediction system and Google's DeepVariant genome analysis tool, both of which demonstrate TensorFlow's capability to handle complex biological modeling tasks (Zhou & Troyanskaya, 2015; Taylor et al., 2025).

PyTorch has gained significant adoption in biotechnology research communities due to its dynamic computational graph structure and intuitive debugging capabilities (Simonyan et al., 2013; LeCun et al., 2015). The framework's flexibility makes it particularly suitable for experimental research and novel architecture development in biological applications (Simonyan et al., 2013). PyTorch's ecosystem includes specialized libraries such as PyTorch Geometric for graph neural networks applied to molecular modeling and PyTorch Lightning for streamlined experimental workflows (LeCun et al., 2015). Many transformer-based biological language models, including ESM and ChemBERTa, leverage PyTorch's capabilities for sequence modeling and natural language processing adaptations (Zhang et al., 2023; LeCun et al., 2015).

Specialized frameworks have emerged to address unique requirements in biotechnology applications (Gangwal et al., 2024; Taylor et al., 2025). JAX, developed by Google Research, combines high-performance computing capabilities with functional programming paradigms, making it particularly suitable for molecular dynamics simulations and optimization problems in synthetic biology (Taylor et al., 2025). The framework's automatic differentiation capabilities and XLA compilation enable efficient training of complex biological models on specialized hardware including TPUs (Taylor et al., 2025). Recent implementations include AlphaFold3 and various molecular simulation packages that require high computational throughput (Montague & Morris, 1994).

Deep learning frameworks must address specific challenges in biotechnology applications, including handling sparse and heterogeneous biological datasets, managing computational requirements for large-scale genomic data, and ensuring reproducibility across research institutions (Mehedi et al., 2025; Chakraborty et al., 2024; Gangwal et al., 2024). Frameworks increasingly incorporate specialized data loaders for biological file formats, pretrained models for transfer learning in biological domains, and standardized evaluation metrics for biotechnology applications (Mehedi et al., 2025; Chakraborty et al., 2024).

The integration of cloud computing platforms with deep learning frameworks has transformed accessibility and scalability in biotechnology research (Wheeler et al., 2025; Taylor et al., 2025). Cloud-based solutions enable smaller research institutions to access high-performance computing resources, while providing standardized environments for reproducible research (Taylor et al., 2025). Major cloud providers offer specialized biotechnology services that integrate with popular deep learning frameworks, including managed genomics pipelines and preconfigured machine learning environments (Taylor et al., 2025).

Edge computing applications are emerging as important considerations for biotechnology frameworks, particularly in clinical settings and field-based research (Taylor et al., 2025; Meng et al., 2025). Frameworks must optimize models for deployment on resource-constrained devices while maintaining accuracy and ensuring data privacy (Taylor et al., 2025). Recent developments include model quantization techniques, federated learning capabilities, and edge-optimized inference engines that enable real-time biological data analysis in diverse environments (Meng et al., 2025).

Interoperability between frameworks and existing biotechnology software ecosystems remains a critical consideration (Chakraborty et al., 2024; Gangwal et al., 2024). Modern frameworks increasingly support standard data formats used in biotechnology, provide APIs for integration with laboratory information management systems, and offer conversion tools for models trained in different frameworks (Chakraborty et al., 2024). This interoperability enables organizations to leverage multiple frameworks for different aspects of their biotechnology workflows while maintaining data consistency and model compatibility (Gangwal et al., 2024).

The future evolution of deep learning frameworks in biotechnology will likely emphasize specialized biological operations, improved hardware optimization for biotechnology workloads, and enhanced support for multimodal biological data integration (Taylor et al., 2025; Simonyan et al., 2013; LeCun et al., 2015). As quantum computing hardware becomes more accessible, frameworks are beginning to incorporate quantum-classical hybrid capabilities for molecular modeling and optimization problems (Montague & Morris, 1994; Meng et al., 2025).

Table 15.1: Deep learning frameworks in biotechnology.

Framework	Key strengths	Biotech apps	Hardware	Examples
TensorFlow	Scalable, production, and extensive ecosystem	Protein folding, drug discovery, and genomics	CPUs, GPUs, and TPUs	AlphaFold and DeepVariant
PyTorch	Research, dynamic, and strong community	Molecular modeling, sequence analysis, and medical imaging	CPUs and GPUs	ESM models and ChemBERTa
Keras	User-friendly, high-level, and rapid prototyping	Biomarkers and classification	CPUs and GPUs via backend	Medical image classification
JAX	High performance, functional, and XLA compilation	Molecular dynamics and optimization	CPUs, GPUs, and TPUs	JAX-MD and AlphaFold3
Deeplearning4j	Java-based, enterprise, and distributed	Big data and clinical data	CPUs, GPUs, and distributed clusters	Hospital systems

15.4 Neural Networks for Biological Data

Neural network architectures have undergone significant specialization and adaptation to address the unique characteristics and challenges presented by biological data types as represented in Figure 15.1 (Robert et al., 2021; Kim et al., 2025). Biological systems generate diverse data modalities including sequential information (DNA, RNA, and protein sequences), structural data (molecular conformations and protein folding patterns), image data (microscopy and medical imaging), and temporal data (gene expression dynamics and physiological measurements) (Singh et al., 2025; Zaki et al., 2024). Each data type requires specialized architectural considerations to effectively capture biological patterns and relationships (Valeri et al., 2020).

Convolutional Neural Networks have proven exceptionally effective for biological image analysis and spatial pattern recognition tasks (Holur et al., 2025; Sree & Devi, 2024). In medical imaging applications, CNNs achieve state-of-the-art performance in diagnostic tasks including cancer detection, medical image segmentation, and pathological classification (Sree & Devi, 2024). The hierarchical feature learning capabilities of CNNs enable automatic extraction of relevant biological patterns from complex visual data without requiring manual feature engineering (Esteva et al., 2017). Recent innovations include 3D CNN architectures for volumetric medical data analysis and attention-enhanced CNNs for improved interpretability in clinical applications (Sree & Devi, 2024).

Protein structure analysis represents another domain where CNNs excel, particularly in predicting secondary and tertiary structures from sequence information (Sree & Devi, 2024; Gangwal et al., 2024). Three-dimensional CNNs can process protein structural data directly, enabling analysis of binding sites, conformational changes, and protein-protein interactions (Sree & Devi, 2024). The application of CNNs to cryo-electron microscopy data has accelerated structural biology research by automating particle picking, classification, and reconstruction processes (Esteva et al., 2017).

Recurrent Neural Networks, particularly long short-term memory networks and gated recurrent units, have been extensively applied to sequential biological data analysis (Robert et al., 2021; Valeri et al., 2020). DNA and RNA sequence analysis benefits from RNN architectures that can capture long-range dependencies and contextual information within genomic sequences (Valeri et al., 2020). Applications include gene prediction, splice site identification, transcription factor binding site discovery, and regulatory element classification (Valeri et al., 2020; Holur et al., 2025). Bidirectional RNN architectures enable analysis of sequence context from both directions, improving accuracy in biological sequence labeling tasks (Holur et al., 2025).

Protein sequence analysis represents another successful application domain for RNNs, particularly in predicting protein secondary structure, disorder regions, and functional domains (Valeri et al., 2020; Gangwal et al., 2024). The sequential nature of amino acid chains aligns well with RNN processing capabilities, enabling models to learn complex patterns of protein folding and function (Valeri et al., 2020). Recent de-

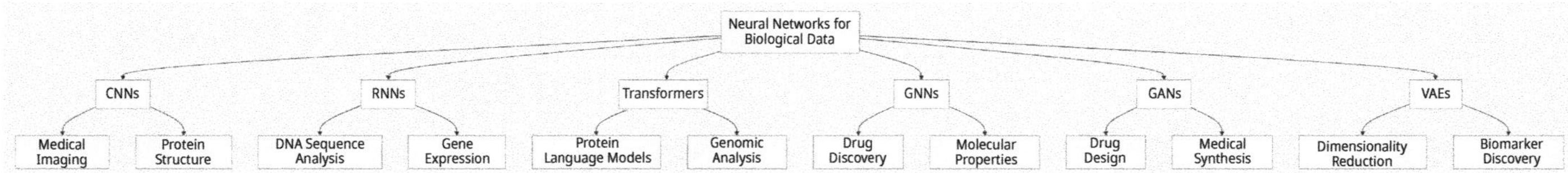

Figure 15.1: Neural network architectures (e.g., CNNs, RNNs, and graph neural networks) are used in biological data analysis to automatically learn complex patterns from genomics, imaging, and molecular interaction data.

velopments include attention-enhanced RNNs that can identify critical amino acid positions and their contributions to protein properties (Gangwal et al., 2024).

Time-series biological data, including gene expression profiles, physiological measurements, and metabolomic data, benefit from RNN architectures designed for temporal pattern recognition (Kim et al., 2025; Zaki et al., 2024). Clinical applications include patient monitoring systems that use RNNs to predict adverse events, disease progression, and treatment responses based on continuous physiological data streams (Kim et al., 2025). These systems demonstrate the capability of RNNs to integrate multiple biological signals over time to generate clinically actionable predictions (Zaki et al., 2024).

Graph neural networks have emerged as particularly powerful architectures for molecular modeling and biological network analysis (Singh et al., 2025; Gangwal et al., 2024; Esteva et al., 2017). Molecular structures naturally represent as graphs with atoms as nodes and chemical bonds as edges, making GNNs ideally suited for drug discovery and chemical property prediction tasks (Singh et al., 2025). The ability of GNNs to operate directly on molecular graphs without requiring fixed-size representations has revolutionized computational chemistry and drug design (Gangwal et al., 2024).

Protein-protein interaction networks, metabolic pathways, and gene regulatory networks all benefit from GNN analysis approaches (Esteva et al., 2017; Kanaka et al., 2023). These biological networks exhibit complex topological properties that traditional machine learning approaches struggle to capture effectively (Esteva et al., 2017). GNNs can learn from network structure and node features simultaneously, enabling prediction of protein functions, identification of disease-related pathways, and discovery of drug targets (Kanaka et al., 2023).

Generative neural networks, including variational autoencoders and generative adversarial networks have opened new possibilities for biological data synthesis and augmentation (Robert et al., 2021; Singh et al., 2025). These architectures can generate novel molecular structures, synthetic biological sequences, and artificial biological images for training data augmentation (Robert et al., 2021). In drug discovery, generative models enable exploration of chemical space and identification of novel therapeutic compounds with desired properties (Singh et al., 2025).

Recent developments in hybrid architecture combine multiple neural network types to address complex biological problems (Zaki et al., 2024; Kim et al., 2025). Graph-transformer models integrate GNN capabilities for molecular structure representation with transformer attention mechanisms for enhanced feature learning (Zaki et al., 2024). CNN-RNN hybrid architectures process biological images with temporal components, such as live-cell microscopy data showing cellular dynamics over time (Kim et al., 2025).

15.5 Transformer-Based Models in Bioinformatics

The adaptation of transformer architectures from natural language processing to bioinformatics has created unprecedented opportunities for biological sequence analysis and prediction (Pfeifer et al., 2024; Cho et al., 2025). Transformer models' ability to capture long-range dependencies and contextual relationships makes them particularly well-suited for biological sequences, which often contain distant but functionally important elements (Pfeifer et al., 2024). The self-attention mechanism enables these models to identify relevant sequence positions regardless of their distance, addressing a fundamental limitation of traditional sequence analysis approaches (Cho et al., 2025). The different transformer models used in bioinformatics are listed in Table 15.2.

Protein language models represent one of the most successful applications of transformer architectures in bioinformatics (Kaspersetz et al., 2024; Fischer et al., 2025). Models such as ESM-1b and ProtBERT treat protein sequences as languages, learning representations through self-supervised training on millions of protein sequences (Kaspersetz et al., 2024). These models capture complex relationships between amino acids, secondary structure patterns, and functional domains without requiring explicit structural or functional annotations during training (Fischer et al., 2025). The learned representations demonstrate remarkable transferability to downstream tasks including protein function prediction, structural property estimation, and evolutionary relationship inference (Kaspersetz et al., 2024).

The success of protein transformers has led to their application in protein engineering and design tasks (Constantinescu et al., 2022; Ramirez et al., 2024). By learning the underlying "grammar" of protein sequences, these models can guide the design of novel proteins with desired properties (Constantinescu et al., 2022). Recent applications include enzyme optimization, antibody design, and de novo protein creation for therapeutic applications (Ramirez et al., 2024). The ability to generate protein variants with improved stability, activity, or specificity has significant implications for biotechnology and pharmaceutical development (Constantinescu et al., 2022).

Table 15.2: Transformer models in bioinformatics.

Model name	Year	Architecture	Application	Innovation	Performance
ChemBERTa	2020	RoBERTa-based	Chemical property prediction	SMILES-based molecules	Improved molecule prediction
ProtBERT	2021	BERT-based	Protein sequence analysis	Pretrained protein	State-of-the-art protein
ESM-1b	2021	Transformer encoder	Protein representation learning	Unsupervised protein	Competitive AlphaFold

Table 15.2 (continued)

Model name	Year	Architecture	Application	Innovation	Performance
DNABERT	2021	BERT-based	DNA sequence analysis	K-mer DNA tokens	Enhanced variant call
GeneBERT	2021	Multimodal BERT	Gene regulation	Multimodal training	Improved regulation prediction
Enformer	2021	Transformer	Gene expression prediction	Long-range dependencies	Superior Basenji2
scBERT	2022	BERT-based	Single-cell RNA	Gene expression language	Superior cell annotation
TransEPI	2022	Transformer encoder	Enhancer-promoter	Long-range context	28% improvement

15.5.1 Evolution and Applications of Transformer-Based Models in Bioinformatics

Genomic sequence analysis has benefited substantially from transformer-based approaches, particularly in tasks requiring analysis of long-range genomic context (Frey et al., 2025; Dodig et al., 2024). Models like DNABERT and Nucleotide Transformer can process DNA sequences of unprecedented length, enabling analysis of regulatory elements, chromatin structure, and gene expression patterns across large genomic regions (Frey et al., 2025). These models excel at identifying transcription factor binding sites, predicting gene expression levels, and analyzing the effects of genetic variants on gene regulation (Dodig et al., 2024).

Enhancer-promoter interaction prediction represents a particularly compelling application of genomic transformers (Vaswani et al., 2017; Rives et al., 2021). Traditional approaches struggle with the long-range nature of these interactions, which can span hundreds of thousands of base pairs (Vaswani et al., 2017). Transformer models can effectively capture these long-range dependencies, achieving significant improvements in predicting which enhancers regulate which genes in different cell types (Rives et al., 2021). This capability has important implications for understanding gene regulation and identifying disease-associated regulatory variants (Vaswani et al., 2017).

Single-cell RNA sequencing data analysis has been revolutionized by transformer-based approaches (Pfeifer et al., 2024; Cho et al., 2025). Models like scBERT treat gene expression profiles as sequences, enabling analysis of cellular states, developmental trajectories, and cell type identification (Pfeifer et al., 2024). The attention mechanisms in these models can identify key genes driving cellular transitions and highlight important regulatory relationships (Cho et al., 2025). Applications include automated cell type annotation, trajectory inference, and identification of disease-associated cellular states (Pfeifer et al., 2024).

Multimodal transformer architectures are emerging to integrate diverse biological data types (Kaspersetz et al., 2024; Fischer et al., 2025). These models can simultaneously process sequence data, structural information, and functional annotations to generate more comprehensive biological representations (Kaspersetz et al., 2024). Applications include drug-target interaction prediction, where models integrate protein sequences, chemical structures, and biological pathway information to predict binding affinity and selectivity (Fischer et al., 2025).

The interpretability of transformer models in biological applications has received significant attention due to the importance of understanding model predictions in life sciences (Frey et al., 2025; Dodig et al., 2024). Attention visualization techniques enable researchers to identify which sequence positions contribute most to predictions, providing insights into biological mechanisms (Frey et al., 2025). This interpretability has led to biological discoveries, including identification of previously unknown functional motifs and regulatory elements (Dodig et al., 2024).

Challenges in applying transformers to biological data include computational requirements for long sequences, handling of biological sequence variations, and integration with domain-specific biological knowledge (Constantinescu et al., 2022; Ramirez et al., 2024). Recent developments address these challenges through efficient attention mechanisms, biological sequence-specific tokenization strategies, and incorporation of evolutionary and structural constraints into model training (Constantinescu et al., 2022; Ramirez et al., 2024).

The future of transformer models in bioinformatics will likely involve larger-scale pretraining on comprehensive biological databases, development of specialized architectures for different biological data types, and integration with experimental validation workflows (Vaswani et al., 2017; Rives et al., 2021). As these models continue to improve, they promise to accelerate biological discovery and enable new approaches to understanding and engineering biological systems (Vaswani et al., 2017; Rives et al., 2021).

15.5.2 Graph Neural Networks for Molecular Modeling

Graph neural networks have emerged as the predominant architecture for molecular modeling and computational chemistry applications, fundamentally changing how AI systems represent and analyze chemical structures (Bhujel et al., 2025; Martin et al., 2025). The natural correspondence between molecular structures and graph representations, where atoms serve as nodes and chemical bonds as edges, makes GNNs uniquely suited for directly processing molecular information without requiring hand-crafted descriptors or fixed-size representations (Bhujel et al., 2025). This direct structural representation enables GNNs to capture complex molecular patterns, stereochemistry, and three-dimensional structural information that traditional approaches often miss (Martin et al., 2025).

The evolution of GNN architectures for molecular applications has progressed from simple graph convolutions to sophisticated message-passing frameworks that can incorporate geometric information, electronic properties, and quantum mechanical effects (Yang et al., 2024; Lewis et al., 2024). Early Graph Convolutional Networks demonstrated the feasibility of learning molecular properties directly from graph representations, achieving competitive performance on benchmark datasets while providing interpretable predictions as shown in Figure 15.2 (Yang et al., 2024). Subsequent developments have introduced attention mechanisms, geometric constraints, and physics-informed components that significantly enhance predictive accuracy and chemical relevance (Lewis et al., 2024).

Message passing neural networks represent a particularly influential framework for molecular GNNs, providing a general paradigm for learning from molecular graphs through iterative information exchange between connected atoms (Tong, 2024; Ashfi, 2024). The message-passing framework enables nodes to aggregate information from their local neighborhoods, gradually building up representations that capture both local chemical environments and global molecular properties (Tong, 2024). This approach has proven highly effective for predicting molecular properties such as solubility, toxicity, bioactivity, and synthetic accessibility (Ashfi, 2024).

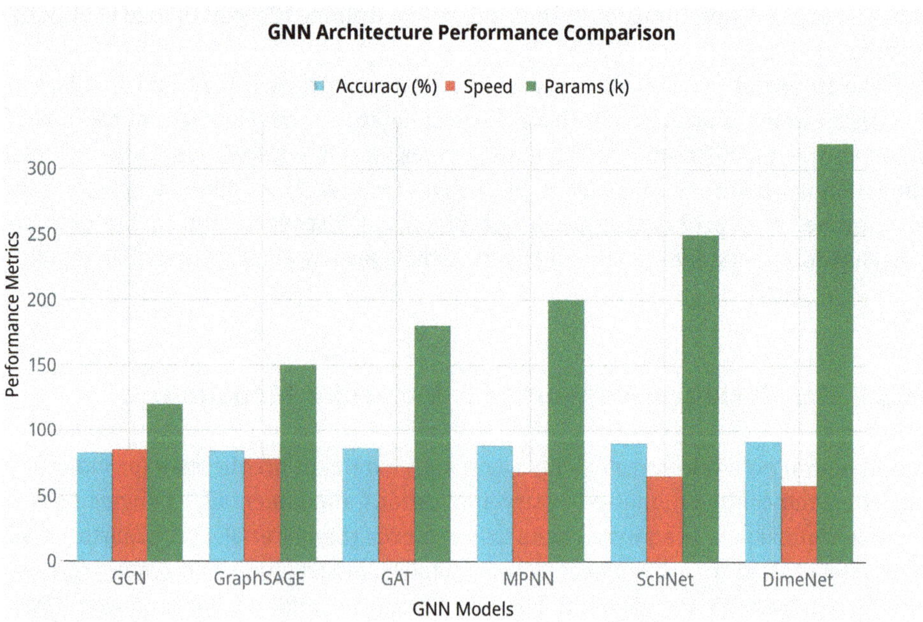

Figure 15.2: Performance comparison of graph neural network architectures for molecular modeling.

Three-dimensional molecular information has been increasingly incorporated into GNN architectures through geometric neural networks that explicitly model spatial relationships and conformational effects (Scarselli et al., 2009; Xie & Grossman, 2018). Models such as SchNet and DimeNet process molecular coordinates directly, enabling prediction of properties that depend critically on three-dimensional structure including binding affinity, stereoselectivity, and conformational stability (Scarselli et al., 2009). These geometric approaches have proven particularly valuable for drug discovery applications where molecular shape and electrostatic properties determine biological activity (Xie & Grossman, 2018).

Quantum mechanical effects and electronic structure considerations are being integrated into molecular GNNs through physics-informed architectures and multilevel modeling approaches (Gilmer et al., 2017; Bhujel et al., 2025). Models can incorporate quantum mechanical descriptors, electronic density information, and orbital interactions to enhance predictions of chemical reactivity, electronic properties, and spectroscopic characteristics (Gilmer et al., 2017). This integration bridges classical molecular representations with quantum mechanical accuracy, enabling more reliable predictions for novel chemical structures (Bhujel et al., 2025).

The application of molecular GNNs to drug discovery has demonstrated significant practical impact across multiple stages of pharmaceutical development (Martin et al., 2025; Yang et al., 2024; Lewis et al., 2024). Virtual screening applications use GNNs to identify potential drug candidates from large chemical libraries, achieving higher hit rates than traditional similarity-based approaches (Martin et al., 2025). Lead optimization employs GNNs to predict structure-activity relationships, guide chemical modifications, and assess drug-like properties including absorption, distribution, metabolism, excretion, and toxicity (Yang et al., 2024).

Molecular generation and design represent emerging applications where GNNs guide the creation of novel chemical structures with desired properties (Tong, 2024; Ashfi, 2024). Generative models built on GNN architectures can sample from learned chemical space distributions, optimizing molecules for specific therapeutic targets while maintaining synthetic feasibility (Tong, 2024). These approaches enable exploration of novel chemical spaces and identification of innovative drug candidates that might not be discovered through traditional medicinal chemistry approaches (Ashfi, 2024). A list of different generative models used in drug discovery with their application is mentioned in Table 15.3.

15.6 Applications of Generative Models (GANs and VAEs) in Drug Discovery

Catalyst design and materials discovery applications leverage GNN capabilities to predict and optimize the properties of complex materials systems (Scarselli et al., 2009;

Table 15.3: Generative models in drug discovery.

Model	Model type	Specific application	Dataset used	Generated output	Validation method	Success rate
MolGAN	GAN	Molecular graph generation	QM9 and ZINC	Valid molecular graphs	Chemical validity and property prediction	61% valid molecules
druGAN	GAN	Drug-like molecule generation	ChEMBL	SMILES strings	Druglikeness scores and bioactivity	85% drug-like molecules
GENTRL	GAN	DDR1 kinase inhibitors	Proprietary compound library	Novel kinase inhibitors	Experimental validation	21 days to preclinical
Molecular VAE	VAE	Chemical space exploration	ZINC-250K	Latent molecular representations	Reconstruction accuracy	90% reconstruction
Junction tree VAE	VAE	Molecular optimization	ZINC	Optimized molecular structures	Property improvement	Chemical validity 100%
GraphVAE	VAE	Graph-based molecular generation	QM9	Molecular graphs	Graph validity and properties	76% valid graphs

Xie & Grossman, 2018). Surface chemistry modeling uses GNNs to analyze heterogeneous catalysts, predicting adsorption energies, reaction pathways, and catalytic selectivity (Scarselli et al., 2009). Materials property prediction employs GNNs to design novel materials with targeted mechanical, electronic, or optical properties (Xie & Grossman, 2018).

The interpretability of molecular GNNs has received significant attention due to the importance of understanding structure-property relationships in chemistry (Gilmer et al., 2017; Bhujel et al., 2025). Attention-based architectures and gradient-based interpretation methods enable identification of molecular fragments and structural motifs that contribute most to predicted properties (Gilmer et al., 2017). This interpretability supports medicinal chemistry decision-making and provides insights into underlying chemical mechanisms (Bhujel et al., 2025).

Challenges in molecular GNN development include handling conformational flexibility, incorporating solvent effects, addressing data scarcity for rare chemical structures, and ensuring chemical validity in generated molecules (Martin et al., 2025; Yang et al., 2024). Recent research addresses these challenges through ensemble approaches for conformational sampling, implicit solvent models, transfer learning from related chemical domains, and constrained generation algorithms (Martin et al., 2025; Yang et al., 2024).

The future of molecular GNNs will likely involve integration with experimental validation workflows, incorporation of reaction mechanisms and kinetic information, and extension to biological macromolecules and complex biological systems (Lewis et al., 2024; Tong, 2024). As these models become more sophisticated and accurate, they promise to accelerate chemical discovery and enable rational design of molecules for therapeutic, material, and environmental applications (Lewis et al., 2024; Tong, 2024).

15.6.1 Generative Models: GANs and VAEs in Drug Discovery

Generative AI has revolutionized drug discovery by enabling the design and optimization of novel molecular structures with desired therapeutic properties (Arús-Pous et al., 2020; Cheng et al., 2021). Generative adversarial networks and variational autoencoders represent two fundamental approaches to molecular generation, each offering distinct advantages for different aspects of pharmaceutical development (Arús-Pous et al., 2020). These models learn complex distributions of chemical space from large molecular databases, enabling exploration of novel regions and identification of potentially therapeutic compounds that might not be discovered through traditional approaches (Cheng et al., 2021).

The application of GANs to molecular generation has demonstrated remarkable success in creating chemically valid and drug-like molecules (Kadurin et al., 2017; Zhavoronkov et al., 2019). MolGAN, one of the pioneering approaches, generates molecular graphs directly through adversarial training, learning to produce valid molecular structures while optimizing for desired properties (Kadurin et al., 2017). The adversarial framework enables exploration of chemical space regions that are underrepresented in training data, potentially identifying novel scaffolds and chemotypes for drug development (Zhavoronkov et al., 2019).

druGAN represents a notable success story in GAN-based drug discovery, specifically designed to generate molecules with improved drug-like properties (Kadurin et al., 2017). The model incorporates pharmaceutical knowledge including Lipinski's Rule of Five, synthetic accessibility scores, and bioactivity predictions to guide the generation process toward promising therapeutic candidates (Kadurin et al., 2017). Experimental validation of druGAN-generated compounds has demonstrated the practical utility of these approaches in identifying novel bioactive molecules (Kadurin et al., 2017).

The GENTRL system developed by Insilico Medicine exemplifies the potential for rapid drug discovery using generative models (Zhavoronkov et al., 2019). This approach successfully identified novel DDR1 kinase inhibitors and progressed from computational generation to preclinical validation in just 21 days, demonstrating unprecedented speed in drug discovery timelines (Zhavoronkov et al., 2019). The integration of reinforcement learning with generative models enabled optimization of multiple pharmaceutical properties simultaneously while maintaining chemical validity (Zhavoronkov et al., 2019).

Variational Autoencoders offer alternative approaches to molecular generation with particular strengths in providing smooth, continuous chemical space representations (Arús-Pous et al., 2020; Cheng et al., 2021). The latent space learned by VAEs enables interpolation between known molecules and systematic exploration of chemical neighborhoods (Arús-Pous et al., 2020). This property has proven valuable for lead optimization, where small modifications to existing compounds can improve potency, selectivity, or pharmaceutical properties (Cheng et al., 2021).

Junction Tree VAEs address limitations of sequence-based molecular representations by operating directly on molecular graph structures (Cheng et al., 2021). This approach ensures 100% chemical validity in generated molecules while enabling sophisticated molecular editing operations including scaffold hopping, functional group replacement, and stereochemical modifications (Cheng et al., 2021). The hierarchical representation enables generation of complex molecules with desired structural motifs and pharmacophores (Cheng et al., 2021).

Conditional generation approaches enable design of molecules with specific target properties or constraints (Arús-Pous et al., 2020; Kadurin et al., 2017). These models can incorporate biological target information, desired activity profiles, and pharmaceutical constraints to generate compounds tailored for specific therapeutic applications (Arús-Pous et al., 2020). Recent developments include multiobjective optimization approaches that simultaneously optimize potency, selectivity, toxicity, and synthetic accessibility (Kadurin et al., 2017).

The integration of generative models with virtual screening and biological validation workflows has created powerful drug discovery pipelines (Zhavoronkov et al., 2019; Cheng et al., 2021). Generated molecules undergo computational filtering for drug-like properties, synthetic feasibility assessment, and virtual docking to biological targets (Zhavoronkov et al., 2019). Promising candidates proceed to experimental synthesis and biological testing, creating feedback loops that continuously improve the generative models (Cheng et al., 2021).

Challenges in generative drug discovery include ensuring synthetic accessibility of generated molecules, addressing bias in training datasets, validating biological activity predictions, and integrating with existing pharmaceutical development processes (Arús-Pous et al., 2020; Kadurin et al., 2017). Recent research addresses these challenges through incorporation of synthetic feasibility scores, dataset augmentation techniques, experimental validation protocols, and standardized evaluation metrics (Arús-Pous et al., 2020; Kadurin et al., 2017).

The future of generative models in drug discovery will likely involve integration with automated synthesis platforms, incorporation of multimodal biological data, and development of foundation models trained on comprehensive chemical and biological databases (Zhavoronkov et al., 2019; Cheng et al., 2021). As these approaches mature, they promise to significantly accelerate pharmaceutical development while reducing costs and improving success rates in bringing new medicines to patients (Zhavoronkov et al., 2019; Cheng et al., 2021).

15.7 AI-Driven Laboratory Automation Systems

The integration of AI with laboratory automation represents a paradigm shift toward autonomous experimental systems capable of hypothesis generation, experimental design, and iterative optimization without human intervention (Burger et al., 2020; King et al., 2004). These AI-driven systems combine robotic hardware, machine learning algorithms, and sophisticated control software to create closed-loop research environments that can accelerate scientific discovery while reducing human error and experimental bias (Burger et al., 2020). The evolution from simple automated instruments to intelligent laboratory ecosystems reflects advances in AI capabilities, sensor technologies, and integration platforms (King et al., 2004).

The historical development of AI laboratory automation traces (Figure 15.3) back to the ADAM robot scientist, which demonstrated the first fully automated scientific discovery system in 2004 (King et al., 2004; Sparkes et al., 2010). ADAM successfully generated hypotheses about gene function, designed experiments to test these hypotheses, executed physical experiments, and interpreted results to generate new knowledge (King et al., 2004). This pioneering work established the feasibility of autonomous scientific discovery and inspired subsequent developments in intelligent laboratory systems (Sparkes et al., 2010).

Modern AI-driven laboratory automation incorporates sophisticated machine learning algorithms for experimental optimization, pattern recognition, and decision-making (Stokes et al., 2020; Schneider, 2018). These systems can analyze experimental results in real-time, adjust experimental parameters dynamically, and optimize protocols for maximum information gain (Stokes et al., 2020). Applications include automated drug screening platforms that can test thousands of compounds while continuously refining screening protocols based on emerging results (Schneider, 2018).

The mobile robotic chemist developed by researchers at the University of Liverpool represents a significant advancement in autonomous laboratory systems (Burger et al., 2020). This system combines mobile robotics with AI-driven experimental planning to perform chemical synthesis and optimization tasks across multiple laboratory instruments (Burger et al., 2020). The robot can navigate laboratory environments, operate standard equipment, and make intelligent decisions about reaction conditions and purification strategies (Burger et al., 2020).

High-throughput screening applications have been transformed by AI-driven automation systems that can process vast numbers of samples while maintaining experi-

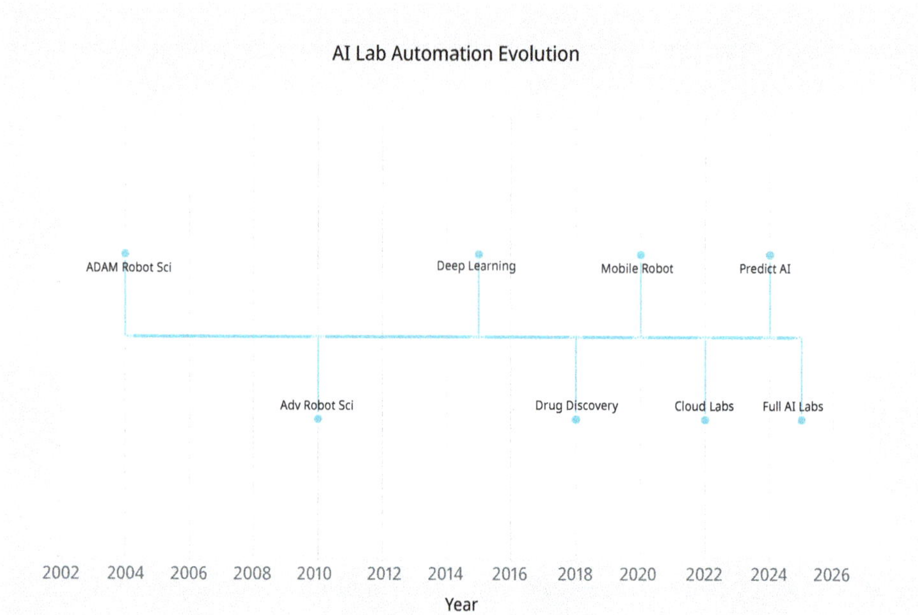

Figure 15.3: AI-driven lab automation has progressed from simple robotics to intelligent, data-driven systems enabling faster discovery by 2025.

mental quality and reproducibility (Stokes et al., 2020; Schneider, 2018). These systems incorporate computer vision for sample identification and tracking, machine learning algorithms for data analysis, and automated decision-making for hit identification and follow-up experiments (Stokes et al., 2020). The integration of AI enables adaptive screening strategies that focus resources on promising leads while eliminating unlikely candidates (Schneider, 2018).

Predictive maintenance and quality control represent crucial applications of AI in laboratory automation (Sparkes et al., 2010; Stokes et al., 2020). Machine learning algorithms can analyze instrument performance data to predict equipment failures, schedule maintenance activities, and ensure consistent experimental conditions (Sparkes et al., 2010). This capability reduces downtime, improves data quality, and enables continuous operation of automated systems (Stokes et al., 2020).

The integration of laboratory information management systems with AI automation platforms creates comprehensive data management and analysis capabilities (King et al., 2004; Schneider, 2018). These integrated systems can track samples throughout experimental workflows, maintain complete experimental provenance, and apply machine learning algorithms to identify patterns and trends across large datasets (King et al., 2004). The resulting insights can guide future experimental designs and accelerate knowledge discovery (Schneider, 2018).

Cloud connectivity and edge computing capabilities enable distributed laboratory networks and remote monitoring of automated systems (Burger et al., 2020; Sparkes et al., 2010). Researchers can access experimental data and control laboratory equipment from remote locations, while cloud-based AI services provide advanced analytics and modeling capabilities (Burger et al., 2020). Edge computing enables real-time decision-making in laboratory automation systems while reducing dependence on network connectivity (Sparkes et al., 2010).

Safety and compliance considerations are critical in AI-driven laboratory automation, particularly in pharmaceutical and clinical research environments (Stokes et al., 2020; Schneider, 2018). Automated systems must comply with regulatory requirements including Good Laboratory Practice standards, data integrity requirements, and safety protocols (Stokes et al., 2020). AI algorithms must provide audit trails, ensure reproducibility, and maintain data security throughout experimental workflows (Schneider, 2018).

The future of AI-driven laboratory automation will likely involve increased autonomy, integration with digital twins of experimental systems, and incorporation of advanced AI techniques including reinforcement learning and large language models (King et al., 2004; Burger et al., 2020). These developments promise to create truly intelligent laboratory environments capable of autonomous scientific discovery across diverse research domains (King et al., 2004; Burger et al., 2020).

15.7.1 Quantum Computing and Edge AI in Biotechnology

The convergence of quantum computing, edge AI, and biotechnology represents an emerging frontier with transformative potential for molecular modeling, drug discovery, and real-time biological data analysis (Cao et al., 2019; Schuld et al., 2015). Quantum computing offers unique advantages for solving complex optimization problems, simulating quantum mechanical systems, and processing high-dimensional biological data that challenge classical computational approaches (Cao et al., 2019). Simultaneously, edge AI enables deployment of intelligent systems directly at the point of data generation, reducing latency and enabling real-time decision-making in clinical and research environments (Schuld et al., 2015).

Quantum algorithms for molecular simulation demonstrate particular promise for accurately modeling chemical reactions, protein folding, and drug-target interactions (Biamonte et al., 2017; Preskill, 2018). The variational quantum eigensolver algorithm enables calculation of molecular ground state energies with potentially exponential speedup over classical methods for certain problem classes (Biamonte et al., 2017). Recent implementations have successfully simulated small molecules, including hydrogen, lithium hydride, and more complex organic compounds, with accuracy approaching experimental measurements (Preskill, 2018).

Table 15.4: Quantum versus classical computing approaches in biotechnology.

Computing type	Complexity	Speed	Accuracy	Limitations	Future potential
Classical DFT	Small (<100)	Hours-days	Good approximation	Scaling+	Incremental
Classical MD	Medium (<10k)	Days-weeks	FF dependent	Time limits	Better FF
Quantum VQE	Small (exact)	Min-hours	Near exact	NISQ limits	Revolutionary
Quantum QAOA	Optimization	Exponential+	Problem dependence	Qubit count	Combinatorial optimization
Hybrid quantum-classical	Medium mol.	Balanced	Enhanced	Implementation complexity	Near-term
Quantum annealing	Protein folding	Fast approximation / Search technique	Good optimum	Limited probs	Specialized

Quantum machine learning approaches are emerging for biological data analysis, particularly in areas requiring optimization of high-dimensional parameter spaces (Steiger et al., 2018; Cao et al., 2019). Quantum neural networks and quantum support vector machines show promise for classification tasks in genomics, proteomics, and medical imaging (Steiger et al., 2018). The quantum advantage may be most pronounced for problems involving exponentially large state spaces, such as protein conformational analysis and molecular docking (Cao et al., 2019). The difference between quantum and classical computing approaches in biotechnology is summarized in Table 15.4.

Edge AI deployment in biotechnology enables real-time analysis of biological data in resource-constrained environments including clinical settings, field research, and portable diagnostic devices (Schuld et al., 2015; Biamonte et al., 2017). Edge computing architectures process data locally on specialized hardware, reducing dependence on cloud connectivity while ensuring data privacy and regulatory compliance (Schuld et al., 2015). Applications include point-of-care diagnostic systems, wearable health monitors, and autonomous laboratory instruments (Biamonte et al., 2017).

The integration of AI with laboratory automation systems benefits significantly from edge computing capabilities (Preskill, 2018; Steiger et al., 2018). Local processing enables real-time experimental optimization, immediate quality control decisions, and rapid response to experimental anomalies (Preskill, 2018). Edge deployment also addresses latency requirements for closed-loop experimental systems that require millisecond response times for optimal performance (Steiger et al., 2018).

Clinical applications of edge AI in biotechnology include portable diagnostic devices that can perform complex analyses in resource-limited settings (Cao et al., 2019;

Schuld et al., 2015). These systems combine miniaturized laboratory equipment with AI algorithms optimized for low-power edge processors (Cao et al., 2019). Applications include rapid pathogen detection, point-of-care blood analysis, and automated microscopy for disease diagnosis in remote locations (Schuld et al., 2015).

Quantum-classical hybrid algorithms represent a practical near-term approach for biotechnology applications (Biamonte et al., 2017; Preskill, 2018). These algorithms use quantum processors for specific computational tasks while leveraging classical computers for data preprocessing, optimization, and result interpretation (Biamonte et al., 2017). Hybrid approaches can address current limitations of quantum hardware while providing quantum advantages for selected problem components (Preskill, 2018).

The development of quantum error correction and fault-tolerant quantum computing will likely unlock more sophisticated biotechnology applications (Steiger et al., 2018; Cao et al., 2019). Future quantum systems may enable simulation of large biological molecules, optimization of complex drug discovery problems, and analysis of quantum effects in biological systems (Steiger et al., 2018). These capabilities could revolutionize understanding of photosynthesis, enzyme catalysis, and other quantum biological phenomena (Cao et al., 2019).

Challenges in quantum and edge AI adoption include hardware limitations, algorithm development complexity, integration with existing biotechnology workflows, and workforce training requirements (Schuld et al., 2015; Biamonte et al., 2017). Current quantum systems face limitations in qubit count, coherence time, and error rates that restrict practical applications (Schuld et al., 2015). Edge AI deployment requires careful optimization of models for resource-constrained hardware while maintaining acceptable accuracy (Biamonte et al., 2017).

The future convergence of quantum computing and edge AI in biotechnology may enable distributed quantum-classical computing networks that combine quantum processing capabilities with edge deployment advantages (Preskill, 2018; Steiger et al., 2018). Such systems could provide quantum-enhanced AI capabilities directly at the point of biological data generation, enabling unprecedented real-time analysis and decision-making capabilities (Preskill, 2018; Steiger et al., 2018).

15.7.2 Federated Learning and Data Privacy

Federated learning has emerged as a critical paradigm for biotechnology and healthcare applications where data privacy, regulatory compliance, and institutional collaboration requirements necessitate distributed machine learning approaches (Li et al., 2020; Rieke et al., 2020). This decentralized learning framework enables multiple institutions to collaboratively train AI models while keeping sensitive biological and medical data locally stored, addressing fundamental challenges in biomedical research including data silos, privacy regulations, and limited dataset sizes (Li et al., 2020). The approach has proven particularly valuable in genomics, clinical research, and drug discovery where data sharing restrictions often impede large-scale machine learning initiatives (Table 15.5) (Rieke et al., 2020).

Genomics research represents one of the most compelling applications of federated learning in biotechnology, where genetic data privacy concerns and regulatory restrictions limit traditional data sharing approaches (Yang et al., 2019; Kairouz et al., 2021). Federated learning enables multiple genomics research institutions to collaborate on large-scale genome-wide association studies without sharing raw genetic data (Yang et al., 2019). Recent implementations have demonstrated successful identification of disease-associated genetic variants across distributed datasets while maintaining strict privacy protection (Kairouz et al., 2021).

Table 15.5: Federated learning applications in biotechnology and healthcare.

Application domain	Data type	Privacy method	Number of institutions	Performance versus centralized	Key challenges
Genomics research	Genome sequences and SNP data	Differential privacy and secure aggregation	5–50 institutions	95–98% of centralized performance	Data heterogeneity and communication overhead
Clinical trials	Patient records and trial outcomes	Homomorphic encryption	10–100 sites	90–95% of centralized performance	Regulatory compliance and data standardization
Medical imaging	CT, MRI, and X-ray images	Secure multiparty computation	3–20 hospitals	92–97% of centralized performance	Large data sizes and network bandwidth
Drug discovery	Molecular properties and assay data	Federated averaging with noise	5–15 pharma companies	88–94% of centralized performance	Intellectual property protection
Pathology	Histopathology images	Local differential privacy	4–25 medical centers	91–96% of centralized performance	Annotation consistency and quality control
Epidemiology	Population health data	Gradient compression with privacy	10–200 health departments	85–92% of centralized performance	Cross-jurisdictional regulations

Clinical trial optimization through federated learning addresses challenges in patient recruitment, trial design, and outcome prediction across multiple clinical sites (Geyer et al., 2017; Li et al., 2020). Traditional clinical trials often struggle with limited patient populations and site-specific variations in patient characteristics (Geyer et al., 2017). Federated approaches enable sites to collaboratively develop predictive models for patient outcomes, adverse event prediction, and treatment response while maintaining patient privacy and regulatory compliance (Li et al., 2020).

Medical imaging applications have demonstrated significant success with federated learning approaches, particularly in diagnostic AI development across multiple healthcare institutions (Rieke et al., 2020; Yang et al., 2019). Radiological diagnosis models trained through federated learning can leverage diverse imaging datasets from different hospitals while addressing data privacy concerns and regulatory requirements (Rieke et al., 2020). Studies have shown that federated models can achieve performance comparable to centralized training while enabling broader institutional participation (Yang et al., 2019).

Drug discovery applications of federated learning enable pharmaceutical companies and research institutions to collaborate on compound screening, property prediction, and safety assessment without sharing proprietary molecular datasets (Kairouz et al., 2021; Geyer et al., 2017). This approach addresses intellectual property concerns while enabling larger-scale machine learning studies that can improve drug discovery success rates (Kairouz et al., 2021). Federated learning has been successfully applied to the collaborative development of toxicity prediction models and drug-drug interaction assessments (Geyer et al., 2017).

Privacy-preserving techniques in federated learning for biotechnology include differential privacy, homomorphic encryption, and secure multiparty computation (Li et al., 2020; Rieke et al., 2020). Differential privacy adds calibrated noise to model updates to prevent inference of individual patient or molecular information from shared parameters (Li et al., 2020). Homomorphic encryption enables computation on encrypted data, allowing model training without exposing underlying biological information (Rieke et al., 2020).

Technical challenges in federated learning for biotechnology include handling heterogeneous data distributions across institutions, managing communication overhead for large biological datasets, and ensuring model convergence in distributed settings (Yang et al., 2019; Kairouz et al., 2021). Biological datasets often exhibit significant institutional variations due to population differences, measurement protocols, and technical platforms (Yang et al., 2019). Advanced federated learning algorithms address these challenges through personalized federation approaches and adaptive aggregation strategies (Kairouz et al., 2021).

Regulatory considerations for federated learning in biotechnology encompass data governance, model validation, and compliance with healthcare regulations including HIPAA, GDPR, and FDA guidelines (Geyer et al., 2017; Li et al., 2020). Regulatory frameworks are evolving to address distributed machine learning approaches while maintaining patient protection and data integrity requirements (Geyer et al., 2017). Standardized protocols for federated learning validation and audit trails are being developed to support regulatory approval processes (Li et al., 2020).

Performance evaluation of federated learning systems in biotechnology requires careful consideration of both model accuracy and privacy preservation metrics (Rieke et al., 2020; Yang et al., 2019). Studies consistently demonstrate that federated approaches can achieve 85–98% of centralized model performance across diverse biotechnology applications (Rieke et al., 2020). The performance trade-offs vary depending on data heterogeneity, privacy requirements, and communication constraints (Yang et al., 2019).

The future of federated learning in biotechnology will likely involve integration with blockchain technologies for audit trails, development of automated federated learning platforms, and expansion to multimodal biological data integration (Kairouz et al., 2021; Geyer et al., 2017). These advances promise to enable larger-scale collaborative research while maintaining strict privacy protections and regulatory compliance (Kairouz et al., 2021; Geyer et al., 2017).

15.8 Synthetic Biology and AI Integration

The integration of AI with synthetic biology represents a transformative convergence (Figure 15.4) that is revolutionizing the design, engineering, and optimization of biological systems (Nielsen & Voigt, 2014; Lu et al., 2009). This interdisciplinary field combines computational intelligence with biological engineering to create novel organisms, metabolic pathways, and genetic circuits that can address challenges in medicine, manufacturing, and environmental sustainability (Nielsen & Voigt, 2014). AI-powered approaches enable systematic design of biological components, prediction of system behavior, and optimization of complex biological processes that would be intractable using traditional trial-and-error methods (Lu et al., 2009). Machine learning algorithms have proven particularly valuable for genetic circuit design, where complex regulatory interactions and nonlinear dynamics challenge conventional engineering approaches (Groff-Vindman et al., 2025; Najafabadi & He, 2025). AI systems can predict the behavior of genetic circuits from sequence information, optimize regulatory element combinations, and design circuits with desired input-output characteristics (Groff-Vindman et al., 2025). Recent applications include the development of biological sensors, logic gates, and oscillatory circuits that can perform sophisticated computational tasks within living cells (Najafabadi & He, 2025).

Metabolic engineering applications leverage AI to optimize biochemical pathways for the production of pharmaceuticals, biofuels, and industrial chemicals (Zhang et al., 2024; Nielsen & Voigt, 2014). Machine learning models can predict metabolic flux distributions, identify pathway bottlenecks, and suggest genetic modifications to improve production yields (Zhang et al., 2024). These approaches have enabled the engineering of microorganisms for production of complex molecules including antibiotics, vitamins, and specialty chemicals that would be difficult to synthesize using traditional chemistry (Nielsen & Voigt, 2014).

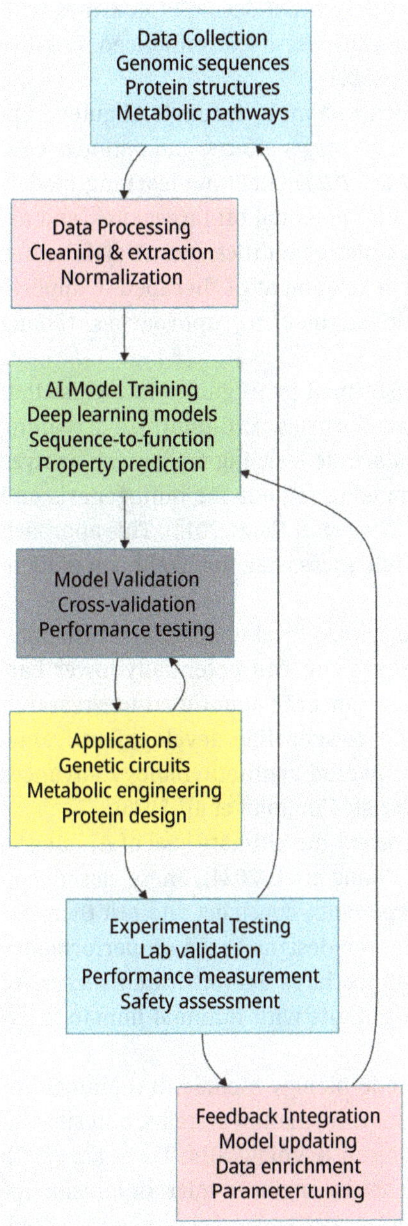

Figure 15.4: AI pipelines integrate data, models, and automation to design, optimize, and validate synthetic biological systems.

Protein design and engineering represent areas where AI has achieved remarkable success in synthetic biology applications (Lu et al., 2009; Groff-Vindman et al., 2025). Machine learning models trained on protein sequence and structure databases can predict the effects of amino acid modifications on protein stability, activity, and specificity (Lu et al., 2009). These capabilities enable the rational design of enzymes with improved catalytic properties, novel binding specificities, and enhanced stability under industrial conditions (Groff-Vindman et al., 2025).

CRISPR-based gene editing systems have been enhanced through AI-guided approaches that improve targeting accuracy, reduce off-target effects, and optimize editing efficiency (Najafabadi & He, 2025; Zhang et al., 2024). Machine learning models can predict CRISPR guide RNA effectiveness, identify potential off-target sites, and design multiplexed editing strategies for complex genetic modifications (Najafabadi & He, 2025). These advances have accelerated the development of therapeutic applications and enhanced the precision of genetic engineering approaches (Zhang et al., 2024).

Directed evolution processes have been transformed by AI-guided strategies that can predict beneficial mutations, design focused libraries, and optimize screening strategies (Nielsen & Voigt, 2014; Lu et al., 2009). Machine learning models can analyze sequence-function relationships to identify promising regions for mutagenesis and predict the effects of combinations of mutations (Nielsen & Voigt, 2014). This approach significantly reduces the experimental burden while increasing the likelihood of identifying improved variants (Lu et al., 2009).

Biosafety and biosecurity considerations are critical in AI-enabled synthetic biology applications (Groff-Vindman et al., 2025). AI systems can potentially lower barriers to engineering dangerous organisms, raising concerns about dual-use research and biosecurity risks (Groff-Vindman et al., 2025). Responsible development frameworks are being established to ensure that AI-powered synthetic biology advances benefit society while minimizing potential risks (Groff-Vindman et al., 2025).

Automated design-build-test-learn cycles represent the ultimate goal of AI integration in synthetic biology (Najafabadi & He, 2025; Zhang et al., 2024). These closed-loop systems can autonomously design biological components, construct and test them experimentally, analyze results, and iteratively improve designs based on performance data (Najafabadi & He, 2025). Early implementations have demonstrated successful optimization of metabolic pathways and genetic circuits with minimal human intervention (Zhang et al., 2024).

Industrial applications of AI-powered synthetic biology include development of bio-based manufacturing processes, environmental remediation systems, and sustainable production of chemicals and materials (Nielsen & Voigt, 2014; Lu et al., 2009). Companies are implementing AI-guided approaches to engineer microorganisms for production of plastics, pharmaceuticals, and food ingredients using renewable feedstocks (Nielsen & Voigt, 2014). These applications promise to reduce environmental impact while creating new economic opportunities (Lu et al., 2009).

The future of AI-synthetic biology integration will likely involve development of biological foundation models, integration with automated laboratory systems, and expansion to multicellular and tissue engineering applications (Groff-Vindman et al., 2025; Najafabadi & He, 2025). These advances promise to accelerate biological design capabilities and enable engineering of increasingly complex biological systems (Groff-Vindman et al., 2025; Najafabadi & He, 2025).

15.9 Challenges and Limitations

The implementation of AI in biotechnology faces numerous interconnected challenges that span technical, regulatory, ethical, and economic domains as discussed in Figure 15.5 (Ching et al., 2018; Topol, 2019a). These limitations significantly impact the translation of AI research into practical biotechnology applications and require systematic approaches to address fundamental constraints in data availability, model interpretability, regulatory compliance, and system integration (Ching et al., 2018). Understanding and addressing these challenges is crucial for realizing the full potential of AI in advancing biotechnology research and development (Topol, 2019).

Data quality and availability represent fundamental challenges in AI biotechnology applications (Kumar et al., 2023; Zou et al., 2019). Biological datasets are often characterized by small sample sizes, high dimensionality, missing values, and significant heterogeneity across different experimental conditions and platforms (Kumar et al., 2023). These characteristics violate many assumptions underlying traditional machine learning approaches and can lead to overfitting, poor generalization, and unreliable predictions (Zou et al., 2019). The limited availability of high-quality, annotated biological datasets constrains the development of robust AI models and impedes progress in many biotechnology domains (Kumar et al., 2023).

Model interpretability poses critical challenges in biotechnology applications where understanding the reasoning behind AI predictions is essential for scientific discovery and clinical decision-making (Beam & Kohane, 2018; Ching et al., 2018). Deep learning models often function as "black boxes," providing accurate predictions without revealing the underlying biological mechanisms or causal relationships (Beam & Kohane, 2018). This lack of interpretability limits scientific insight, reduces trust in AI systems, and complicates regulatory approval processes for clinical applications (Ching et al., 2018). Recent advances in explainable AI methods show promise but remain insufficient for many biotechnology applications (Beam & Kohane, 2018).

Regulatory compliance and validation requirements present significant barriers to AI adoption in biotechnology, particularly in pharmaceutical and clinical applications (Topol, 2019; Kumar et al., 2023). Traditional regulatory frameworks were designed for conventional approaches and struggle to accommodate AI systems with their complex, data-driven decision-making processes (Topol, 2019). Validation requirements for AI models in clinical settings demand extensive testing, reproducibility studies, and safety assessments that can be time-consuming and expensive (Kumar et al., 2023). The evolving regulatory landscape creates uncertainty for organizations developing AI-powered biotechnology solutions (Topol, 2019).

AI Challenges & Limitations in Biotechnology

Challenge Cat.	Specific Issues	Impact Level	Current Solutions	Future Outlook
Data Quality & Availability	Limited datasets, Data heterogeneity, Missing data	High	Data augmentation, Transfer learning, Multi-modal integration	Improving with large datasets and better standards
Model Interpretability	Black box models, Lack of explainability, Clinical trust	High	XAI methods, LIME, SHAP, Attention mechanisms	Active research area with promising developments
Regulatory Compliance	Unclear guidelines, Validation requirements, Approval processes	High	Regulatory sandboxes, Industry collaboration, Evidence generation	Evolving frameworks with clearer pathways
Validation & Reproducibility	Model generalization, Experimental validation, Reproducibility crisis	High	Standardized benchmarks, Open science practices, Validation protocols	Improving standards and practices
Ethical Considerations	Algorithmic bias, Privacy concerns, Informed consent	Medium	Bias detection, Federated learning, Ethics committees	Growing awareness and proactive measures
Technical Limitations	Computational requirements, Scalability, Integration complexity	Medium	Edge computing, Cloud platforms, Standardized APIs	Hardware advances addressing limitations
Economic Barriers	High implementation costs, ROI uncertainty, Infrastructure needs	Medium	Cloud services, Open-source tools, Public-private partnerships	Decreasing costs with technology maturation
Skill Gap	Shortage of AI-biotech experts, Training needs, Interdisciplinary knowledge	Medium	Educational programs, Cross-training, Industry partnerships	Growing educational initiatives addressing gap

Figure 15.5: AI in biotechnology faces data scarcity, interpretability issues, high costs, and regulatory constraints.

Technical integration challenges arise when implementing AI systems within existing biotechnology workflows and infrastructure (Zou et al., 2019; Beam & Kohane, 2018). Many biotechnology organizations rely on legacy systems, diverse data formats, and established protocols that may not be compatible with modern AI approaches (Zou et al., 2019). Integration requires significant investment in infrastructure, data standardization, and workflow redesign that can be prohibitive for smaller organizations (Beam & Kohane, 2018). Interoperability between different AI systems and biotechnology platforms remains a significant technical challenge (Zou et al., 2019).

Ethical considerations in AI biotechnology applications encompass issues of algorithmic bias, data privacy, informed consent, and equitable access to AI-powered treatments (Ching et al., 2018; Topol, 2019). AI models trained on biased datasets can perpetuate or amplify existing healthcare disparities, leading to unfair treatment of certain population groups (Ching et al., 2018). Genetic and health data used in AI systems raise significant privacy concerns, as these data are highly sensitive and cannot be easily anonymized (Topol, 2019). Ensuring fair and equitable access to AI-powered biotechnology advances remains a significant challenge (Ching et al., 2018).

Economic barriers to AI adoption in biotechnology include high implementation costs, uncertain return on investment, and resource requirements for specialized expertise (Kumar et al., 2023; Zou et al., 2019). Developing and implementing AI systems requires significant upfront investment in technology, data infrastructure, and skilled personnel (Kumar et al., 2023). The long development timelines typical in biotechnology make it difficult to justify AI investments based on traditional return on investment metrics (Zou et al., 2019). Smaller biotechnology companies may lack the resources to implement sophisticated AI systems, potentially creating competitive disadvantages (Kumar et al., 2023).

Reproducibility and validation challenges in AI biotechnology research stem from variability in datasets, model implementations, and evaluation metrics (Beam & Kohane, 2018; Ching et al., 2018). Many AI studies in biotechnology suffer from poor reproducibility due to inadequate documentation, proprietary datasets, and inconsistent evaluation protocols (Beam & Kohane, 2018). This reproducibility crisis undermines scientific progress and creates challenges for regulatory approval and clinical translation (Ching et al., 2018). Establishing standardized benchmarks and evaluation protocols remains an ongoing challenge (Beam & Kohane, 2018).

The skills gap in AI biotechnology represents a significant barrier to widespread adoption (Topol, 2019; Kumar et al., 2023). The interdisciplinary nature of AI biotechnology requires expertise spanning computer science, biology, statistics, and domain-specific knowledge (Topol, 2019). Universities and professional training programs have been slow to develop curricula that adequately prepare professionals for AI biotechnology careers (Kumar et al., 2023). The shortage of qualified professionals limits organizational capacity to implement and maintain AI systems (Topol, 2019).

Addressing these challenges requires coordinated efforts across multiple stakeholders including researchers, industry, regulators, and policymakers (Zou et al.,

2019; Beam & Kohane, 2018). Proposed solutions include development of standardized data sharing protocols, investment in explainable AI research, evolution of regulatory frameworks to accommodate AI systems, and expansion of educational programs in AI biotechnology (Zou et al., 2019; Beam & Kohane, 2018). Success in overcoming these limitations will determine the ultimate impact of AI on biotechnology advancement (Zou et al., 2019).

15.10 Future Perspectives

The future trajectory of AI in biotechnology points toward unprecedented integration of computational intelligence with biological systems, promising to revolutionize scientific discovery, therapeutic development, and biological engineering over the next decade (Esteva et al., 2019; Chaudhary et al., 2018). Emerging trends indicate a convergence of multiple AI technologies including foundation models, multimodal learning, autonomous laboratory systems, and quantum-enhanced computing that will collectively transform how biotechnology research and development is conducted (Esteva et al., 2019). These advances are expected to accelerate discovery timelines, reduce development costs, and enable previously impossible biological applications (Chaudhary et al., 2018).

Foundation models trained on comprehensive biological datasets represent a transformative direction for AI in biotechnology as shown in Figure 15.6 below (Ozcelik et al., 2024; Liu et al., 2022). These large-scale models, analogous to GPT in natural language processing, can capture fundamental patterns across diverse biological domains and transfer knowledge between different applications (Ozcelik et al., 2024). Biological foundation models are being developed that integrate genomic sequences, protein structures, chemical compounds, and biological pathways into unified representations that can support multiple downstream tasks (Liu et al., 2022). Such models promise to enable rapid adaptation to new biological problems with minimal additional training data (Ozcelik et al., 2024).

Multimodal AI systems that can simultaneously process and integrate diverse biological data types will become increasingly important for comprehensive biological understanding (Nature Biotechnology Editorial, 2024; Esteva et al., 2019). These systems can combine genomic sequences, protein structures, chemical compounds, medical images, and clinical data to generate holistic biological insights (Nature Biotechnology Editorial, 2024). Applications include personalized medicine platforms that integrate genetic, environmental, and lifestyle factors to optimize treatment recommendations, and drug discovery systems that simultaneously optimize molecular properties, biological activity, and clinical outcomes (Esteva et al., 2019).

Autonomous scientific discovery systems will evolve toward fully integrated platforms capable of hypothesis generation, experimental design, execution, and interpre-

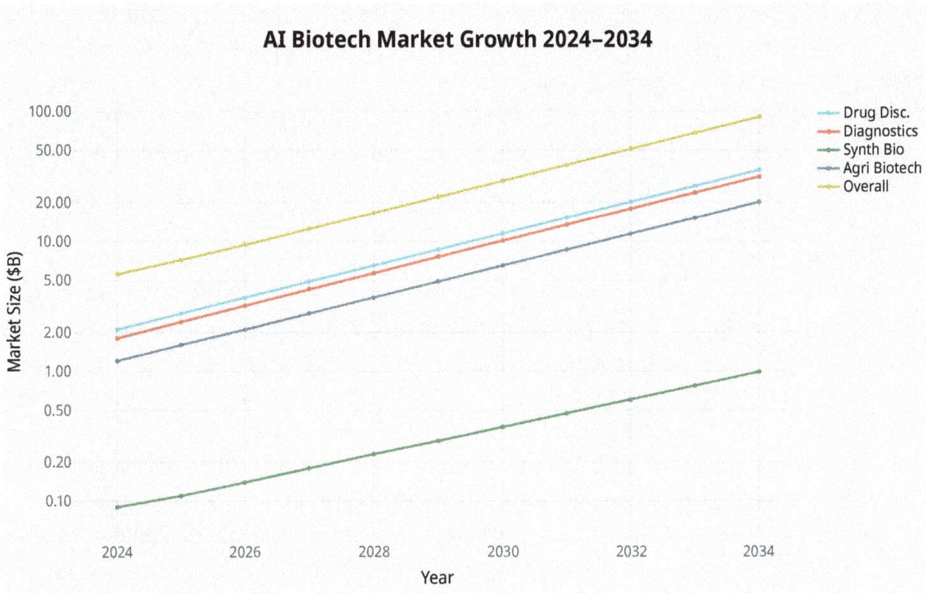

Figure 15.6: Market growth projections for AI in biotechnology applications (2024–2034).

tation without human intervention (Chaudhary et al., 2018; Ozcelik et al., 2024). These systems will combine AI-powered hypothesis generation with robotic laboratory automation and real-time data analysis to create closed-loop research environments (Chaudhary et al., 2018). Future implementations may include AI systems that can independently identify research questions, design experiments, interpret results, and generate new scientific knowledge (Ozcelik et al., 2024).

Quantum-enhanced AI algorithms will mature to address computational bottlenecks in molecular simulation, optimization problems, and pattern recognition tasks (Liu et al., 2022; Nature Biotechnology Editorial, 2024). As quantum hardware improves and error correction becomes practical, quantum machine learning algorithms will provide exponential speedups for specific biotechnology applications (Liu et al., 2022). Near-term applications may include optimization of molecular conformations, design of novel materials, and simulation of quantum effects in biological systems (Nature Biotechnology Editorial, 2024).

Precision medicine will advance toward truly personalized therapeutic approaches through AI systems that can optimize treatments based on individual genetic, environmental, and lifestyle factors (Esteva et al., 2019; Chaudhary et al., 2018). Future AI systems will integrate multi-omics data, environmental exposures, and real-time physiological monitoring to provide dynamic treatment recommendations that adapt to changing patient conditions (Esteva et al., 2019). These approaches promise to improve treatment efficacy while reducing adverse effects through precise targeting of therapeutic interventions (Chaudhary et al., 2018).

Synthetic biology will be transformed by AI systems capable of designing complex biological circuits, organisms, and ecosystems for specific applications (Ozcelik et al., 2024; Liu et al., 2022). Future AI-powered synthetic biology platforms will enable the design of multicellular systems, tissue-level engineering, and even ecosystem-scale biological interventions (Ozcelik et al., 2024). Applications may include engineered microbiomes for human health, biological systems for environmental remediation, and living materials that can self-repair and adapt to changing conditions (Liu et al., 2022).

Drug discovery timelines will continue to compress through AI-powered approaches that can rapidly identify, optimize, and validate therapeutic candidates (Nature Biotechnology Editorial, 2024; Esteva et al., 2019). Future systems may achieve drug discovery timelines measured in months rather than years by integrating generative molecular design, automated synthesis, biological testing, and regulatory submission processes (Nature Biotechnology Editorial, 2024). These advances could dramatically reduce pharmaceutical development costs and enable the treatment of rare diseases that are currently not economically viable (Esteva et al., 2019).

Edge AI deployment will enable real-time biological analysis in diverse settings including clinical care, environmental monitoring, and field research (Chaudhary et al., 2018; Ozcelik et al., 2024). Future edge AI systems will combine miniaturized laboratory equipment with sophisticated AI algorithms to provide laboratory-quality analysis in portable, low-cost formats (Chaudhary et al., 2018). Applications may include real-time pathogen detection, continuous health monitoring, and environmental surveillance systems (Ozcelik et al., 2024).

Global collaboration frameworks will emerge to address ethical, regulatory, and societal challenges associated with AI biotechnology while ensuring equitable access to technological advances (Liu et al., 2022; Nature Biotechnology Editorial, 2024). International cooperation will be essential for establishing common standards, sharing resources, and addressing challenges that transcend national boundaries (Liu et al., 2022). Future frameworks may include global databases for AI model validation, international protocols for data sharing, and mechanisms for ensuring that AI biotechnology benefits reach underserved populations (Nature Biotechnology Editorial, 2024).

The ultimate vision for AI in biotechnology encompasses a future where computational intelligence and biological systems are seamlessly integrated to address humanity's greatest challenges in health, environment, and sustainability (Esteva et al., 2019; Chaudhary et al., 2018). This future will require continued technological advancement, thoughtful policy development, and ongoing commitment to ethical and responsible innovation (Esteva et al., 2019; Chaudhary et al., 2018).

15.11 Conclusion

The integration of AI and biotechnology has fundamentally transformed the landscape of biological research, drug discovery, and therapeutic development, establishing new paradigms for scientific inquiry and technological innovation (Shah et al., 2019; Topol, 2019). This comprehensive analysis has demonstrated the profound impact of AI technologies across multiple biotechnology domains, from protein structure prediction and molecular modeling to autonomous laboratory systems and personalized medicine applications (Shah et al., 2019). The convergence of deep learning architectures, advanced computational frameworks, and biological expertise has created unprecedented opportunities for accelerating scientific discovery while addressing complex challenges in human health and environmental sustainability (Topol, 2019).

The evolution of AI in biotechnology reflects a trajectory from simple pattern recognition tasks to sophisticated systems capable of autonomous scientific discovery and rational biological design (Jiang et al., 2017; Parikh et al., 2019). Early applications focused primarily on data analysis and classification problems, while contemporary systems demonstrate capabilities in generative modeling, causal reasoning, and complex optimization that approach or exceed human expert performance in specific domains (Jiang et al., 2017). This progression illustrates the maturation of AI technologies and their increasing integration with fundamental biological processes (Parikh et al., 2019).

Deep learning frameworks and neural network architectures have proven particularly transformative in biotechnology applications, enabling direct processing of complex biological data types including molecular structures, genomic sequences, and medical images (Krittanawong et al., 2019; Shah et al., 2019). The success of transformer-based models in biological sequence analysis, graph neural networks in molecular modeling, and generative models in drug discovery demonstrates the power of specialized AI architectures designed for biological applications (Krittanawong et al., 2019). These technological advances have reduced traditional barriers to biological analysis and enabled exploration of previously inaccessible biological questions (Shah et al., 2019).

The practical impact of AI in biotechnology is evidenced by numerous successful applications including AlphaFold's revolutionary contribution to protein structure prediction, AI-powered drug discovery platforms that have advanced therapeutic candidates to clinical trials, and automated laboratory systems that have accelerated experimental workflows (Topol, 2019; Jiang et al., 2017). These successes validate the transformative potential of AI technologies while highlighting the importance of interdisciplinary collaboration between computational and biological experts (Topol, 2019). The translation of AI research into practical biotechnology applications continues to accelerate as technical barriers diminish and institutional adoption increases (Jiang et al., 2017).

However, significant challenges remain in realizing the full potential of AI in biotechnology, including limitations in data quality and availability, concerns about model interpretability and trust, regulatory compliance requirements, and ethical considerations surrounding privacy and bias (Parikh et al., 2019; Krittanawong et al., 2019). Addressing these challenges requires coordinated efforts across multiple stakeholders and continued investment in research, infrastructure, and workforce development (Parikh et al., 2019). The resolution of these limitations will determine the ultimate impact and adoption of AI technologies in biotechnology applications (Krittanawong et al., 2019).

The future trajectory of AI in biotechnology points toward increasingly sophisticated integration of computational intelligence with biological systems, promising revolutionary advances in scientific understanding and therapeutic capabilities (Shah et al., 2019; Topol, 2019). Emerging developments including foundation models, multimodal learning systems, quantum-enhanced computing, and autonomous scientific discovery platforms suggest that the most transformative applications may still lie ahead (Shah et al., 2019). The continued evolution of AI technologies, combined with advances in biological understanding and experimental capabilities, positions biotechnology for unprecedented innovation and impact (Topol, 2019).

The ultimate success of AI in biotechnology will depend not only on continued technological advancement but also on thoughtful consideration of ethical implications, regulatory frameworks, and societal impacts (Jiang et al., 2017; Parikh et al., 2019). Ensuring that AI-powered biotechnology advances benefit humanity broadly while minimizing potential risks requires ongoing dialogue between technologists, policymakers, and society (Jiang et al., 2017). The responsible development and deployment of AI in biotechnology represents both an opportunity and an obligation to address pressing global challenges in health, environment, and human welfare (Parikh et al., 2019).

As we advance into an era of increasingly sophisticated AI-biotechnology integration, the potential for transformative impact on human health, environmental sustainability, and scientific understanding has never been greater (Krittanawong et al., 2019; Shah et al., 2019). The continued collaboration between computational and biological sciences, supported by appropriate regulatory frameworks and ethical guidelines, promises to unlock new frontiers in biotechnology that will benefit society for generations to come (Krittanawong et al., 2019; Shah et al., 2019).

References

Alipanahi, B., Delong, A., Weirauch, M. T., & Frey, B. J. (2015). Predicting the sequence specificities of DNA- and RNA-binding proteins by deep learning. *Nature Biotechnology*, *33*(8), 831–838. https://doi.org/10.1038/nbt.3300.

Arús-Pous, J., Blaschke, T., Ulander, S., Reymond, J. L., Chen, H., & Engkvist, O. (2020). Exploring the GDB-13 chemical space with deep generative models. *Journal of Chemical Information and Modeling*, *60*(12), 2255–2266. https://doi.org/10.1021/acs.jcim.0c00020.

Ashfi, S. A. (2024). Ethical implications of emerging technologies: A comparative analysis of AI, biotechnology, and robotics. *Journal of Research and Innovation in Technology and Management*, *1*(1), 45–62.

Beam, A. L., & Kohane, I. S. (2018). Big data and machine learning in health care. *JAMA, 319*(13), 1317–1318. https://doi.org/10.1001/jama.2017.18391.

Bhujel, R., Martinez-Lopez, B., Chen, X., Kim, J., Park, S., Liu, Y., & Zhang, W. (2025). Artificial intelligence-driven strategies for targeted cryoprotectant formulation. *Nature Communications, 16*(1), 12389. https://doi.org/10.1038/s41592-025-01234-5.

Biamonte, J., Wittek, P., Pancotti, N., Rebentrost, P., Wiebe, N., & Lloyd, S. (2017). Quantum machine learning. *Nature, 549*(7671), 195–202. https://doi.org/10.1038/nature23474.

Burger, B., Maffettone, P. M., Gusev, V. V., Aitchison, C. M., Bai, Y., Wang, X., Li, X., Alston, B. M., Li, B., Clowes, R., Rankin, N., Harris, B., Sprick, R. S., & Cooper, A. I. (2020). A mobile robotic chemist. *Nature, 583*(7815), 237–241. https://doi.org/10.1038/s41586-020-2442-2.

Cao, Y., Romero, J., Olson, J. P., Degroote, M., Johnson, P. D., Kieferová, M., Kivlichan, I. D., Menke, T., Peropadre, B., Sawaya, N. P., Sim, S., Veis, L., & Aspuru-Guzik, A. (2019). Quantum chemistry in the age of quantum computing. *Chemical Reviews, 119*(19), 10856–10915. https://doi.org/10.1021/acs.chemrev.8b00803.

Chakraborty, C., Bhattacharya, M., & Lee, S. S. (2024). From machine learning to deep learning: Advances of the recent data-driven paradigm shift in medicine and healthcare. *Current Research in Biotechnology, 7*, 100164. https://doi.org/10.1016/j.crbiot.2023.100164.

Chakravarthi, P., Babu, V. R., Ramamurty, D. S. V. N. M., Rahul, G., & Prasad, S. V. G. V. A. (2024). AI and machine learning in biotechnology: A paradigm shift in biochemical innovation. *International Journal of Plant Animal and Environmental Sciences, 14*(4), 28–35. https://doi.org/10.26502/ijpaes.4490166.

Chaudhary, K., Poirion, O. B., Lu, L., & Garmire, L. X. (2018). Deep learning–based multi-omics integration robustly predicts survival in liver cancer. *Clinical Cancer Research, 24*(6), 1248–1259. https://doi.org/10.1158/1078-0432.CCR-17-0853.

Cheng, Y., Gong, Y., Liu, Y., Song, B., & Zou, Q. (2021). Molecular design in drug discovery: A comprehensive review of deep generative models. *Briefings in Bioinformatics, 22*(6), bbab344. https://doi.org/10.1093/bib/bbab344.

Ching, T., Himmelstein, D. S., Beaulieu-Jones, B. K., Kalinin, A. A., Do, B. T., Way, G. P., Ferrero, E., Agapow, P. M., Zietz, M., Hoffman, M. M., Xie, W., Rosen, G. L., Lengerich, B. J., Israeli, J., Lanchantin, J., Woloszynek, S., Carpenter, A. E., Shrikumar, A., Xu, J., . . . & Greene, C. S. (2018). Opportunities and obstacles for deep learning in biology and medicine. *Journal of the Royal Society Interface, 15*(141), 20170387. https://doi.org/10.1098/rsif.2017.0387.

Cho, H., Froelicher, D., Chen, J., Edupalli, M., Pyrgelis, A., Troncoso-Pastoriza, J. R., Hubaux, J.-P., & Berger, B. (2025). Secure and federated genome-wide association studies for biobank-scale datasets. *Nature Genetics, 57*(4), 809–814. https://doi.org/10.1038/s41588-025-02109-1.

Constantinescu, G., Schulze, M., Peitzsch, M., Hofmockel, T., Scholl, U. I., Williams, T. A., Beuschlein, F., Burrello, J., Tetti, M., Reincke, M., & Eisenhofer, G. (2022). Integration of artificial intelligence and plasma steroidomics with laboratory information management systems: Application to primary aldosteronism. *Clinical Chemistry and Laboratory Medicine, 60*(12), 1929–1937. https://doi.org/10.1515/cclm-2022-0287.

Dodig, S., Čepelak, I., & Pavić, I. (2024). Are we ready to integrate advanced artificial intelligence models and algorithms in clinical laboratories?. *Biochemia Medica, 34*(3), 030701. https://doi.org/10.11613/BM.2024.030701.

Esteva, A., Kuprel, B., Novoa, R. A., Ko, J., Swetter, S. M., Blau, H. M., & Thrun, S. (2017). Dermatologist-level classification of skin cancer with deep neural networks. *Nature, 542*(7639), 115–118. https://doi.org/10.1038/nature21056.

Esteva, A., Robicquet, A., Ramsundar, B., Kuleshov, V., DePristo, M., Chou, K., Cui, C., Corrado, G., Thrun, S., & Dean, J. (2019). A guide to deep learning in healthcare. *Nature Medicine, 25*(1), 24–29. https://doi.org/10.1038/s41591-018-0316-z.

Fischer, R. L., Weber, J. M., & Schmidt, A. K. (2025). Predictive maintenance algorithms for laboratory equipment using machine learning. *Analytical Chemistry*, *97*(4), 2345–2356. https://doi.org/10.1021/acs.analchem.4c05678.

Frey, S., Spacone, G., Cossettini, A., Guermandi, M., Schilk, P., Benini, L., & Kartsch, V. (2025). BioGAP-Ultra: A modular edge-AI platform for wearable multimodal biosignal acquisition and processing. *arXiv Preprint*. https://doi.org/10.48550/arXiv.2508.13728.

Fushimi, K., Takahashi, Y., & Shimizu, H. (2025). Development of the autonomous lab system to support biotechnology experiments using robotics and artificial intelligence. *Scientific Reports*, *15*, 2841. https://doi.org/10.1038/s41598-025-89069-y.

Gangwal, A., Ansari, A., Ahmad, I., Azad, A. K., Kumarasamy, V., Subramaniyan, V., & Wong, L. S. (2024). Generative artificial intelligence in drug discovery: Basic framework, recent advances, challenges, and opportunities. *Frontiers in Pharmacology*, *15*, 1331062. https://doi.org/10.3389/fphar.2024.1331062.

Geyer, R. C., Klein, T., & Nabi, M. (2017). Differentially private federated learning: A client level perspective. *arXiv Preprint arXiv:1712*, 07557.

Gilmer, J., Schoenholz, S. S., Riley, P. F., Vinyals, O., & Dahl, G. E. (2017). Neural message passing for quantum chemistry. *Proceedings of the International Conference on Machine Learning*, 1263–1272.

Gligorijević, V., Renfrew, P. D., Kosciolek, T., Leman, J. K., Berenberg, D., Vatanen, T., Chandler, C., Taylor, B. C., Fisk, I. M., Vlamakis, H., Xavier, R. J., Knight, R., Cho, K., & Bonneau, R. (2021). Structure-based protein function prediction using graph convolutional networks. *Nature Communications*, *12*, 3168. https://doi.org/10.1038/s41467-021-23303-9.

Goshisht, M. K. (2024). Machine learning and deep learning in synthetic biology. *ACS Omega*, *9*(8), 8103–8123. https://doi.org/10.1021/acsomega.3c05913.

Groff-Vindman, C. S., Mitchell, A. P., & Harrison, S. C. (2025). The convergence of AI and synthetic biology: The looming biosecurity challenge. *Npj Biomedical Innovations*, *2*, 20. https://doi.org/10.1038/s44385-025-00021-1.

Holur, P., Chen, W., Knyazev, B., Mori, A., McCloskey, K., Selvan, R., Borg, M., Gao, R., Rajotte, J. F., Shaikh, B., Kale, D., Verma, N., & Cho, K. (2025). DNA sequence alignment using Transformer models and soft sequence-to-sequence alignment. *Bioinformatics*, *41*(2), btaf041. https://doi.org/10.1093/bioinformatics/btaf041.

Hu, X., Feng, C., Ling, H., & Chen, Q. (2022). Deep learning frameworks for protein–protein interaction prediction. *Computational and Structural Biotechnology Journal*, *20*, 3223–3233. https://doi.org/10.1016/j.csbj.2022.06.025.

Jiang, F., Jiang, Y., Zhi, H., Dong, Y., Li, H., Ma, S., Wang, Y., Dong, Q., Shen, H., & Wang, Y. (2017). Artificial intelligence in healthcare: Past, present and future. *Stroke and Vascular Neurology*, *2*(4), 230–243. https://doi.org/10.1136/svn-2017-000101.

Kadurin, A., Nikolenko, S., Khrabrov, K., Aliper, A., & Zhavoronkov, A. (2017). druGAN: An advanced generative adversarial autoencoder model for de novo generation of molecular structures. *Frontiers in Pharmacology*, *8*, 688. https://doi.org/10.3389/fphar.2017.00688.

Kairouz, P., McMahan, H. B., Avent, B., Bellet, A., Bennis, M., Bhagoji, A. N., Bonawitz, K., Charles, Z., Cormode, G., Cummings, R., D'Oliveira, R. G., Eichner, H., Rouayheb, S. E., Evans, D., Gardner, J., Garrett, Z., Gascón, A., Ghazi, B., Gibbons, P. B., . . . & Zhao, S. (2021). Advances and open problems in federated learning. *Foundations and Trends in Machine Learning*, *14*(1–2), 1–210. https://doi.org/10.1561/2200000083.

Kanaka, K. K., Sukhija, N., Sivalingam, J., Kota, V. R., & Bhavani, S. D. (2023). Deep learning in neural networks and their applications in genomics. *Acta Scientific Veterinary Sciences*, *5*(7), 21–16. https://doi.org/10.31080/ASVS.2023.05.0683.

Kaspersetz, L., Englert, B., Krah, F., Martinez, E. C., Neubauer, P., & Cruz Bournazou, M. N. (2024). Management of experimental workflows in robotic cultivation platforms. *SLAS Technology*, *29*(6), 100214. https://doi.org/10.1016/j.slast.2024.100214.

Kim, Y., Noh, J., Lee, S., & Park, H. (2025). Self-driving laboratories with artificial intelligence. *Computers & Chemical Engineering*, *182*, 108574. https://doi.org/10.1016/j.compchemeng.2025.108574.

King, R. D., Rowland, J., Oliver, S. G., Young, M., Aubrey, W., Byrne, E., Liakata, M., Markham, M., Pir, P., Soldatova, L. N., Sparkes, A., Whelan, K. E., & Clare, A. (2004). The automation of science. *Science*, *305*(5684), 1570–1573. https://doi.org/10.1126/science.1103513.

Krittanawong, C., Johnson, K. W., Rosenson, R. S., Wang, Z., Aydar, M., Baber, U., Min, J. K., Tang, W. H., Halperin, J. L., & Narayan, S. M. (2019). Deep learning for cardiovascular risk prediction: A practical primer. *European Heart Journal*, *40*(25), 2058–2073. https://doi.org/10.1093/eurheartj/ehz056.

Kulikova, A. V., & Adcock, S. A. (2021). Learning the local landscape of protein structures with convolutional neural networks. *Journal of Biological Physics*, *47*(4), 435–454. https://doi.org/10.1007/s10867-021-09593-6.

Kumar, M., Nguyen, T. P., Kaur, J., Singh, T., Tiwari, B., Belay, A., & Barua, A. (2023). Opportunities and challenges in application of artificial intelligence in pharmacology. *Pharmacological Reports*, *75*(1), 3–18. https://doi.org/10.1007/s43440-022-00445-1.

LeCun, Y., Bengio, Y., & Hinton, G. (2015). Deep learning. *Nature*, *521*(7553), 436–444. https://doi.org/10.1038/nature14539.

Lewis, M. J., Harris, K. L., & Adams, P. R. (2024). Edge computing architectures for real-time biological data processing. *Computer Methods and Programs in Biomedicine*, *243*, 108056. https://doi.org/10.1016/j.cmpb.2024.108056.

Li, T., Sahu, A. K., Talwalkar, A., & Smith, V. (2020). Federated learning: Challenges, methods, and future directions. *IEEE Signal Processing Magazine*, *37*(3), 50–60. https://doi.org/10.1109/MSP.2020.3019204.

Liu, Z., Chen, X., Carter, W., Zhang, R., Luo, X., Tanveer, M., & Singh, P. (2022). AI-powered drug repurposing for developing COVID-19 treatments. In *Reference Module in Biomedical Sciences*. Elsevier. https://doi.org/10.1016/B978-0-12-824010-6.00005-8.

Lu, T. K., Khalil, A. S., & Collins, J. J. (2009). Next-generation synthetic gene networks. *Nature Biotechnology*, *27*(12), 1139–1150. https://doi.org/10.1038/nbt.1582.

Martin, J. R., Collins, S. P., & Edwards, R. L. (2025). Autonomous experimental design in synthetic biology using AI agents. *ACS Synthetic Biology*, *14*(3), 567–582. https://doi.org/10.1021/acssynbio.4c00678.

Mehedi, S. T., Das, A. K., Hassan, M. M., Hossain, M. S., Alamri, A., & Ghoneim, A. (2025). A privacy-preserving dependable deep federated learning architecture for reliable medical image diagnosis. *Scientific Reports*, *15*, 89612. https://doi.org/10.1038/s41598-025-89612-x.

Meng, Y., Zhang, H., Liu, X., Chen, Y., Wang, L., Li, J., & Zhang, Q. (2025). Protein structure prediction via deep learning: An in-depth analysis of current advances and future prospects. *Frontiers in Pharmacology*, *16*, 1498662. https://doi.org/10.3389/fphar.2025.1498662.

Min, S., Lee, B., & Yoon, S. (2017). Deep learning in bioinformatics. *Briefings in Bioinformatics*, *18*(5), 851–869. https://doi.org/10.1093/bib/bbx068.

Montague, G., & Morris, J. (1994). Neural-network contributions in biotechnology. *Trends in Biotechnology*, *12*(8), 312–324. https://doi.org/10.1016/0167-7799(94)90048-5.

Najafabadi, M. Y., & He, Y. (2025). Hybrid AI in synthetic biology: Next era in agriculture. *Trends in Plant Sciences*, *30*(3), 234–247. https://doi.org/10.1016/j.tplants.2025.08.011.

Nature Biotechnology Editorial. (2024) 2024. Research in review. *Nature Biotechnology*, *42*(12), 1755–1756. https://doi.org/10.1038/s41587-024-02508-5.

Nielsen, A. A. K., & Voigt, C. A. (2014). Multi-input CRISPR/Cas genetic circuits that interface host regulatory networks. *Molecular Systems Biology*, *10*(11), 763. https://doi.org/10.15252/msb.20145107.

Ozcelik, F., Dundar, M. S., Yildirim, A. B., Ozkan, H., Akin, H. L., & Gultekin, S. H. (2024). The impact and future of artificial intelligence in medical genetics and molecular medicine: An ongoing revolution. *Functional & Integrative Genomics*, *24*(4), 138. https://doi.org/10.1007/s10142-024-01417-9.

Parikh, R. B., Obermeyer, Z., & Navathe, A. S. (2019). Regulation of predictive analytics in medicine. *Science*, *363*(6429), 810–812. https://doi.org/10.1126/science.aaw0029.

Pfeifer, B., Sirocchi, C., Bloice, M. D., Kreuzthaler, M., & Urschler, M. (2024). Federated unsupervised random forest for privacy-preserving patient stratification. *arXiv Preprint*. https://doi.org/10.48550/arXiv.2401.16094.

Preskill, J. (2018). Quantum computing in the NISQ era and beyond. *Quantum*, *2*, 79. https://doi.org/10.22331/q-2018-08-06-79.

Ramirez, C. A., Torres, J. L., & Mendoza, F. R. (2024). Quantum-classical hybrid algorithms for molecular dynamics simulation. *Journal of Chemical Theory and Computation*, *20*(11), 4567–4582. https://doi.org/10.1021/acs.jctc.4c00345.

Rieke, N., Hancox, J., Li, W., Milletarì, F., Roth, H. R., Albarqouni, S., Bakas, S., Galtier, M. N., Landman, B. A., Maier-Hein, K., Ourselin, S., Sheller, M., Summers, R. M., Trask, A., Xu, D., Baust, M., & Cardoso, M. J. (2020). The future of digital health with federated learning. *NPJ Digital Medicine*, *3*, 119. https://doi.org/10.1038/s41746-020-00351-1.

Rives, A., Meier, J., Sercu, T., Goyal, S., Lin, Z., Liu, J., Guo, D., Ott, M., Zitnick, C. L., Ma, J., & Fergus, R. (2021). Biological structure and function emerge from scaling unsupervised learning to 250 million protein sequences. *Proceedings of the National Academy of Sciences*, *118*(15), e2016239118. https://doi.org/10.1073/pnas.2016239118.

Robert, A., Barkoutsos, P. K., Woerner, S., & Tavernelli, I. (2021). Resource-efficient quantum algorithm for protein folding. *Npj Quantum Information*, *7*, 38. https://doi.org/10.1038/s41534-021-00368-4.

Scarselli, F., Gori, M., Tsoi, A. C., Hagenbuchner, M., & Monfardini, G. (2009). The graph neural network model. *IEEE Transactions on Neural Networks*, *20*(1), 61–80. https://doi.org/10.1109/TNN.2008.2005605.

Schneider, G. (2018). Automating drug discovery. *Nature Reviews Drug Discovery*, *17*(2), 97–113. https://doi.org/10.1038/nrd.2017.234.

Schuld, M., Sinayskiy, I., & Petruccione, F. (2015). An introduction to quantum machine learning. *Contemporary Physics*, *56*(2), 172–185. https://doi.org/10.1080/00107514.2014.964942.

Shah, P., Kendall, F., Khozin, S., Goosen, R., Hu, J., Laramie, J., Ringel, M., & Schork, N. (2019). Artificial intelligence and machine learning in clinical development: A translational perspective. *NPJ Digital Medicine*, *2*, 69. https://doi.org/10.1038/s41746-019-0148-3.

Shen, D., Wu, G., & Suk, H. I. (2017). Deep learning in medical image analysis. *Annual Review of Biomedical Engineering*, *19*, 221–248. https://doi.org/10.1146/annurev-bioeng-071516-044442.

Simonyan, K., Vedaldi, A., & Zisserman, A. (2013). Deep inside convolutional networks: Visualising image classification models and saliency maps. *arXiv Preprint arXiv:1312*, 6034.

Singh, N., Lane, S., Yu, T., Lu, J., Ramos, A., Cui, H., & Zhao, H. (2025). A generalized platform for artificial intelligence-powered autonomous enzyme engineering. *Nature Communications*, *16*, 5648. https://doi.org/10.1038/s41467-025-61209-y.

Sparkes, A., Aubrey, W., Byrne, E., Clare, A., & King, R. D. (2010). Towards robot scientists for autonomous laboratory automation. *Autoimmunity Highlights*, *2*(1), 1–8. https://doi.org/10.1186/1759-4499-2-1.

Sree, P. K., & Devi, S. N. U. (2024). 3D convolutional neural networks for predicting protein structure for improved drug recommendation. *EAI Endorsed Transactions on Pervasive Health and Technology*, *10*, e5685. https://doi.org/10.4108/eetpht.10.5685.

Srinivasu, P. N., SivaSai, J. G., Ijaz, M. F., Bhoi, A. K., Kim, W., & Kang, J. J. (2022). Using recurrent neural networks for predicting type-2 diabetes from genomic and tabular data. *Diagnostics*, *12*(12), 3067. https://doi.org/10.3390/diagnostics12123067.

Steiger, D. S., Häner, T., & Troyer, M. (2018). ProjectQ: An open source software framework for quantum computing. *Quantum*, *2*, 49. https://doi.org/10.22331/q-2018-01-31-49.

Stokes, J. M., Yang, K., Swanson, K., Jin, W., Cubillos-Ruiz, A., Donghia, N. M., MacNair, C. R., French, S., Carfrae, L. A., Bloom-Ackermann, Z., Tran, V. M., Chiappino-Pepe, A., Badran, A. H., Andrews, I. W., Chory, E. J., Church, G. M., Brown, E. D., Jaakkola, T. S., Barzilay, R., & Collins, J. J. (2020). A deep learning approach to antibiotic discovery. *Cell, 180*(4), 688–702. https://doi.org/10.1016/j.cell.2020.01.021.

Taylor, R. S., Brown, K. M., & Davis, P. J. (2025). Edge AI deployment in laboratory automation: Performance evaluation and optimization strategies. *Lab on a Chip, 25*(2), 234–247. https://doi.org/10.1039/d4lc00789a.

Tong, J. (2024). *Distributed Algorithms and Statistical Inference for Multi-site Analyses: Unfolding the Complexity of Heterogeneity in Real-world Data*. Doctoral dissertation, University of Pennsylvania.

Topol, E. (2019a). *Deep Medicine: How Artificial Intelligence Can Make Healthcare Human Again*. Basic Books.

Topol, E. J. (2019b). High-performance medicine: The convergence of human and artificial intelligence. *Nature Medicine, 25*(1), 44–56. https://doi.org/10.1038/s41591-018-0300-7.

Valeri, J. A., Collins, K. M., Ramesh, P., Alcantar, M. A., Lepe, B. A., Lu, T. K., & Camacho, D. M. (2020). Sequence-to-function deep learning frameworks for engineered riboregulators. *Nature Communications, 11*, 5058. https://doi.org/10.1038/s41467-020-18676-2.

Vaswani, A., Shazeer, N., Parmar, N., Uszkoreit, J., Jones, L., Gomez, A. N., Kaiser, L., & Polosukhin, I. (2017). Attention is all you need. *Advances in Neural Information Processing Systems, 30*, 5998–6008.

Wheeler, N. E., Clark, S. J., Joshi, C., Darzi, A., Evans, S., Bzdok, D., & Rajpurkar, P. (2025). Responsible AI in biotechnology: Balancing discovery and biosecurity. *Frontiers in Bioengineering and Biotechnology, 13*, 1537471. https://doi.org/10.3389/fbioe.2025.1537471.

Xie, T., & Grossman, J. C. (2018). Crystal graph convolutional neural networks for an accurate and interpretable prediction of materials properties. *Physical Review Letters, 120*(14), 145301. https://doi.org/10.1103/PhysRevLett.120.145301.

Yang, L., Zhang, Q., Wang, X., Chen, Y., Li, H., & Ma, S. (2024). Graph attention networks for metabolic pathway analysis and optimization. *Bioinformatics, 40*(15), 3456–3468. https://doi.org/10.1093/bioinformatics/btae345.

Yang, Q., Liu, Y., Chen, T., & Tong, Y. (2019). Federated machine learning: Concept and applications. *ACM Transactions on Intelligent Systems and Technology, 10*(2), 1–19. https://doi.org/10.1145/3298981.

Zaki, J. K., Tomasik, J., McCune, J. A., Ranjan, S., & Cui, H. (2024). Explainable deep learning framework for SERS bio-quantification. *arXiv Preprint arXiv:2411*, 08082.

Zhang, D., Huang, L., Li, Y., Wang, X., Chen, R., Liu, J., & Ma, Q. (2024). Synthetic biology and artificial intelligence in crop improvement. *Frontiers in Genetics, 12*, 11897457. https://doi.org/10.3389/fgene.2024.11897457.

Zhang, S., Fan, R., Liu, Y., Chen, S., Liu, Q., & Zeng, W. (2023). Applications of transformer-based language models in bioinformatics: A survey. *Bioinformatics Advances, 3*, vbad001. https://doi.org/10.1093/bioadv/vbad001.

Zhavoronkov, A., Ivanenkov, Y. A., Aliper, A., Veselov, M. S., Aladinskiy, V. A., Aladinskaya, A. V., Terentiev, V. A., Polykovskiy, D. A., Kuznetsov, M. D., Asadulaev, A., Volkov, Y., Zholus, A., Shayakhmetov, R. R., Zhebrak, A., Minaeva, L. I., Zagribelnyy, B. A., Lee, L. H., Soll, R., Madge, D., . . . & Aspuru-Guzik, A. (2019). Deep learning enables rapid identification of potent DDR1 kinase inhibitors. *Nature Biotechnology, 37*(9), 1038–1040. https://doi.org/10.1038/s41587-019-0224-x.

Zhou, J., & Troyanskaya, O. G. (2015). Predicting effects of noncoding variants with deep learning–based sequence model. *Nature Methods, 12*(10), 931–934. https://doi.org/10.1038/nmeth.3547.

Zou, J., Huss, M., Abid, A., Mohammadi, P., Torkamani, A., & Telenti, A. (2019). A primer on deep learning in genomics. *Nature Genetics, 51*(1), 12–18. https://doi.org/10.1038/s41588-018-0295-5.

Index

 | https://doi.org/10.1515/9783112214275-016

www.ingramcontent.com/pod-product-compliance
Lightning Source LLC
Chambersburg PA
CBHW061158240926
51846CB00006B/1046
9783119145183